CHEMISTRY 142

EXAMINATIONS

&

SOLUTIONS

SPRING SEMESTER 2006

to

SPRING SEMESTER 2011

edited by

Wendy Whitford & Steve Poulios

Chemistry 142 Examinations and Solutions

(Volume 2 of a series of Examination books)

Published by the Okemos Press
Scientific and Technical Books
Okemos MI 48805-0085

Printed Summer 2011
Printed by
Sheridan Books, Inc., Ann Arbor, MI 48103

ISBN 0-9630471-6-7

CONTENTS

EXAMS

SOLUTIONS

EXAMINATION 1

CHEMISTRY 142 SPRING 2006

Wednesday February 15th 2006

1. Dinitrogen pentoxide decomposes to nitrogen dioxide and oxygen according to the reaction:

$$2N_2O_5(g) \rightarrow 4NO_2(g) + O_2(g)$$

In one particular experiment, four moles of dinitrogen pentoxide was placed in a 1.0 L flask and the concentrations of the reactant and products were plotted as time progressed.

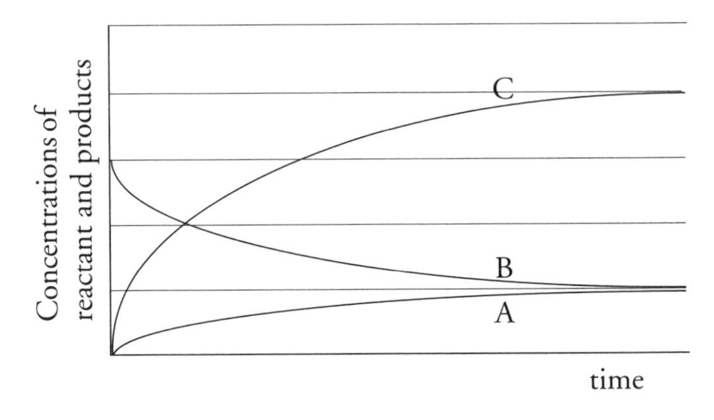

The graph shown on the right was obtained. Identify the components A, B, and C. *Choose the line where all components are identified properly.*

	A	B	C
a.	N_2O_5	NO_2	O_2
b.	N_2O_5	O_2	NO_2
c.	NO_2	N_2O_5	O_2
d.	NO_2	O_2	N_2O_5
e.	O_2	N_2O_5	NO_2
f.	O_2	NO_2	N_2O_5

2. Phosgene $COCl_2$ is a toxic gas prepared by the reaction of carbon monoxide and chlorine:

$$CO(g) + Cl_2(g) \rightarrow COCl_2(g)$$

The rate of the reaction was investigated and the data tabulated below were collected. Concentrations are expressed in mol/L. What is the overall reaction order?

	Initial [CO]	*Initial* [Cl_2]	*Initial rate of formation of* $COCl_2$
Experiment 1	1.00	0.100	1.29×10^{-29} mol L^{-1} s^{-1}
Experiment 2	0.100	0.100	1.30×10^{-30} mol L^{-1} s^{-1}
Experiment 3	0.100	1.00	1.29×10^{-29} mol L^{-1} s^{-1}
Experiment 4	0.100	0.010	1.30×10^{-31} mol L^{-1} s^{-1}

a. –4 c. –2 e. 0 g. 2 i. 4
b. –3 d. –1 f. 1 h. 3 j. 5

3. An initial concentration of a reactant (0.240 M) was found to drop to one-half of this value (0.120 M) in 4 minutes in a *second-order* reaction. How long does the reaction take for the concentration of this reactant to drop from its initial value (0.240 M) to one quarter of this value (0.060 M)?

a. 1 min c. 4 min e. 8 min g. 12 min
b. 2 min d. 6 min f. 10 min h. 16 min

4. The mechanism for the catalytic decomposition of hydrogen peroxide is:

$$H_2O_2(aq) + I^-(aq) \rightarrow OI^-(aq) + H_2O(l)$$
$$H_2O_2(aq) + OI^-(aq) \rightarrow O_2(g) + H_2O(l) + I^-(aq)$$

Label the species involved in this reaction appropriately.

	reactant	*intermediate*	*catalyst*	*product*
a.	H_2O_2	I^-	OI^-	H_2O
b.	H_2O_2	OI^-	I^-	O_2
c.	H_2O	I^-	OI^-	O_2
d.	I^-	H_2O_2	OI^-	H_2O
e.	I^-	OI^-	O_2	H_2O
f.	H_2O	OI^-	I^-	O_2
g.	I^-	H_2O	OI^-	O_2

5. Carbon–14 (^{14}C) has a half-life of 5270 years. If a piece of wood found in an American Indian burial mound near Marietta, Ohio, contains 77% of the amount of ^{14}C that you would expect to find in a modern piece of wood, what is the approximate age of the burial mound?

a. 23 years
b. 2,000 years
c. 500 years
d. 4,057 years
e. 1,212 years
f. 10,000 years

6. The *principal* reason why the rates of reactions increase when the temperature is increased is that

a. more of the collisions have sufficient energy
b. the required orientation in the collision becomes less important
c. the probability of termolecular collisions increases
d. the activation energy for the reaction decreases
e. the rate at which the molecules collide increases

7. Which of the following statements about a catalyst is *not* true?

a. A catalyst can speed up a reaction
b. A catalyst can cause a reaction to proceed by a different pathway
c. A catalyst takes part in the reaction but is not used up
d. A catalyst can be involved in an intermediate step in a reaction mechanism
e. A catalyst can decrease the activation energy for a reaction
f. A catalyst can increase the yield of product at equilibrium

8. In a 20 L flask *at equilibrium*, there are 4.0 moles of nitrous oxide N_2O, 2.0 moles of oxygen O_2, and an unknown concentration of nitrogen dioxide NO_2. The value of the equilibrium constant for the system at this temperature is 1.60×10^6. What is the concentration of the NO_2 in the flask?

$$2\,N_2O(g) + 3\,O_2(g) \rightleftharpoons 4\,NO_2(g)$$

a. 0.0054 mol L^{-1}
b. 1.4 mol L^{-1}
c. 2.8 mol L^{-1}
d. 4.0 mol L^{-1}
e. 8.0 mol L^{-1}
f. 16 mol L^{-1}
g. 57 mol L^{-1}
h. 120 mol L^{-1}

9. For which gas-phase equilibrium is K_p numerically larger than K_c?

a. $PCl_5 \rightleftharpoons PCl_3 + Cl_2$
b. $2SO_2 + O_2 \rightleftharpoons 2SO_3$
c. $CO + Cl_2 \rightleftharpoons COCl_2$
d. $C_2H_2 + 3H_2 \rightleftharpoons 2CH_4$
e. $N_2 + 3H_2 \rightleftharpoons 2NH_3$
f. $2NH_3 + Cl_2 \rightleftharpoons N_2H_4 + 2HCl$

10. As indicated above for the following equilibrium, the equilibrium constant is

$$2\,N_2O(g) + 3\,O_2(g) \rightleftharpoons 4\,NO_2(g) \qquad K = 1.60 \times 10^6$$

What is the equilibrium constant for the reaction:

$$2\,NO_2(g) \rightleftharpoons N_2O(g) + 1\tfrac{1}{2}\,O_2(g) \qquad K = ?$$

a. 8.00×10^5
b. 1.27×10^3
c. -8.00×10^5
d. 3.13×10^{-7}
e. 7.91×10^{-4}
f. 2.81×10^{-2}
g. 3.91×10^{-13}
h. 2.56×10^{12}

11. Which of the following acids is the *strongest* acid?

a. HIO
b. HClO
c. $HClO_2$
d. $HClO_4$
e. HNO_2
f. HCN
g. CH_3CO_2H
h. HF

12. As seen in the demonstration in class, the juice of a red cabbage can be used as an acid-base indicator. Which one of the following substances would cause the indicator to turn green?

a. HCl
b. ethanol C_2H_5OH
c. oxalic acid (HO_2CCO_2H)
d. window cleaner (NH_3*(aq)*)
e. vinegar (CH_3CO_2H*(aq)*)
f. lemon juice (citric acid)

13. Label the species in the following aqueous equilibrium as acid or base according to the definitions of Brønsted and Lowry. *It is recommended that you check the acid–base partnerships.*

	O^{2-} +	H_2O ⇌	OH^- +	OH^-
a.	acid	base	acid	base
b.	acid	acid	base	base
c.	base	acid	acid	base
d.	acid	base	acid	base
e.	base	base	base	base
f.	acid	base	base	acid

Use the distribution diagram and acid ionization constants shown below for the triprotic H_3PO_4 to answer the following questions (14, 15, and 16). *It is recommended that you label the curves.*

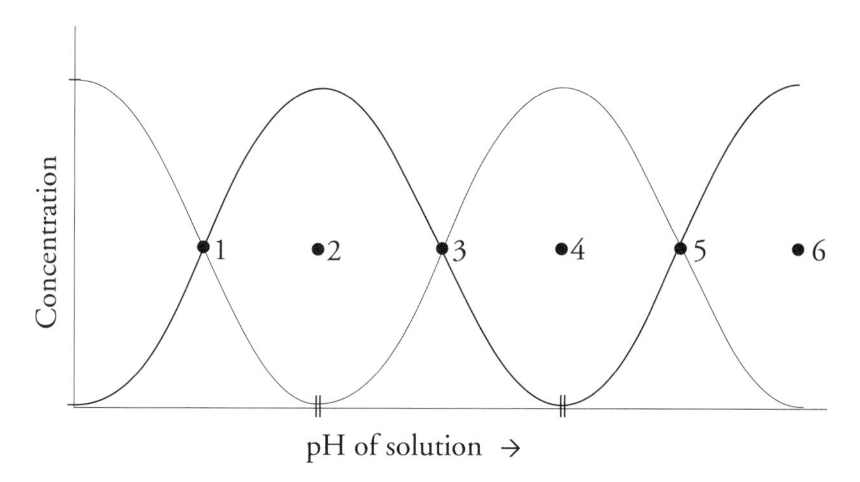

$H_3PO_4 + H_2O \rightleftharpoons H_3O^+ + H_2PO_4^-$ $K_{a1} = 7.5 \times 10^{-3}$ $pK_{a1} = 2.13$

$H_2PO_4^- + H_2O \rightleftharpoons H_3O^+ + HPO_4^{2-}$ $K_{a2} = 6.2 \times 10^{-8}$ $pK_{a2} = 7.21$

$HPO_4^{2-} + H_2O \rightleftharpoons H_3O^+ + PO_4^{3-}$ $K_{a3} = 3.6 \times 10^{-13}$ $pK_{a3} = 12.44$

14. What is (are) the predominant phosphorus-containing species in solution at point 3?

a. H_3PO_4
b. H_3PO_4 and $H_2PO_4^-$
c. $H_2PO_4^-$
d. $H_2PO_4^-$ and HPO_4^{2-}
e. HPO_4^{2-}
f. HPO_4^{2-} and PO_4^{3-}
g. PO_4^{3-}

15. What is the pH at point 3?

a. 2.13
b. 4.67
c. 6.79
d. 7.21
e. 9.82
f. 10.42
g. 11.31
h. 12.44

16. What is the pH at point 2?

a. 2.13
b. 4.67
c. 6.79
d. 7.21
e. 9.82
f. 10.42
g. 11.31
h. 12.44

17. What is the value of K_b for the hydrolysis of rubidium acetate (RbAc) at 25°C? K_a for acetic acid = 1.8×10^{-5} at 25°C.

a. 1.8×10^{-5}
b. 5.6×10^{-10}
c. 1.8×10^{-9}
d. 1.8×10^{-19}
e. 5.6×10^{18}
f. 3.2×10^{-10}

18. The pH of a 0.20 M solution of acetic acid is 2.72. What would the pH be if enough sodium acetate ($NaCH_3CO_2$) was added to make the solution 0.30 M in sodium acetate? Assume the volume doesn't change when the sodium acetate is added.

a. 2.72
b. 3.52
c. 4.92
d. 7.00
e. 7.52
f. 8.64

19. Given the following acid ionization constants, K_a, determine which salt will produce the most basic solution.

Acid	K_a	*Acid*	K_a
Acetic acid	1.8×10^{-5}	Hydrocyanic acid	4.0×10^{-10}
Hydrofluoric acid	3.5×10^{-4}	Nitrous acid	4.3×10^{-4}
Hypochlorous acid	3.0×10^{-8}	Benzoic acid	6.5×10^{-5}
Hydrogen sulfate	1.1×10^{-2}		

a. NaF
b. $NaC_6H_5CO_2$
c. KCH_3CO_2
d. KNO_2
e. NaOCl
f. NaCN
g. NH_4F
h. $KHSO_4$

20. Vitamin C is ascorbic acid (a weak diprotic acid with $K_{a1} = 7.9 \times 10^{-5}$ and $K_{a2} = 1.6 \times 10^{-12}$). What is the approximate pH of a buffer solution consisting of equal concentrations of ascorbic acid and sodium hydrogen ascorbate?

a. 1
b. 2
c. 3
d. 4
e. 5
f. 6
g. 7
h. 8
i. 10
j. 12

21. Which statement is applicable to the Lewis theory of acids and bases?

a. Acids produce hydronium ions in aqueous solution
b. The hydroxide ion is the strongest base than can exist in water
c. Bases donate electron pairs
d. Every acid has a conjugate base
e. All acid-base reactions involve the transfer of a hydrogen ion

22. Hydrazine NH_2NH_2 is a weak base like ammonia. Its K_b at 25°C is equal to 1.7×10^{-6}. What is the value of the neutralization constant K_{neut} in a titration of hydrazine with hydrochloric acid HCl?

a. 1.7×10^{-6}
b. 5.1×10^{7}
c. 5.9×10^{5}
d. 1.0×10^{-14}
e. 5.9×10^{-9}
f. 6.4×10^{-11}
g. 1.7×10^{8}
h. 5.9×10^{-1}
i. 6.4×10^{-8}
j. 3.2×10^{10}

EXAMINATION 1

CHEMISTRY 142 FALL 2006

Thursday October 5th 2006

1. Fluorine F_2 reacts with water H_2O to produce hydrofluoric acid HF, oxygen difluoride OF_2 and oxygen O_2:

$$4\,F_2(g) \;+\; 3\,H_2O(l) \;\rightarrow\; 6\,HF(aq) \;+\; OF_2(g) \;+\; O_2(g)$$

If fluorine is used at a rate of 24.0 mol L^{-1} s^{-1}, what are the rates at which hydrofluoric acid and oxygen are produced? *Choose the row in which both answers are correct.*

	rate of HF production	rate of O_2 production
a.	60.2 mol L^{-1} s^{-1}	22.3 mol L^{-1} s^{-1}
b.	42.0 mol L^{-1} s^{-1}	12.5 mol L^{-1} s^{-1}
c.	5.55 mol L^{-1} s^{-1}	16.7 mol L^{-1} s^{-1}
d.	36.0 mol L^{-1} s^{-1}	6.00 mol L^{-1} s^{-1}
e.	84.0 mol L^{-1} s^{-1}	21.0 mol L^{-1} s^{-1}
f.	42.0 mol L^{-1} s^{-1}	42.0 mol L^{-1} s^{-1}
g.	21.0 mol L^{-1} s^{-1}	21.0 mol L^{-1} s^{-1}
h.	14.9 mol L^{-1} s^{-1}	29.9 mol L^{-1} s^{-1}

2. Ethene C_2H_4 reacts with hydrogen H_2 to produce ethane C_2H_6:

$$C_2H_4(g) \;+\; H_2(g) \;\rightarrow\; C_2H_6(g)$$

The rate of the reaction was investigated and the data tabulated below were collected. Concentrations are expressed in mol L^{-1}. What is the order of the reaction with respect to the concentration of hydrogen?

	Initial $[C_2H_4]$	Initial $[H_2]$	Initial rate of formation of C_2H_6
Experiment 1	0.100	0.200	6.20×10^{-5} mol L^{-1} s^{-1}
Experiment 2	0.050	0.200	3.10×10^{-5} mol L^{-1} s^{-1}
Experiment 3	0.100	0.100	1.55×10^{-5} mol L^{-1} s^{-1}

a. −4 c. −2 e. 0 g. 2 i. 4
b. −3 d. −1 f. 1 h. 3 j. 5

3. Dinitrogen pentoxide N_2O_5 decays to nitrogen dioxide NO_2 and oxygen O_2 in a first-order process. If it takes 5.2 minutes for the N_2O_5 concentration to fall from 0.60 M to 0.15 M, what is the rate constant, k?

a. 10.2 c. 5.60 e. 20.5 g. 78.1
b. 4.44 d. 6.60 f. 100 h. 0.27

4. The rate constant for a reaction increases by a factor of two when the temperature increases from 298 K to 377 K. What is the activation energy E_A (in J mol^{-1}) for this reaction?

a. 2,200 c. 8,200 e. 800 g. 4,600
b. 3,000 d. 9,700 f. 1,000 h. 900

5. Hydrogen bromide HBr reacts with 2-methylpropene $(CH_3)_2C{=}CH_2$ to produce 2–bromo–2–methylpropane $(CH_3)_3C{-}Br$:

$$HBr + (CH_3)_2C{=}CH_2 \rightarrow (CH_3)_3C{-}Br$$

The mechanism for this reaction includes two steps, which are:

$$(CH_3)_2C{=}CH_2 + HBr \rightarrow (CH_3)_3C^+ + Br^- \quad \text{SLOW}$$

$$(CH_3)_3C^+ + Br^- \rightleftharpoons (CH_3)_3C{-}Br \quad \text{FAST}$$

How many intermediates are involved in this mechanism for the reaction?

a. 0
b. 1
c. 2
d. 3
e. 4
f. 5

6. Which is the rate law for the reaction described in question 5?

a. Rate = k $[(CH_3)_3C{-}Br]$
b. Rate = k $[(CH_3)_3C^+]\,[Br^-]$
c. Rate = k $[(CH_3)_2C{=}CH_2][HBr]$
d. Rate = k $[(CH_3)_3C{-}Br]^2$
e. Rate = k $[(CH_3)_3C^+]^2[Br^-]^2$
f. Rate = k $[(CH_3)_2C{=}CH_2]^2[HBr]^2$
g. Rate = k $[(CH_3)_2C{=}CH_2]\,[HBr]^2$
h. Rate = k

7. Which of the following statements best explains the difference between a heterogeneous catalyst and a homogeneous catalyst?

a. A heterogeneous catalyst is in a different phase than the reactants, whereas a homogeneous catalyst is in the same phase as the reactants.
b. A homogeneous catalyst is in a different phase than the reactants, whereas a heterogeneous catalyst is in the same phase as the reactants.
c. A heterogeneous catalyst lowers the activation energy of a reaction, but a homogeneous catalyst does not.
d. A homogeneous catalyst lowers the activation energy of a reaction, but a heterogeneous catalyst does not.
e. A heterogeneous catalyst is an intermediate in a reaction, but a homogeneous catalyst is not.
f. A homogeneous catalyst is completely used up in a reaction, but a heterogeneous catalyst is not.

8. Which of the following compounds is a weak acid in aqueous solution?

a. $HClO_4$
b. $HClO_3$
c. KSCN
d. $MgCrO_4$
e. HCN
f. $MgSO_4$
g. NaOH
h. HI

9. Calculate the equilibrium constant for the following reaction at 25°C given the equilibrium concentrations shown:

$$2\,NO(g) + Br_2(g) \rightleftharpoons 2\,NOBr(g) \qquad [NO] = 1.80\text{ M};\ [Br_2] = 3.0\text{ M};\ [NOBr] = 5.0\text{ M}$$

a. 0.6
b. 1.5
c. 2.6
d. 4.5
e. 6.0
f. 6.5
g. 7.0
h. 5.7
i. 4.3

10. Calculate the pH of a 1.75 M solution of the weak base sodium hypochlorite, NaClO. (K_a for HClO is 3.2×10^{-8})

a. 1.30
b. 2.50
c. 3.50
d. 4.00
e. 7.80
f. 9.90
g. 10.9
h. 13.1

11. Consider the following Brønsted-Lowry equilibrium:

$$HSCN + H_2O \rightleftharpoons H_3O^+ + \_\_\_\_$$

Identify the missing product, and decide whether that product is acting as a base or an acid in the equilibrium.

a. SCN^-, acid
b. SCN^-, base
c. HSCN, acid
d. HSCN, base
e. H_2SCN, acid
f. H_2SCN, base
g. OH^-, acid
h. OH^-, base

12. Which one of the following compounds *cannot* act as a Lewis base?

a. NH_3
b. H_2O
c. OH^-
d. CH_4
e. Cl^-
f. F^-
g. Br^-
h. CO

13. In which direction will the equilibrium of the following *endothermic* reaction shift if the indicated changes are made? *Choose the row in which all answers are correct.*

$$TiCl_4(g) + O_2(g) \rightleftharpoons TiO_2(s) + 2\ Cl_2(g)$$

	add O_2	add Cl_2	add a catalyst	increase T
a.	left	right	left	right
b.	left	right	right	right
c.	right	right	right	right
d.	left	left	left	no change
e.	right	left	right	left
f.	no change	left	right	left
g.	right	left	no change	right
h.	left	no change	no change	right

14. What is K_b for HSO_3^-? Assume that the temperature is 25°C. (K_a for $H_2SO_3 = 1.2 \times 10^{-2}$)

a. 3.4×10^2
b. 1.0×10^1
c. 1.6×10^{-2}
d. 2.4×10^{-4}
e. 8.7×10^{-7}
f. 5.1×10^{-8}
g. 5.5×10^{-10}
h. 8.3×10^{-13}

15. Methane CH_4 and chlorine Cl_2 react to produce methyl chloride CH_3Cl and hydrogen chloride HCl according to the following reaction:

$$CH_4(g) + Cl_2(g) \rightleftharpoons CH_3Cl(g) + HCl(g)$$

At equilibrium, the partial pressures of CH_4 and Cl_2 are 0.5 atm, and the partial pressures of CH_3Cl and HCl are 0.75 atm. Assuming the temperature is 298 K, what is K_c?

a. 2.25
b. 4.75
c. 10.2
d. 5.65
e. 8.90
f. 12.2
g. 16.8
h. 20.7

16. As indicated below for the following equilibria, the equilibrium constants are:

$$C(s) + CO_2(g) \rightleftharpoons 2\ CO(g) \qquad K = 0.64 \qquad (T = 1200\ K)$$
$$C(s) + \tfrac{1}{2}\ O_2(g) \rightleftharpoons CO(g) \qquad K = 1 \times 10^3 \qquad (T = 1200\ K)$$

What is the equilibrium constant for the reaction:

$$CO_2(g) \rightleftharpoons \tfrac{1}{2}\ O_2(g) + CO(g) \qquad K = ? \qquad (T = 1200\ K)$$

a. 1.5×10^4
b. 8.9×10^2
c. 4.6×10^1
d. 2.2×10^{-2}
e. 6.4×10^{-4}
f. 3.7×10^{-8}
g. 7.3×10^{-12}
h. 7.9×10^{-20}

17. Which of the following salts produces a *neutral* solution when dissolved in water?

a. NH_4ClO_4 c. NaF e. $NaClO_4$ g. $Mg(ClO)_2$
b. KCN d. $Mg(CH_3CO_2)_2$ f. LiCN h. NH_4Cl

18. What is the pH of a 0.25 M solution of sodium hydrogen sulfide NaHS?
Hydrosulfuric acid H_2S is a weak diprotic acid with $K_{a1} = 9.5 \times 10^{-8}$ and $K_{a2} = 1 \times 10^{-19}$.

a. 13.0 c. 9.80 e. 7.80 g. 3.70
b. 10.2 d. 8.80 f. 5.30 h. 0.50

19. Calculate the pH of a 0.8 M hydrocyanic acid (HCN) and 0.25 M cyanide (CN^-) solution.
$K_a = 4.9 \times 10^{-10}$ for HCN.

a. 0.4 c. 3.7 e. 8.8 g. 13.1
b. 2.2 d. 5.8 f. 9.3 h. 13.9

20. What is the value of the neutralization constant K_{neut} in a titration of hypoiodous acid with ammonia NH_3?
Hypoiodous acid, HIO, is a weak acid with $K_a = 2 \times 10^{-11}$.
For ammonia NH_3, $K_b = 1.8 \times 10^{-5}$.

a. 7.6×10^{-19} c. 3.6×10^{-15} e. 6.8×10^{-6} g. 3.6×10^{-2}
b. 3.9×10^{-16} d. 2.8×10^{-9} f. 5.2×10^{-4} h. 9.7×10^{2}

EXAMINATION 1

CHEMISTRY 142 SPRING 2007

Wednesday February 14th 2007

1. Ammonia is oxidized by oxygen to form nitrogen dioxide and water according to the equation:

$$4\,NH_3(g) \;+\; O_2(g) \;\rightarrow\; 4\,NO_2(g) \;+\; 6\,H_2O(g)$$

In one particular experiment, five moles of ammonia were placed in a 1.0 L flask in the presence of excess oxygen and the concentrations of the ammonia and the products (nitrogen dioxide and water) were plotted as time progressed.

The graph shown on the right was obtained.

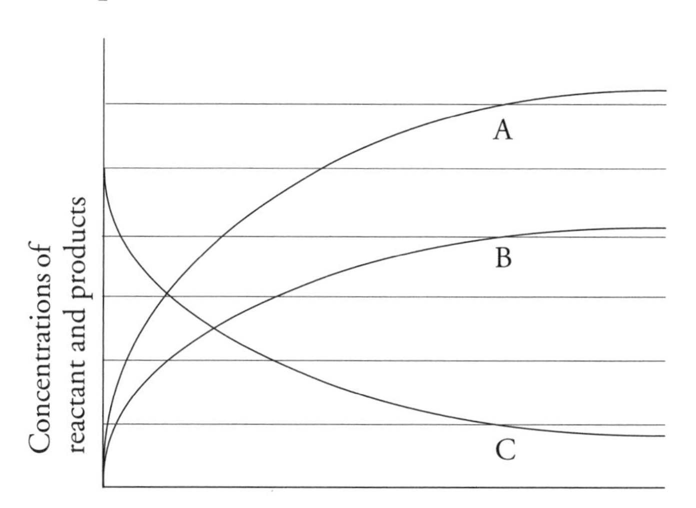

Identify the components A, B, and C:

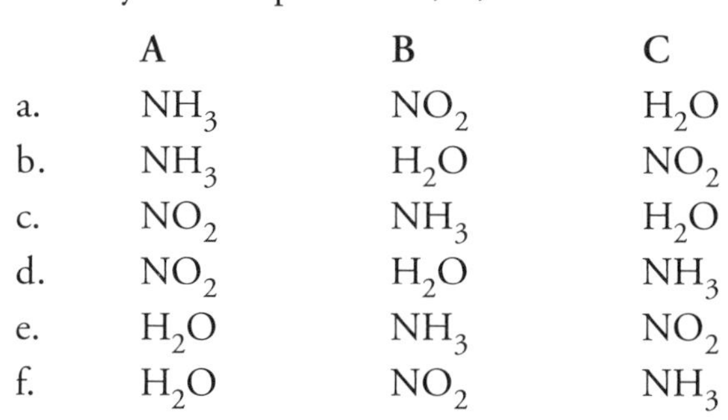

	A	B	C
a.	NH_3	NO_2	H_2O
b.	NH_3	H_2O	NO_2
c.	NO_2	NH_3	H_2O
d.	NO_2	H_2O	NH_3
e.	H_2O	NH_3	NO_2
f.	H_2O	NO_2	NH_3

2. In the combustion of the hydrocarbon pentane C_5H_{12} shown below, carbon dioxide is produced at a rate of 60 mol L^{-1} min^{-1}. At what rate is oxygen being used up ?

$$C_5H_{12} \;+\; 8\,O_2 \;\rightarrow\; 5\,CO_2 \;+\; 6\,H_2O$$

a. 38 mol $L^{-1}min^{-1}$ c. 60 mol $L^{-1}min^{-1}$ e. 96 mol $L^{-1}min^{-1}$
b. 48 mol $L^{-1}min^{-1}$ d. 64 mol $L^{-1}min^{-1}$ f. 108 mol $L^{-1}min^{-1}$

3. In a decomposition reaction which is zero-order with respect to the reactant, the concentration of the reactant decreases from 24 mol L^{-1} to 16 mol L^{-1} in just 30 minutes. How long does it take for the reactant to be completely used up, i.e. to decrease from 24 mol L^{-1} to zero concentration?

a. 10 min c. 30 min e. 60 min g. 90 min i. 180 min
b. 20 min d. 45 min f. 75 min h. 120 min j. 210 min

4. A compound undergoing a second-order decomposition decreases in concentration from 56 mol L^{-1} to 14 mol L^{-1} in 12 minutes. How long does it take for the 14 mol L^{-1} to decrease to 3.5 mol L^{-1}?

a. 12 min c. 20 min e. 28 min g. 36 min i. 60 min
b. 16 min d. 24 min f. 32 min h. 48 min j. 90 min

5. The wrappings from a mummy discovered in an Egyptian tomb in the Valley of Kings was dated by counting its ^{14}C content. It was found to have 75% of the ^{14}C activity/gram that one might expect to find in a modern artifact. What is the approximate age of the mummy? (^{14}C decays by a first-order mechanism with a half-life of 5730 years.)

a. somewhat less than 2500 years
b. somewhat greater than 2500 years
c. about 4300 years
d. about 5700 years
e. somewhat less than 1500 years

6. In the energy profile for a reaction shown, _____ is the energy of the activated complex, _____ is the energy of the reaction, _____ is the energy of the products, and _____ is the activation energy for the reverse reaction. *(Choose your answers in the correct order.)*

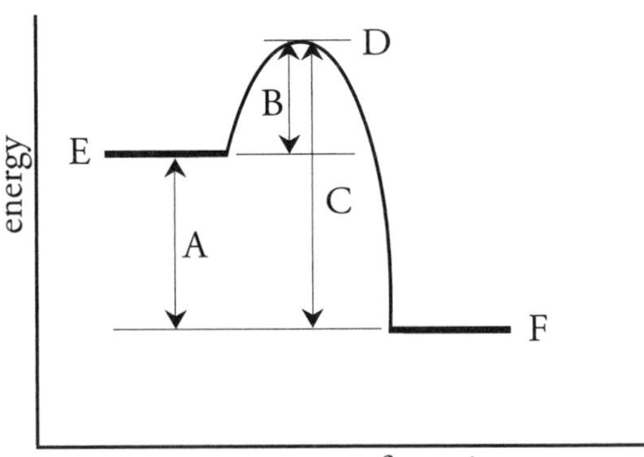

a. C D F C
b. B C D A
c. C D E F
d. D C F A
e. B A F A
f. D A F C
g. B A E D
h. C F D E
i. D C E A
j. B C F A

7. The following kinetic data were obtained for the reaction:

$2\ ICl(g) + H_2(g) \rightarrow I_2(g) + 2\ HCl(g)$

Concentrations are expressed in *m*mol L^{-1}. What is the overall reaction order?

	Initial [ICl]	*Initial* [H_2]	*Initial rate of formation of* I_2
Experiment 1	1.5	1.5	3.7×10^{-7} mol L^{-1} s^{-1}
Experiment 2	3.0	1.5	7.4×10^{-7} mol L^{-1} s^{-1}
Experiment 3	3.0	4.5	2.2×10^{-6} mol L^{-1} s^{-1}
Experiment 4	4.7	2.7	*see the following question*

a. −4
b. −3
c. −2
d. −1
e. 0
f. 1
g. 2
h. 3
i. 4
j. 5

8. What would the initial rate of formation of iodine I_2 be for the Experiment 4 described in the previous question, where $[ICl]_{init}$ = 4.7 *m*mol L^{-1} and $[H_2]_{init}$ = 2.7 *m*mol L^{-1} as shown?

a. 6.5×10^{-6} mol L^{-1} s^{-1}
b. 5.6×10^{-6} mol L^{-1} s^{-1}
c. 8.2×10^{-7} mol L^{-1} s^{-1}
d. 2.1×10^{-6} mol L^{-1} s^{-1}
e. 3.5×10^{-6} mol L^{-1} s^{-1}
f. 5.6×10^{-7} mol L^{-1} s^{-1}
g. 1.7×10^{-7} mol L^{-1} s^{-1}
h. 2.2×10^{-7} mol L^{-1} s^{-1}

9. The industrial synthesis of ammonia by the reaction of hydrogen and nitrogen involves the use of catalysts. A simplified mechanism for one method is:

$$N_2(g) + Fe(s) \rightarrow FeN_2(s)$$
$$3\ H_2(g) + 3\ Fe(s) \rightarrow 3\ FeH_2(s)$$
$$FeN_2(s) + 3\ FeH_2(s) \rightarrow 4\ Fe(s) + 2\ NH_3(g)$$

Label the species involved in this reaction appropriately.

	reactant	*intermediate*	*catalyst*	*product*
a.	H_2	FeH_2	Fe	NH_3
b.	H_2	Fe	FeN_2	NH_3
c.	Fe	FeN_2	H_2	NH_3
d.	Fe	FeH_2	FeN_2	Fe
e.	N_2	FeN_2	Fe	Fe
f.	N_2	Fe	H_2	Fe
g.	H_2	FeN_2	NH_3	Fe
h.	Fe	FeH_2	H_2	NH_3

10. The rate law for the disproportionation of the hypochlorite ion in aqueous solution is Rate = $k[ClO^-]^2$.

$$3\ ClO^- \rightarrow ClO_3^- + 2\ Cl^-$$

The fact that the rate equation (specifically the exponent 2 of the concentration term) does not match the stoichiometry of the balanced chemical equation for the reaction indicates that...

a. the rate equation cannot be correct
b. the mechanism of the reaction is incorrect
c. the reaction occurs by a single termolecular collision of three hypochlorite ions
d. doubling the initial concentration of hypochlorite ion should triple the rate of the reaction
e. the reaction does not occur in a single step
f. some error was made in determining the rate equation

11. The *principal* reason why the rates of reactions increase when the temperature is increased is that

a. a greater fraction of the collisions have sufficient energy to reach the activated complex
b. the required orientation in the collision becomes less critical
c. the molecules collide more frequently
d. the reaction becomes more favored thermodynamically
e. the simultaneous collision of more than two molecules becomes more frequent
f. other mechanisms for the reaction become available
g. the activation energy for the reaction changes

12. An enzyme reduces the activation energy of a reaction at 298 K from 55 kJ mol^{-1} to 15 kJ mol^{-1} by changing the route (the mechanism) of the reaction. Assuming all factors other than the activation energy remain the same, by how much does the rate of the reaction increase?

a. 10 times faster
b. 100 times faster
c. 1000 times faster
d. 10,000 times faster
e. 100,000 times faster
f. one million times faster
g. 10 million times faster
h. 100 million times faster
i. one billion times faster

13. Classify the following solutes in aqueous solution.
Choose the row where all responses are correct.

	Acid	*Base*	*Salt*
a.	NaCN	KOH	NH_4Cl
b.	HF	$NaNO_3$	KCH_3CO_2
c.	NH_3	KF	Na_2HPO_4
d.	H_2SO_4	NaOH	KF
e.	$NaNH_2$	NH_3	$NaHCO_3$
f.	$NaCH_3CO_2$	HCN	NaCl

14. What is the pH of a solution prepared by diluting 10.0 mL of a 0.220 *M* KOH solution to 0.250 L?

a. 1.09
b. 2.06
c. 3.76
d. 4.22
e. 5.21
f. 6.37
g. 7.63
h. 11.94
i. 12.94
j. 13.34

15. The element At (#85 in Group 7) is not very well characterized chemically because its longest-lived isotope, ^{210}At, has a half-life of only 8.3 hours. Nevertheless, we can make definite predictions about the strength of its acids. Which of the following acids would be the weakest?

a. HAt b. HAtO c. $HAtO_2$ d. $HAtO_3$ e. $HAtO_4$

16. Which one of the following equations does *not* represent a Lewis acid-base reaction?

a. $H_2O + H^+ \rightarrow H_3O^+$
b. $NH_3 + BF_3 \rightarrow H_3NBF_3$
c. $PF_3 + F_2 \rightarrow PF_5$
d. $Al(OH)_3 + OH^- \rightarrow Al(OH)_4^-$
e. $Co^{3+} + 6\,NH_3 \rightarrow [Co(NH_3)_6]^{3+}$
f. $NH_3 + H_2O \rightarrow NH_4^+ + OH^-$
g. $CH_3CO_2H + H_2O \rightarrow H_3O^+ + CH_3CO_2^-$

17. In a 2.0 L flask *at equilibrium*, there are 2.0 moles of ammonia, 2.0 moles of chlorine, 6.0 moles of hydrazine N_2H_4, and an unknown concentration of hydrogen chloride HCl. The value of the equilibrium constant K_c for the system at this temperature is 12.0. How many *moles* of HCl are in the 2.0 L flask at equilibrium?

$$2\,NH_3(g) + Cl_2(g) \rightleftharpoons N_2H_4(g) + 2\,HCl(g)$$

a. 0.50 mol
b. 1.0 mol
c. 1.5 mol
d. 2.0 mol
e. 2.5 mol
f. 4.0 mol
g. 5.0 mol
h. 8.0 mol

18. According to LeChatelier's Principle, determine the influence the following changes would have upon the system:

$$2\,N_2O(g) + 3\,O_2(g) \rightleftharpoons 4\,NO_2(g) \qquad \Delta H° \text{ negative}$$

	decrease in P	*increase in T*	*addition of O_2*	*removal of N_2O*
a.	shift left	shift left	shift left	shift left
b.	shift right	shift left	shift right	shift left
c.	shift left	shift right	shift left	shift right
d.	shift right	shift right	shift right	shift right
e.	shift left	shift left	shift right	shift left
f.	shift right	shift right	shift left	shift left

19. Label the species in the following aqueous equilibrium as acid or base according to the definitions of Brønsted and Lowry. *It is recommended that you check the acid–base partnerships.*

	S^{2-} +	H_2O ⇌	SH^- +	OH^-
a.	acid	base	acid	base
b.	acid	acid	base	base
c.	base	acid	acid	base
d.	acid	base	acid	base
e.	base	base	base	base
f.	acid	base	base	acid

20. Arsenic acid H_3AsO_4 is a weak triprotic acid:

$$H_3AsO_4 + H_2O \rightleftharpoons H_3O^+ + H_2AsO_4^- \qquad K_{a1} = 2.5 \times 10^{-4} \qquad pK_{a1} = 3.60$$
$$H_2AsO_4^- + H_2O \rightleftharpoons H_3O^+ + HAsO_4^{2-} \qquad K_{a2} = 5.6 \times 10^{-8} \qquad pK_{a2} = 7.25$$
$$HAsO_4^{2-} + H_2O \rightleftharpoons H_3O^+ + AsO_4^{3-} \qquad K_{a3} = 3.0 \times 10^{-13} \qquad pK_{a3} = 12.52$$

Determine the acidity or basicity of solutions of the following salts:

	NaH_2AsO_4	Na_2HAsO_4	Na_3AsO_4
a.	acidic	acidic	acidic
b.	acidic	acidic	basic
c.	acidic	basic	basic
d.	acidic	basic	acidic
e.	basic	basic	basic
f.	basic	acidic	basic
g.	basic	basic	acidic

21. Consider again the data for the triprotic acid H_3AsO_4 presented in the previous question. Salts of the dihydrogenarsenate ion $H_2AsO_4^-$, such as NaH_2AsO_4, undergo autoprotolysis in aqueous solution, a reaction in which the dihydrogenarsenate ion acts as *both* acid *and* base:

$$H_2AsO_4^- + H_2AsO_4^- \rightleftharpoons H_3AsO_4 + HAsO_4^{2-} \qquad K = ??$$

What is the value of the equilibrium constant for this autoprotolysis reaction?

a. 2.5×10^{-4} c. 3.0×10^{-13} e. 2.2×10^{-4} g. 5.4×10^{-6}

b. 5.6×10^{-8} d. 4.5×10^{3} f. 1.9×10^{5} h. 1.2×10^{-9}

22. Given the following distribution diagram for the titration of arsenic acid H_3AsO_4 against sodium hydroxide NaOH,

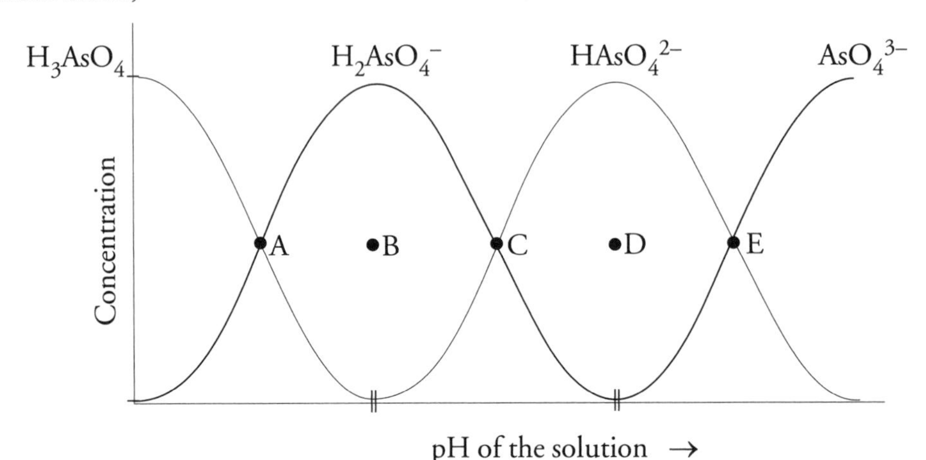

what is/are the predominant species, other than H_2O, in the solution at a pH of 7.25? Appropriate data are provided in question 20.

a. H_3AsO_4
b. H_3AsO_4 and $H_2AsO_4^-$
c. $H_2AsO_4^-$
d. $H_2AsO_4^-$ and $HAsO_4^{2-}$
e. $HAsO_4^{2-}$
f. $HAsO_4^{2-}$ and AsO_4^{3-}

23. Which two solutes mixed together in aqueous solution would *not* act as a buffer solution?

a. HCN and NaCN
b. NH_3 and NH_4NO_3
c. KH_2PO_4 and K_2HPO_4
d. KF and HF
e. NH_4Cl and NH_3
f. H_2CO_3 and $NaHCO_3$
g. HI and NaI

24. A buffer solution has a pH = pK_a whenever

a. the weak acid and its conjugate base have equal values for their ionization constants.
b. the solution of the weak acid and its conjugate base is neutral.
c. the solution contains concentrations of the weak acid and conjugate base inversely proportional to their pK values.
d. the solution contains no strong acid or base.
e. the solution contains equal concentrations of the weak acid and its conjugate base.

25. Calculate the equilibrium constant K at 25°C for the reaction between hydrofluoric acid and potassium hydroxide. The acid ionization constant K_a for hydrofluoric acid = 3.50×10^{-4}.

a. 1.80×10^{-5}
b. 2.86×10^{-11}
c. 5.56×10^{4}
d. 3.13×10^{-7}
e. 5.56×10^{-10}
f. 2.86×10^{3}
g. 3.50×10^{10}
h. 2.56×10^{12}

EXAMINATION 1

CHEMISTRY 142 FALL 2007

Thursday October 4th 2007

1. The reaction between methylhydrazine and dinitrogen tetroxide in a rocket engine is represented by the equation shown below. At one stage in this reaction, the gaseous products were produced at a rate of 5,000 liters per second. At what rate was the methylhydrazine being used at this stage?

$$4\ CH_3NHNH_2(g) \ + \ 5\ N_2O_4(g) \ \rightarrow \ 9\ N_2(g) \ + \ 12\ H_2O(g) \ + \ 4\ CO_2(g)$$

a. $200\ L\ s^{-1}$ c. $800\ L\ s^{-1}$ e. $1250\ L\ s^{-1}$ g. $2500\ L\ s^{-1}$
b. $500\ L\ s^{-1}$ d. $1000\ L\ s^{-1}$ f. $2000\ L\ s^{-1}$ h. $6250\ L\ s^{-1}$

2. The rate law for the decomposition of ozone is found to be: Rate = $k[O_3]^2[O_2]^{-1}$
The overall order for this reaction is:

a. −3   c. −1   e. 1   g. 3
b. −2   d. 0   f. 2   h. 4

3. A radioactive isotope decays by a first order mechanism and has a half-life of 12.0 days. If you initially have 264 mg of the isotope, how many milligrams would you have 36.0 days later?

a. 33 mg   c. 66 mg   e. 112 mg   g. 178 mg
b. 45 mg   d. 88 mg   f. 132 mg   h. 225 mg

4. A compound undergoing a second-order decomposition decreases in concentration from 144 mol L^{-1} to 36 mol L^{-1} in 30 minutes. How long does it take for the 36 mol L^{-1} to decrease by another factor of four to 9.0 mol L^{-1}?

a. 10 min   c. 20 min   e. 40 min   g. 80 min   i. 100 min
b. 15 min   d. 30 min   f. 1 hour   h. 90 min   j. 2 hours

5. The reaction of a reactant A with another reactant B to form a product C was investigated in a series of experiments with different initial concentrations of the two reactants. The following initial rates were observed. What is the overall reaction order?

	Initial [A]	*Initial [B]*	*Initial rate of formation of C*
Experiment 1	0.20 M	0.10 M	8.00×10^{-3} mol L^{-1} s^{-1}
Experiment 2	0.40 M	0.10 M	1.60×10^{-2} mol L^{-1} s^{-1}
Experiment 3	0.60 M	0.20 M	9.60×10^{-2} mol L^{-1} s^{-1}
Experiment 4	0.40 M	0.20 M	6.40×10^{-2} mol L^{-1} s^{-1}

a. −4   c. −2   e. 0   g. 2   i. 4
b. −3   d. −1   f. 1   h. 3   j. 5

6. The principal reason why the rates of reactions increase when the temperature is increased is that

a. the rate of collisions increases
b. catalysts are more effective at higher temperatures
c. the molecules collide with greater energy
d. the molecules move faster
e. the activation energy for the reaction decreases
f. molecules decompose at higher temperatures

7. In the energy profile for a reaction shown,

_____ is the energy of the reaction,
_____ is the energy of the activated complex,
_____ is the energy of the products, and
_____ is the activation energy for the forward reaction.

(Choose your answers in the correct order.)

a. E B F C
b. E B C A
c. A B E F
d. A C F B
e. C E F A
f. C D F A
g. A D F B
h. D B E C
i. D B E F
j. B D F C

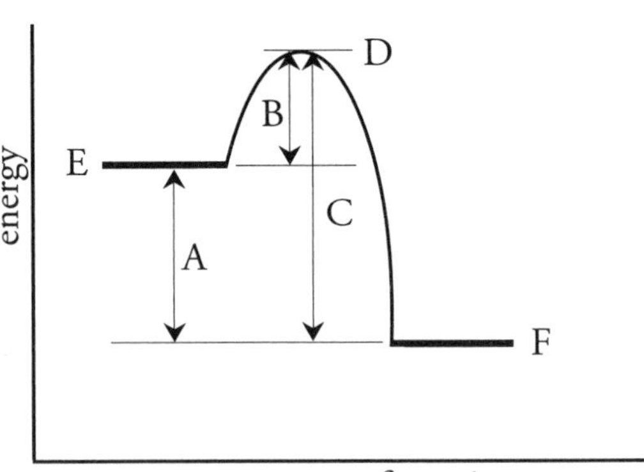

8. The reaction between bromine and hydrogen occurs by the mechanism illustrated below:

$Br_2(g)$	$\rightleftharpoons$	$2\,Br(g)$				*fast*
$H_2(g)$	+ $Br(g)$	$\rightarrow$	$HBr(g)$	+	$H(g)$	*slow*
$H(g)$	+ $Br_2(g)$	$\rightarrow$	$HBr(g)$	+	$Br(g)$	*fast*
$H(g)$	+ $Br(g)$	$\rightarrow$	$HBr(g)$			*very fast*

Label the species involved in this mechanism appropriately.

	reactant	*intermediate*	*product*
a.	H_2	Br	Br_2
b.	Br_2	H_2	HBr
c.	H	H	H_2
d.	Br	H_2	HBr
e.	H_2	Br	HBr
f.	Br_2	HBr	H_2
g.	H	Br	HBr
h.	Br	H_2	Br_2

9. A catalyst changes the rate of a chemical reaction because...

a. it changes the equilibrium constant K
b. it changes the coefficients in the balanced reaction
c. it changes the temperature
d. it changes the route of the reaction
e. it changes the average speed of the molecules involved

10. Calculate the equilibrium constant K_c for the following equilibrium if the concentrations of the various components are as shown:

$$2\ SO_2(g) + O_2(g) \rightleftharpoons 2\ SO_3(g)$$

Concentrations at equilibrium: $[SO_2(g)] = 0.5$ mole/L
$[O_2(g)] = 2.0$ mole/L
$[SO_3(g)] = 1.5$ mole/L

a. 1.0 c. 2.0 e. 3.0 g. 4.0
b. 1.5 d. 2.5 f. 3.5 h. 4.5

11. 2.0 moles of hydrogen, 2.0 moles of chlorine, and 24.0 moles of hydrogen chloride are placed in a 1.0 liter flask and the system is allowed to reach equilibrium. The value of the equilibrium constant for the system at this temperature is 25.0. What is the concentration of chlorine in the flask at equilibrium?

$$H_2(g) + Cl_2(g) \rightleftharpoons 2HCl(g)$$

a. 0 mol L^{-1} c. 1.0 mol L^{-1} e. 2.0 mol L^{-1} g. 3.0 mol L^{-1} i. 5.0 mol L^{-1}
b. 0.5 mol L^{-1} d. 1.5 mol L^{-1} f. 2.5 mol L^{-1} h. 4.0 mol L^{-1} j. 6.0 mol L^{-1}

12. For which gas-phase equilibrium does $K_p = K_c \times RT$?

a. $PCl_5 \rightleftharpoons PCl_3 + Cl_2$
b. $2SO_2 + O_2 \rightleftharpoons 2SO_3$
c. $CO + Cl_2 \rightleftharpoons COCl_2$
d. $C_2H_2 + 3H_2 \rightleftharpoons 2CH_4$
e. $N_2 + 3H_2 \rightleftharpoons 2NH_3$
f. $2NH_3 + Cl_2 \rightleftharpoons N_2H_4 + 2HCl$

13. Which one of the following electrolytes is classified as weak?

a. $HClO_4$ c. KCl e. KCN g. LiI i. NaOH
b. Na_2SO_3 d. NH_4Cl f. $Ca(OH)_2$ h. H_2SO_4 j. HNO_2

14. Calculate the equilibrium constant for the system: $P + 2\ Q \rightleftharpoons R + S$

given the chemical data: $2\ R + 2\ S \rightleftharpoons 4\ V + 2\ W$ $K_1 = 25$
$W + 2\ V \rightleftharpoons 2\ Q + P$ $K_2 = 0.040$

a. 2.0 c. 10 e. 40 g. 250 i. 500
b. 5.0 d. 25 f. 100 h. 400 j. 1000

15. The net ionic equation representing the equilibrium established when potassium sulfide dissolves in water is shown. Label the species present as either acid or base according to Brønsted-Lowry theory.

	S^{2-}	+ H_2O	$\rightleftharpoons$ HS^-	+ OH^-
a.	acid	base	acid	base
b.	acid	acid	base	base
c.	acid	base	base	acid
d.	base	acid	base	acid
e.	base	acid	acid	base
f.	base	base	acid	acid

16. Which of the following choices represents the net ionic equation for the reaction between lithium hydroxide and hydrofluoric acid in aqueous solution?

a. $LiOH + HF \rightleftharpoons LiF + H_2O$
b. $Li^+ + OH^- + HF \rightleftharpoons Li^+ + F^- + H_2O$
c. $LiOH + H^+ + F^- \rightleftharpoons Li^+ + F^- + H_2O$
d. $Li^+ + OH^- + H^+ + F^- \rightleftharpoons Li^+ + F^- + H_2O$
e. $LiOH + HF \rightleftharpoons Li^+ + F^- + H_2O$
f. $OH^- + HF \rightleftharpoons F^- + H_2O$

17. In the neutralization reaction between 0.20 M aqueous ammonia and 0.30 M aqueous hydrofluoric acid illustrated by the equation:

$NH_3 + HF \rightleftharpoons NH_4^+ + F^-$ $K = ?$

what is the value for the equilibrium constant?

K_b for ammonia = 1.8×10^{-5}
K_a for hydrofluoric acid = 6.8×10^{-4}

a. 1.2×10^{-8} c. 3.8×10^{1} e. 1.2×10^{6} g. 6.8×10^{10} i. 5.6×10^{-10}
b. 2.6×10^{-2} d. 8.2×10^{-7} f. 1.8×10^{9} h. 1.5×10^{-11} j. 3.7×10^{8}

18. Which statement distinguishes the Lewis theory of acid–base behavior from the theories of Arrhenius and Brønsted–Lowry?

a. Acids produce hydronium ions in aqueous solution
b. Strong acids are those acids stronger than the hydronium ion in aqueous soltion
c. The hydroxide ion is the strongest base that can exist in water
d. Acids are electron pair acceptors
e. Every base in aqueous solution has a conjugate acid
f. All acid-base reactions involve the transfer of a hydrogen ion

19. What is the approximate pH of an aqueous 10^{-8} M solution of nitric acid (a very dilute solution of a strong acid)?

a. 5.0 b. 6.0 c. 6.9 d. 7.1 e. 8.0 f. 9.0

Use the distribution diagram and acid ionization constants shown below for the diprotic H_2SO_3 to assist in answering the following questions (20 to 25).

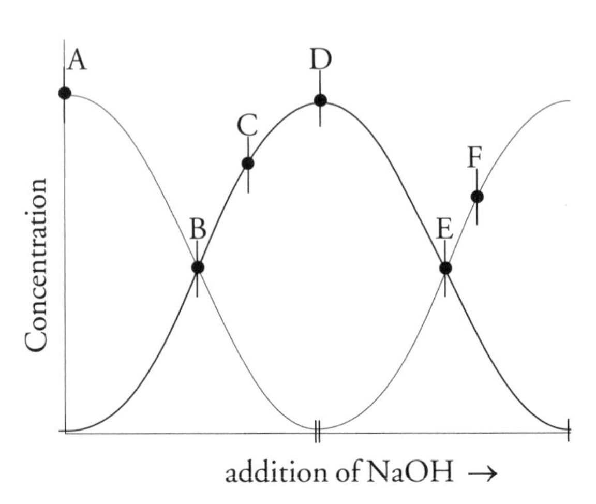

$H_2SO_3 + H_2O \rightleftharpoons H_3O^+ + HSO_3^-$ $K_{a1} = 1.3 \times 10^{-2}$ $pK_{a1} = 1.89$

$HSO_3^- + H_2O \rightleftharpoons H_3O^+ + SO_3^{2-}$ $K_{a2} = 6.3 \times 10^{-8}$ $pK_{a2} = 7.20$

20. 50 mL of a 0.1 M solution of sulfurous acid is titrated against a 0.1 M solution of sodium hydroxide. What is the pH of the solution at point A before any NaOH has been added?

a. 1.44 b. 1.52 c. 1.63 d. 1.89 e. 2.11 f. 2.37 g. 3.42 h. 3.89 i. 4.08 j. 4.55

21. What is the pH of the solution at point B halfway to the first equivalence point?

a. 1.44 b. 1.52 c. 1.63 d. 1.89 e. 2.11 f. 2.37 g. 3.42 h. 3.89 i. 4.08 j. 4.55

22. What is the pH of the solution at point C three quarters of the way to the first equivalence point?

a. 1.63 b. 1.89 c. 2.11 d. 2.37 e. 3.42 f. 3.89 g. 4.08 h. 4.55 i. 5.67 j. 6.13

23. What is the pH of the solution at the first equivalence point (point D)?

a. 1.63 b. 1.89 c. 2.11 d. 2.37 e. 3.42 f. 3.89 g. 4.08 h. 4.55 i. 5.67 j. 6.13

24. What is the pH of the solution at point E halfway to the second equivalence point when 75 mL of NaOH has been added?

a. 3.42 b. 3.89 c. 4.08 d. 4.55 e. 5.67 f. 6.13 g. 7.20 h. 7.50 i. 8.19 j. 8.67

25. What is the pH of the solution at point F two-thirds of the way to the second equivalence point when 83.3 mL of NaOH has been added?

a. 3.42 b. 3.89 c. 4.08 d. 4.55 e. 5.67 f. 6.13 g. 7.20 h. 7.50 i. 8.19 j. 8.67

EXAMINATION 1

CHEMISTRY 142 SPRING 2008

Wednesday February 13th 2008

1. Consider the rection

$$4\,PH_3(g) \rightarrow P_4(g) + 6\,H_2(g)$$

If $P_4(g)$ is formed at a rate of x Ms^{-1}, what are the corresponding rates for $PH_3(g)$ and $H_2(g)$?

	$\Delta[PH_3]/\Delta t$	$\Delta[H_2]/\Delta t$
a.	$4\,x$	$6\,x$
b.	$-4\,x$	$-6\,x$
c.	$-4\,x$	$6\,x$
d.	$(1/4)\,x$	$(1/6)\,x$
e.	$(1/4)\,x$	$-(1/6)\,x$
f.	$-(1/4)\,x$	$(1/6)\,x$

2. Which of the following statements about chemical reaction rates are true?

I Reactions are faster at higher temperature because the activation energies are lower.
II Rates increase with increasing concentrations of reactants because there are more collisions between reactant molecules.
III At higher temperature a larger fraction of the molecules have sufficient energy to get over the activation energy barrier.
IV Catalyzed and uncatalyzed reaction have identical mechanisms.

a. I and II
b. I and III
c. III and IV
d. II and III
e. II and IV
f. I, II, and III
g. All the statements are true

3. The following diagram represents the change in the concentration of the reactant as time passes for a zero-order, a first-order, and a second-order reaction. Which answer correctly identifies the graphs?

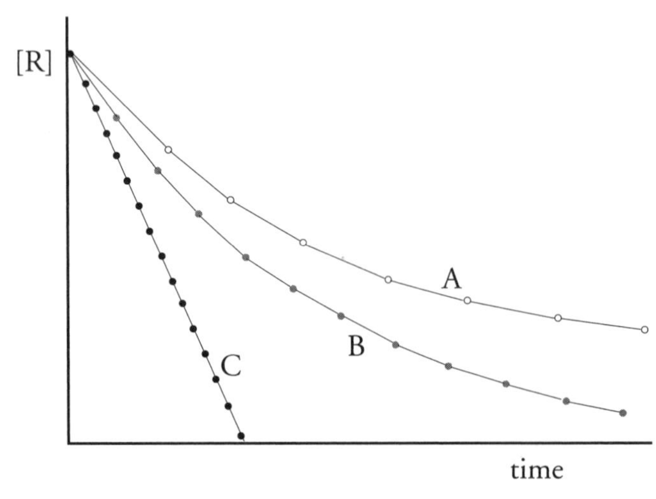

	zero-order	*first-order*	*second-order*
a.	A	B	C
b.	A	C	B
c.	B	A	C
d.	B	C	A
e.	C	A	B
f.	C	B	A

4. A reactant involved in a reaction in which its order is 2 (second-order) decreases in concentration from 16 mol L^{-1} to 4.0 mol L^{-1} in 300 seconds. How long does it take for its concentration to decrease from 4.0 mol L^{-1} to 1.0 mol L^{-1}?

   a. 100 s
   b. 200 s
   c. 300 s
   d. 6 min
   e. 8 min
   f. 10 min
   g. 20 min
   h. 30 min
   i. 45 min
   j. 1 hour

5. Initial rate data are given below for the reaction

$$NH_4^+(aq) + NO_2^-(aq) \rightarrow N_2(g) + 2\,H_2O(l)$$

Experiment	*Initial* $[NH_4^+]$	*Initial* $[NO_2^-]$	*Initial rate* $M\ s^{-1}$
1	0.24 M	0.10 M	7.2×10^{-6}
2	0.12 M	0.10 M	3.6×10^{-6}
3	0.12 M	0.15 M	5.4×10^{-6}

The rate equation for this reaction is

   a. Rate = $k[NH_4^+][NO_2^-]^{-1}$
   b. Rate = $k[NH_4^+][NO_2^-]$
   c. Rate = $k[NH_4^+]$
   d. Rate = $k[NH_4^+]^2[NO_2^-]$
   e. Rate = $k[NH_4^+]^2$
   f. Rate = $k[NH_4^+]^{-1}[NO_2^-]$
   g. Rate = $k[NO_2^-]$

6. Consider the following mechanism for a gas-phase reaction:

$$2\,NO \underset{k_{-1}}{\overset{k_1}{\rightleftharpoons}} N_2O_2 \qquad \textit{fast}$$

$$N_2O_2 + H_2 \xrightarrow{k_2} N_2O + H_2O \qquad \textit{slow}$$

$$N_2O + H_2 \xrightarrow{k_3} N_2 + H_2O \qquad \textit{fast}$$

Identify the reactants, products, and intermediates in this reaction.

	Reactants	*Products*	*Intermediates*
a.	NO, H_2, N_2O	N_2, H_2O	N_2O_2
b.	NO, H_2, N_2O, N_2O_2	N_2, H_2O	none
c.	NO, H_2	N_2, H_2O, N_2O	N_2O_2
d.	NO, H_2	N_2, H_2O	N_2O_2, N_2O
e.	NO, H_2, N_2O_2	N_2, H_2O	N_2O
f.	NO, H_2	N_2	N_2O_2, N_2O, H_2O

7. Consider again the reaction mechanism described in the previous question. The rate law based upon this mechanism is

   a. Rate = $k[N_2O][H_2]$
   b. Rate = $k[N_2O_2][H_2]$
   c. Rate = $k[NO][N_2O_2]$
   d. Rate = $k[NO]^2[H_2]$
   e. Rate = $k[NO]^{1/2}[H_2]$
   f. Rate = $k[NO][H_2]$
   g. Rate = $k[NO]^2$

8. The decomposition of formic acid (HCO_2H) at 550°C obeys the rate law: Rate = k[HCO_2H]. If a 5.00 atm sample of formic acid decays to 1.25 atm in 72 seconds, what is the half-life $t_{1/2}$ for the reaction?

a. 0.028 s
b. 6.0 s
c. 12 s
d. 24 s
e. 36 s
f. 48 s
g. 72 s
h. 144 s

9. The progress of a reaction from reactants to products is often represented by a reaction profile such as that shown on the right. Some of the labels on the diagram indicate thermodynamic properties of the system whereas other labels describe the kinetics of the reaction. Which are which?

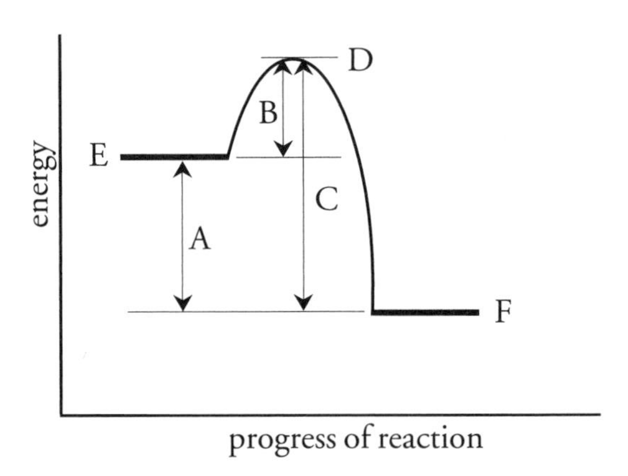

	thermodynamic	*kinetic*
a.	A B C	D E F
b.	E D F	A B C
c.	A E D F	B C
d.	E F	A B C D
e.	A E F	B C D
f.	A C E F	B D

10. Classify the solutes given below based on the type of aqueous solution that they form when dissolved in water. Choose the row where all the responses are correct.

	Acidic	*Basic*	*Neutral*
a.	NH_4Cl	KCN	LiBr
b.	KCN	$NaNO_3$	KCl
c.	NaF	KOH	Na_3PO_4
d.	CH_3NH_2	$NaNO_3$	NH_4I
e.	CH_3CO_2H	NaI	KCN
f.	KOH	CH_3NH_2	$NaNO_3$
g.	H_2SO_4	KCH_3CO_2	CH_3NH_2

11. Consider the reaction:

$$H_2O(g) + CH_4(g) \rightleftharpoons CO(g) + 3\,H_2(g) \qquad K = 4.7 \text{ at } 1400\text{ K}$$

If a mixture of reactants and products at 1400 K contains 0.035 M H_2O, 0.050 M CH_4, 0.15 M CO, and 0.30 M H_2, in which direction must the reaction proceed to reach equilibrium?

a. Toward products
b. Toward reactants
c. The system is already at equilibrium

12. What is the pH of a 1.5×10^{-4} M solution of potassium hydroxide at 25°C?

a. 3.82
b. 4.00
c. 4.18
d. 4.82
e. 9.18
f. 10.00
g. 10.18
h. 11.00

13. Methanol (CH_3OH) is manufactured from carbon monoxide and hydrogen in the presence of a ZnO/ Cr_2O_3 catalyst

$$CO(g) + 2H_2(g) \xrightleftharpoons{ZnO/Cr_2O_3} CH_3OH(g) \qquad \Delta H° = -91 \text{ kJ}$$

Suppose that you want to increase the yield of methanol and the following options are available:

I add CO
II increase temperature
III increase volume

Which of these changes to the system would increase the yield of methanol?

a. I
b. II
c. III
d. I and II
e. I and III
f. II and III
g. I, II, and III
h. none of them

14. The net ionic equation for equilibrium established when HN_3 is added to water is shown below. Classify the participants in the reaction based on the Brønsted-Lowry definitions for acids and bases.

	$HN_3(aq)$ +	$H_2O(l)$ ⇌	$N_3^-(aq)$ +	$H_3O^+(aq)$
a.	acid	base	acid	base
b.	acid	acid	base	base
c.	base	acid	acid	base
d.	base	base	acid	acid
e.	base	acid	base	acid
f.	acid	base	base	acid

15. What is the pH of a 0.20 M solution of the weak monoprotic acid formic acid (HCO_2H)?

$K_a = 1.8 \times 10^{-4}$

a. 2.23
b. 3.34
c. 4.45
d. 5.56
e. 6.67
f. 9.56
g. 10.10
h. 11.77

16. Consider the weak acids HSO_3^- ($K_a = 6.3 \times 10^{-8}$) and HNO_2 ($K_a = 4.5 \times 10^{-4}$). Identify the corresponding conjugate bases and rank them according to their base strength.

	stronger		*weaker*
a.	H_2SO_3	>	HNO_2^+
b.	H_2SO_3	>	NO_2^-
c.	SO_3^{2-}	>	NO_2^-
d.	NO_2^-	>	H_2SO_3
e.	NO_2^-	>	SO_3^{2-}
f.	SO_3^{2-}	>	HNO_2^+

17. At one point during the titration of formic acid (HCO_2H) *vs.* NaOH, it was found that the concentration of the acid was 0.066 M and the concentration of its conjugate base (HCO_2^-) was 0.022 M. What is the pH of the solution at this point?

    K_a for $HCO_2H = 1.8 \times 10^{-4}$

    a. 1.43   b. 1.87   c. 2.11   d. 2.89   e. 3.27   f. 3.75   g. 4.22   h. 4.56   i. 4.97   j. 5.61

18. Given the following information, calculate the equilibrium constant K for the reaction shown, illustrating the hydrolysis of the salt ammonium benzoate in aqueous solution at 25°C.

    $$NH_4^+ + Bz^- \rightleftharpoons NH_3 + HBz \qquad K = ?$$

    K_a for benzoic acid $= 6.5 \times 10^{-5}$
    K_b for ammonia $= 1.8 \times 10^{-5}$
    K_w for water $= 1.0 \times 10^{-14}$

    Bz^- = benzoate anion
    HBz = benzoic acid

    a. 1.17×10^{-23} b. 1.17×10^{-9} c. 2.77×10^{-1} d. 8.55×10^{-6} e. 3.61   f. 1.8×10^{9} g. 1.17×10^{5} h. 8.55×10^{-34}

19. Which two solutes mixed together in aqueous solution would *not* act as a buffer solution?

    a. H_2SO_3 and $KHSO_3$
    b. HF and NaF
    c. NH_3 and NH_4NO_3
    d. KBr and HBr
    e. KCN and HCN
    f. NH_4Cl and NH_3
    g. K_2CO_3 and $NaHCO_3$
    h. HNO_2 and $NaNO_2$

20. The distribution diagram and acid ionization constants are shown below for the diprotic weak acid selenous acid H_2SeO_3.

    $$H_2SeO_3 + H_2O \rightleftharpoons H_3O^+ + HSeO_3^- \qquad K_{a1} = 3.5 \times 10^{-2} \qquad pK_{a1} = 1.46$$
    $$HSeO_3^- + H_2O \rightleftharpoons H_3O^+ + SeO_3^{2-} \qquad K_{a2} = 5.0 \times 10^{-8} \qquad pK_{a2} = 7.30$$

    What is the predominant selenium-containing species in solution at point 2, and what is the pH?

	species present	*pH*
a.	H_2SeO_3	1.46
b.	H_2SeO_3	4.38
c.	H_2SeO_3	7.30
d.	$HSeO_3^-$	1.46
e.	$HSeO_3^-$	4.38
f.	$HSeO_3^-$	7.30
g.	SeO_3^{2-}	1.46
h.	SeO_3^{2-}	4.38
i.	SeO_3^{2-}	7.30

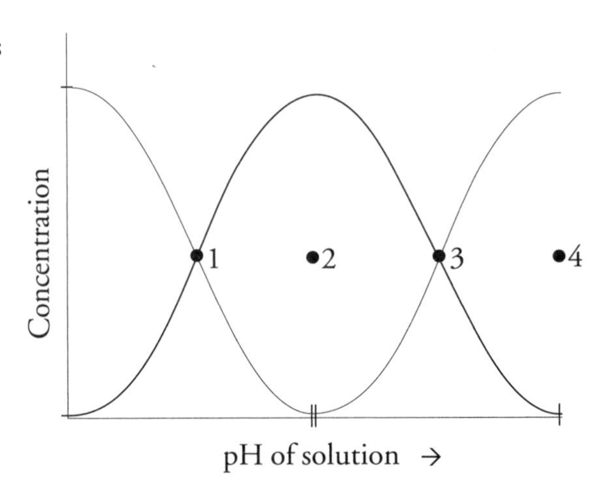

EXAMINATION 1

CHEMISTRY 142 FALL 2008

Thursday October 2nd 2008

1. Consider the unbalanced equation of A reacting with B to produce C:

$$\_\_A + \_\_B \rightarrow \_\_C$$

When C is formed at the rate of 0.086 M s^{-1}, A is consumed at a rate of 0.172 M s^{-1}, and B is consumed at a rate of 0.258 M s^{-1}. What is the balanced equation for this reaction?

a. $A + B \rightarrow C$
b. $2A + B \rightarrow C$
c. $2A + 3B \rightarrow 2C$
d. $A + 3B \rightarrow C$
e. $3A + 2B \rightarrow C$
f. $A + B \rightarrow 2C$
g. $A + 2B \rightarrow 3C$
h. $2A + 3B \rightarrow C$
i. $2A + 3B \rightarrow 4C$

2. Mercury(II) chloride reacts with the oxalate ion to produce the chloride ion, carbon dioxide and mercury(I) chloride:

$$2\,HgCl_2 + C_2O_4^{2-} \rightarrow 2\,Cl^- + 2\,CO_2 + Hg_2Cl_2$$

This reaction was studied in a series of experiments with different initial concentrations of reactant and the following initial rates were observed. What is the overall reaction order?

	$[HgCl_2]$	$[C_2O_4^{2-}]$	*Initial Rate*
Experiment 1	0.105 M	0.15 M	1.8×10^{-5} M min^{-1}
Experiment 2	0.105 M	0.30 M	7.1×10^{-5} M min^{-1}
Experiment 3	0.052 M	0.30 M	3.5×10^{-5} M min^{-1}

a. −4
b. −3
c. −2
d. −1
e. 0
f. 1
g. 2
h. 3
i. 4
j. 5

3. ^{32}P decomposes by a first-order mechanism and has a half-life of 14.3 days. How many milligrams of a 298 mg sample remains unreacted after 23.0 days?

a. 6.35 mg
b. 32.3 mg
c. 48.5 mg
d. 97.8 mg
e. 112 mg
f. 149 mg
g. 200 mg
h. 328 mg

4. A compound undergoing a zero-order decomposition decreases in concentration from 12.0 M to 3.0 M in one hour. How long does it take for the 3.0 M sample to decrease to 0.75 M?

a. 5 min
b. 15 min
c. 20 min
d. 40 min
e. 60 min
f. 75 min
g. 90 min
h. 120 min

5. Which of the following statements regarding a catalyst is/are true?

I. A catalyst must be in the same phase as the reactants.
II. A catalyst lowers the activation energy by changing the reaction mechanism.
III. A catalyst may be included in the rate equation.
IV. A catalyst is not consumed or changed in the reaction.

a. I only
b. II only
c. III only
d. IV only
e. II and IV
f. I and III
g. II, III, and IV
h. I, II, and IV
i. I, II, and III
j. all are true

6. The rate constant for the decomposition of nitrogen dioxide at 838 K is 0.0011 L mol^{-1} s^{-1}. At 1053 K, the same reaction has a rate constant of 1.67 L mol^{-1} s^{-1}. What is the activation energy for this reaction?

a. 0.28 kJ
b. 2.5 kJ
c. 17 kJ
d. 108 kJ
e. 250 kJ
f. 267 kJ
g. 283 kJ
h. 300 kJ

7. The mechanism for the reaction of chlorine and chloroform $CHCl_3$ is:

$Cl_2(g) \rightleftharpoons 2\,Cl(g)$ *fast*
$Cl(g) + CHCl_3(g) \rightarrow HCl(g) + CCl_3(g)$ *slow*
$CCl_3(g) + Cl(g) \rightarrow CCl_4(g)$ *fast*

What is the correct rate law for this reaction according to the mechanism?

a. Rate = $k[Cl_2]^{1/2}[CHCl_3]$
b. Rate = $k[Cl_2][CHCl_3]$
c. Rate = $k[Cl][CHCl_3]$
d. Rate = $k[Cl_2][Cl]^2$
e. Rate = $k[Cl_2][CCl_4]$
f. Rate = $k[HCl][CCl_4][Cl_2]^{-1}[CHCl_3]^{-1}$

8. What is (are) the intermediate(s) in the mechanism given in question 7?

a. Cl_2
b. Cl and CCl_3
c. Cl and $CHCl_3$
d. CCl_3 and Cl_2
e. HCl
f. CCl_4
g. HCl and CCl_4
h. Cl_2 and $CHCl_3$

9. What is the value of K_c for the following equilibrium at 25°C given the equilibrium concentrations shown?

$3\,F_2(g) + Cl_2(g) \rightleftharpoons 2\,ClF_3(g)$

$[F_2] = 2.0$ M
$[Cl_2] = 1.5$ M
$[ClF_3] = 3.0$ M

a. 0.00125
b. 0.00167
c. 1.22×10^{-7}
d. 3.62
e. 0.0312
f. 0.75
g. 449
h. 1.0

10. Calculate the equilibrium constant for:

$2\,NO(g) + 2\,BrCl(g) \rightleftharpoons 2\,NOBr(g) + Cl_2(g)$ K = ???

Given the following information:

$2\,NOBr(g) \rightleftharpoons 2\,NO(g) + Br_2(g)$ K = 0.014
$Br_2(g) + Cl_2(g) \rightleftharpoons 2\,BrCl(g)$ K = 7.2

a. 0.101
b. 6.34
c. 7.21
d. 9.92
e. 71.6
f. 98.4
g. 362
h. 514

11. 2.0 mol $H_2(g)$, 2.0 mol $I_2(g)$, and 12.0 mol $HI(g)$ were placed in a 2.0 L flask. The system was allowed to reach equilibrium. The value of the equilibrium constant at this temperature is 54.3. What is the concentration of H_2 in the flask at equilibrium?

$H_2(g) + I_2(g) \rightleftharpoons 2\,HI(g)$ $K_c = 54.3$

a. 0.146 M
b. 0.164 M
c. 0.362 M
d. 0.547 M
e. 0.700 M
f. 0.854 M
g. 0.962 M
h. 1.0 M

12. Which one of the following would you classify as a weak acid?

a. H_2SO_4 c. HI e. NaBr g. CdI_2 i. KF
b. NH_3 d. $Mg(OH)_2$ f. KNO_3 h. $HClO_4$ j. CH_3CO_2H

13. What is the pH of a 0.00213 M $Sr(OH)_2$ solution?

a. 11.63 c. 10.96 e. 8.23 g. 6.67
b. 11.33 d. 9.73 f. 7.42 h. 2.37

14. How many of the following molecules / ions can act as a Lewis acid?

H^+ NH_3 SbF_5 Ag^+
F^- Co^{3+} BF_3 Cl^-

a. 0 c. 2 e. 4 g. 6 i. 8
b. 1 d. 3 f. 5 h. 7

15. Consider the following Brønsted-Lowry equilibrium:

$$HOBr + H_2O \rightleftharpoons H_3O^+ + \_\_\_\_$$

Identify the missing product and decide if that product is acting as an acid or a base in this equilibrium.

a. HOBr base d. H_2OBr acid g. OH^- base
b. HOBr acid e. OBr^- base h. OH^- acid
c. H_2OBr base f. OBr^- acid

16. What is the K_b for $H_2PO_4^-$ at 25°C?

For H_3PO_4: $K_{a1} = 7.1 \times 10^{-3}$ $K_{a2} = 6.3 \times 10^{-8}$ $K_{a3} = 4.2 \times 10^{-13}$

a. 4.2×10^{-13} c. 6.3×10^{-8} e. 5.8×10^{-5} g. 0.013
b. 1.4×10^{-12} d. 1.6×10^{-7} f. 7.1×10^{-3} h. 0.024

17. What is the pH of a 0.150 M solution of sodium acetate? (K_a for acetic acid = 1.8×10^{-5})

a. 5.04 c. 8.96 e. 10.1 g. 11.4
b. 6.39 d. 9.37 f. 10.8 h. 12.7

18. A 0.1 M solution of glycine, a diprotic weak acid, was titrated with 0.1 M NaOH. What is the pH of the solution at the following points in the titration? *(Hint: Sketch the titration curve or the distribution curve!)*

$$^+H_3NCH_2CO_2H + H_2O \rightleftharpoons {}^+H_3NCH_2CO_2^- + H_3O^+ \quad K_{a1} = 4.57 \times 10^{-3}$$
$$^+H_3NCH_2CO_2^- + H_2O \rightleftharpoons H_2NCH_2CO_2^- + H_3O^+ \quad K_{a2} = 2.51 \times 10^{-10}$$

	$[^+H_3NCH_2CO_2^-] = [H_2NCH_2CO_2^-]$	$[^+H_3NCH_2CO_2^-] = 0.1M$
a.	pH = 2.34	pH = 5.97
b.	pH = 2.34	pH = 7.00
c.	pH = 5.97	pH = 7.00
d.	pH = 5.97	pH = 2.34
e.	pH = 9.60	pH = 5.97
f.	pH = 9.60	pH = 2.34

19. Referring to the titration of glycine described in question 18, what is the pH of the solution when $[H_3^+NCH_2COO^-] = 0.075$ M and $[H_2NCH_2COO^-] = 0.025$ M?

a. 1.86
b. 2.34
c. 4.27
d. 5.97
e. 7.00
f. 9.12
g. 10.07
h. 11.56

20. A 1.0 L solution contains 0.15 M NH_3 and 0.35 M NH_4Cl. What is the pH of the solution? (K_b for NH_3 = 1.8×10^{-5})

a. 13.7
b. 9.26
c. 8.89
d. 7.71
e. 5.11
f. 4.74
g. 3.50
h. 1.80

21. A 1.0 L solution contains 0.391 M HF and 0.293 M KF. What will happen if 0.147 moles HI are added to this solution? Assume the volume does not change upon the addition of HI.

a. The pH will increase slightly.
b. The pH will decrease slightly.
c. The buffer capacity is exceeded causing the pH to increase by several units.
d. The buffer capacity is exceeded causing the pH to decrease by several units.
e. There is no change in the pH.

22. What is the value of the neutralization constant K_{neut} in a titration of hydrocyanic acid with ammonia?

K_a for HCN $= 6.2 \times 10^{-10}$
K_b for NH_3 $= 1.8 \times 10^{-5}$

a. 1.12×10^{-4}
b. 2.90×10^{-10}
c. 3.44×10^{-5}
d. 1.81×10^{-5}
e. 4.40×10^{-4}
f. 8.01×10^{-3}
g. 1.12
h. 6.20×10^{4}

EXAMINATION 1

CHEMISTRY 142 SPRING 2009

Thursday February 19th 2009

1. Nitrogen dioxide decomposes to form nitric oxide and oxygen according to the equation:

$$2\,NO_2(g) \rightarrow 2\,NO(g) + O_2(g)$$

After 4 moles of NO_2 were placed in a 1.0 L flask, the concentrations of the reactants and products were monitored as a function of time. The graph shown on the right was obtained.
Identify the components A, B, and C:

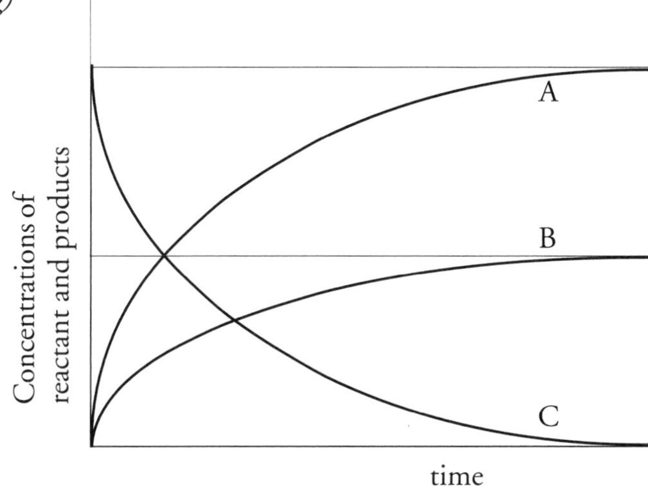

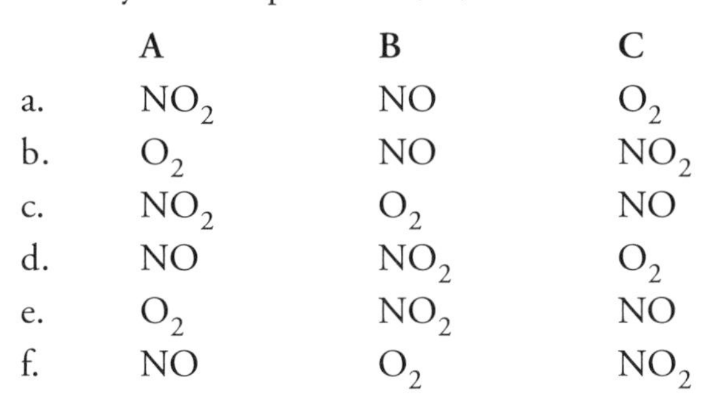

	A	B	C
a.	NO_2	NO	O_2
b.	O_2	NO	NO_2
c.	NO_2	O_2	NO
d.	NO	NO_2	O_2
e.	O_2	NO_2	NO
f.	NO	O_2	NO_2

2. Consider the following data for the decomposition of dinitrogen pentoxide N_2O_5 in the reaction:

$$2\,N_2O_5(g) \rightarrow 4\,NO_2(g) + O_2(g)$$

time, s	*concentration of dinitrogen pentoxide [$N_2O_5(g)$], M*
0	0.80
315	0.40
630	0.20
945	0.10

The reaction is ______ order and the rate constant k is ____________

a.	zeroth	315 s
b.	zeroth	1.3×10^{-3} M s^{-1}
c.	zeroth	7.4×10^{-4} M s^{-1}
d.	first	315 s
e.	first	2.2×10^{-3} s^{-1}
f.	first	945 s
g.	second	315 s
h.	second	4.0×10^{-3} M^{-1} s^{-1}
i.	second	1.6×10^{-2} M^{-1} s^{-1}

3. In the decomposition reaction of ozone

$$2\,O_3(g) \rightarrow 3\,O_2(g)$$

the experimentally determined rate law is: Rate = $k\,[O_3]^2\,[O_2]^{-1}$.
What is the overall reaction order, and what affect will the formation of O_2 have on the rate of reaction?

a. 2, increase c. 2, no effect e. 1, decrease g. 0, increase
b. 2, decrease d. 1, increase f. 1, no effect h. 0, no effect

4. Based upon the potential energy profile shown on the right, complete the following statements:

____ is the activation energy E_A.
____ is an intermediate.
Step ____ is the rate determining step.
The overall reaction is ____.

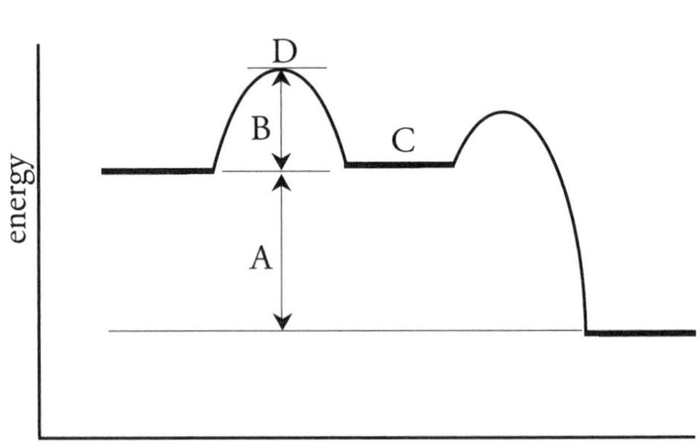

a. B C 2 exothermic
b. A D 1 exothermic
c. B C 2 endothermic
d. A D 1 endothermic
e. B C 1 exothermic
f. B D 2 endothermic
g. A D 2 endothermic
h. A C 1 exothermic

5. A compound undergoing a second-order decomposition decreases in concentration from 96 mol L^{-1} to 24 mol L^{-1} in 30 minutes. How long does it take for the 96 mol L^{-1} to decrease to 3.0 mol L^{-1}?

a. 10 min c. 30 min e. 80 min g. 160 min i. 310 min
b. 20 min d. 70 min f. 150 min h. 200 min j. 630 min

6. The decomposition of hydroxylamine NH_2OH in the presence of oxygen follows the rate law:

Rate = $k\,[NH_2OH][O_2]$ where $k = 2.64 \times 10^{-4}$ L mol^{-1} s^{-1} at 25°C
and $k = 20.1 \times 10^{-4}$ L mol^{-1} s^{-1} at 50°C.

Calculate the activation energy E_A.

a. –0.642 kJ mol^{-1} c. 0.642 kJ mol^{-1} e. 844 kJ mol^{-1} g. –65.0 kJ mol^{-1}
b. 0.844 kJ mol^{-1} d. 65.0 kJ mol^{-1} f. 8.33 kJ mol^{-1}

7. A solution made by dissolving 0.60 mol of chlorous acid $HClO_2$ in sufficient water to make 1.0 L has a pH of 1.102. What is the value of K_a for chlorous acid?

a. 1.0×10^{-2} c. 7.9×10^{-2} e. 1.5×10^{-1} g. 1.3×10^{-3}
b. 1.2×10^{-2} d. 1.3×10^{-1} f. 6.0×10^{-1} h. 1.5×10^{-2}

8. Consider the following two-step reaction mechanism:

$$Br_2(aq) + H_2O_2(aq) \rightarrow 2\,Br^-(aq) + 2\,H^+(aq) + O_2(g)$$

$$2\,Br^-(aq) + H_2O_2(aq) + 2H^+(aq) \rightarrow Br_2(aq) + 2\,H_2O(l)$$

Identify the reactant(s), product(s), intermediate(s), and catalyst(s).

	Reactant(s)	*Product(s)*	*Intermediate(s)*	*Catalyst(s)*
a.	Br_2, H_2O_2	Br_2, H_2O, O_2	Br^-, H^+	none
b.	H_2O_2	O_2, H_2O	Br^-, H^+	Br_2
c.	Br_2, H_2O_2, Br^-, H^+	Br_2, H_2O_2, Br^-, H^+, O_2, H_2O	none	none
d.	H_2O_2	O_2, H_2O	Br_2	Br^-, H^+
e.	H_2O_2	Br_2, H_2O, O_2	Br^-, H^+	none
f.	Br_2, H_2O_2	Br_2, H_2O	Br^-, H^+, O_2	none
g.	H_2O_2	H_2O	Br^-, H^+, O_2	Br

9. Label each of the species in the following aqueous equilibrium as either acid or base according to the Brønsted-Lowry theory.

	C_6H_5OH	+ H_2O	⇌ $C_6H_5O^-$	+ H_3O^+
a.	base	acid	acid	base
b.	base	acid	base	acid
c.	acid	base	base	acid
d.	acid	acid	base	base
e.	acid	base	acid	base
f.	base	base	acid	acid

10. Predict the pH of aqueous solutions containing the compounds given below.
Choose the row where all the responses are correct.

	LiCl	$NaCH_3CO_2$	NH_4I	KSCN
a.	7.0	> 7.0	< 7.0	7.0
b.	> 7.0	< 7.0	> 7.0	7.0
c.	7.0	> 7.0	< 7.0	> 7.0
d.	7.0	> 7.0	> 7.0	> 7.0
e.	> 7.0	7.0	7.0	7.0
f.	< 7.0	> 7.0	< 7.0	7.0
g.	< 7.0	> 7.0	< 7.0	> 7.0

11. All the following aqueous reactions have an equilibrium constant K > 1.

$$IO^- + CH_3CH_2CO_2H \rightleftharpoons HIO + CH_3CH_2CO_2^-$$

$$IO^- + C_6H_5NH_3^+ \rightleftharpoons HIO + C_6H_5NH_2$$

$$C_6H_5NH_2 + CH_3CH_2CO_2H \rightleftharpoons C_6H_5NH_3^+ + CH_3CH_2CO_2^-$$

Identify the three bases in these equilibria and then order them as the strongest, intermediate, and weakest base. Choose the row in which all responses are correct.

	Strongest Base	*Intermediate Base*	*Weakest Base*
a.	$CH_3CH_2CO_2^-$	$C_6H_5NH_2$	IO^-
b.	IO^-	$CH_3CH_2CO_2^-$	$C_6H_5NH_2$
c.	$C_6H_5NH_2$	IO^-	$CH_3CH_2CO_2^-$
d.	HIO	$C_6H_5NH_3^+$	$CH_3CH_2CO_2H$
e.	$CH_3CH_2CO_2H$	$C_6H_5NH_3^+$	HIO
f.	$C_6H_5NH_3^+$	IO^-	$CH_3CH_2CO_2^-$
g.	$CH_3CH_2CO_2H$	HIO	$C_6H_5NH_3^+$
h.	IO^-	$C_6H_5NH_2$	$CH_3CH_2CO_2^-$

12. Which of the following would be the best choice for preparing a pH 5.0 buffer?

a. NH_4Cl and NH_3 ($K_b = 1.8 \times 10^{-5}$)
b. H_2CO_3 and $NaHCO_3$ ($K_{a1} = 4.3 \times 10^{-7}$)
c. KH_2PO_4 and K_2HPO_4 ($K_{a1} = 7.52 \times 10^{-3}$, $K_{a2} = 6.23 \times 10^{-8}$, $K_{a3} = 2.2 \times 10^{-13}$)
d. HF and LiF ($K_a = 6.6 \times 10^{-4}$)
e. $NaCH_3CO_2$ and CH_3CO_2H ($K_a = 1.76 \times 10^{-5}$)

Questions 13 and 14 refer to the following equilibrium:

$$N_2(g) + 3\ H_2(g) \rightleftharpoons 2\ NH_3(g) \qquad K_p = 1.9 \times 10^{-4} \text{ at } 400°C$$

13. Suppose that a reaction mixture contains 65 atm of N_2, 36 atm H_2, and 22 atm NH_3. In which direction must the reaction proceed in order to reach equilibrium?

a. toward reactants
b. toward products
c. neither direction, the system is already at equilibrium

14. According to LeChatelier's Principle, in which direction will the equilibrium shift if the specified changes are made? Choose the row in which all answers are correct.

	add N_2	*reduce the volume*	*add a catalyst*	*remove H_2*
a.	right	right	right	left
b.	right	left	right	left
c.	left	left	left	right
d.	left	right	right	right
e.	right	right	no effect	left
f.	left	left	no effect	right
g.	right	left	no effect	left
h.	right	right	left	right

15. The K_a for HCO_3^- is 5.6×10^{-11}. Identify the conjugate base of HCO_3^-, and determine its K_b.

a. CO_3^{2-} 1.0×10^{-7}
b. CO_3^{2-} 1.8×10^{-4}
c. CO_3^{2-} 5.6×10^{4}
d. H_2CO_3 1.0×10^{-7}
e. H_2CO_3 1.8×10^{-4}
f. H_2CO_3 5.6×10^{4}

16. The ionization constant K_a for hydrazoic acid HN_3 is 2.2×10^{-5}.
What is the equilibrium constant for the reaction:

$$HN_3(aq) + OH^-(aq) \rightleftharpoons N_3^-(aq) + H_2O(l)$$

a. 2.2×10^{-19}
b. 4.5×10^{-10}
c. 2.2×10^{-5}
d. 4.5×10^{4}
e. 2.2×10^{9}
f. 4.5×10^{18}
g. 7.2×10^{-5}
h. 6.8×10^{8}

17. Consider the reaction: $2\,SO_2(g) + O_2(g) \rightleftharpoons 2\,SO_3(g)$ $K_p = 2.5 \times 10^{10}$ (T = 500 K)
What is the equilibrium constant for the reaction:

$$SO_3(g) \rightleftharpoons SO_2(g) + \tfrac{1}{2}\,O_2(g) \qquad K_p = ?? \qquad (T = 500\text{ K})$$

a. 1.6×10^{-21}
b. 4.0×10^{-11}
c. 6.3×10^{-6}
d. 1.6×10^{5}
e. 2.5×10^{10}
f. 6.25×10^{20}
g. 5.0×10^{2}
h. 2.0×10^{2}

18. For which of the following gas-phase reactions does $K_p = K_c(RT)^{-1}$

a. $2\,NH_3 + Cl_2 \rightleftharpoons N_2H_4 + 2\,HCl$
b. $C_2H_2 + 3\,H_2 \rightleftharpoons 2\,CH_4$
c. $CO + Cl_2 \rightleftharpoons COCl_2$
d. $PCl_5 \rightleftharpoons PCl_3 + Cl_2$
e. $H_2 + CO_2 \rightleftharpoons H_2O + CO$
f. $3\,F_2 + Cl_2 \rightleftharpoons 2\,ClF_3$

19. How many of the following substances can act as a Lewis base?

H_2O H^+ NH_3 F^- BF_3 Fe^{2+}

a. 0 b. 1 c. 2 d. 3 e. 4 f. 5 g. 6

20. A buffer solution is prepared by combining equal volumes of 0.50 M HClO and 0.40 M NaClO. The acid ionization constant K_a of HClO is 3.5×10^{-8}. What is the pH of the solution?

a. 6.44 c. 6.64 e. 7.46
b. 6.54 d. 7.36 f. 7.56

Questions 21 and 22 refer to the distribution diagram and acid ionization constants provided below for oxalic acid $H_2C_2O_4$, a weak diprotic acid.

$H_2C_2O_4 + H_2O \rightleftharpoons HC_2O_4^- + H_3O^+$ $K_{a1} = 6.5 \times 10^{-2}$ $pK_{a1} = 1.19$
$HC_2O_4^- + H_2O \rightleftharpoons C_2O_4^{2-} + H_3O^+$ $K_{a2} = 6.1 \times 10^{-5}$ $pK_{a2} = 4.21$

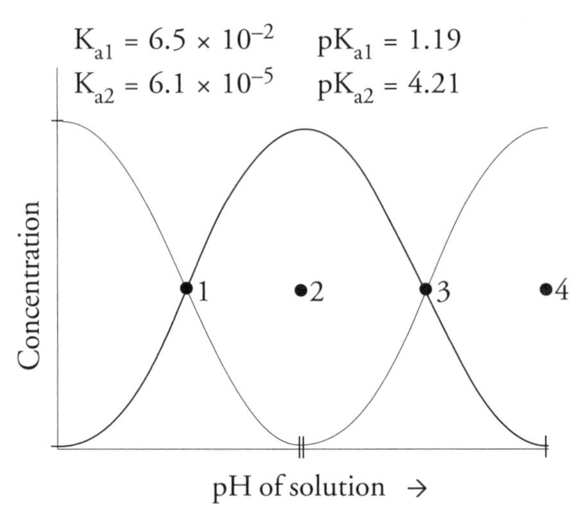

21. At point 1 in the distribution diagram the dominant carbon-containing specie(s) present in solution is/are ____, and the pH is ____.

	Species present	*pH*		*Species present*	*pH*
a.	$H_2C_2O_4$	1.19	f.	$H_2C_2O_4$, $HC_2O_4^-$	4.21
b.	$H_2C_2O_4$	2.70	g.	$H_2C_2O_4$, $HC_2O_4^-$, $C_2O_4^{2-}$	1.19
c.	$H_2C_2O_4$	4.21	h.	$H_2C_2O_4$, $HC_2O_4^-$, $C_2O_4^{2-}$	2.70
d.	$H_2C_2O_4$, $HC_2O_4^-$	1.19	i.	$H_2C_2O_4$, $HC_2O_4^-$, $C_2O_4^{2-}$	4.21
e.	$H_2C_2O_4$, $HC_2O_4^-$	2.70			

22. 50 mL of a 0.1 M solution of oxalic acid is titrated with 0.1 M NaOH. What is(are) the predominant carbon containing specie(s) present after 75 mL of NaOH has been added?

a. $H_2C_2O_4$ d. $HC_2O_4^-$ g. $C_2O_4^{2-}$
b. $H_2C_2O_4$, $HC_2O_4^-$ e. $HC_2O_4^-$, $C_2O_4^{2-}$
c. $H_2C_2O_4$, $HC_2O_4^-$, $C_2O_4^{2-}$ f. $H_2C_2O_4$, $C_2O_4^{2-}$

EXAMINATION 1

CHEMISTRY 142 FALL 2009

Tuesday October 13th 2009

1. The first step in the synthesis of nitric acid involves the oxidation of ammonia at high temperature in the gas phase:

$$4\,NH_3 + 5\,O_2 \rightarrow 4\,NO + 6\,H_2O$$

At one stage in this reaction, water is produced at a rate of 30 mol $L^{-1}min^{-1}$. At what rate is ammonia used up at this stage?

a. 12 mol $L^{-1}min^{-1}$
b. 18 mol $L^{-1}min^{-1}$
c. 20 mol $L^{-1}min^{-1}$
d. 25 mol $L^{-1}min^{-1}$
e. 36 mol $L^{-1}min^{-1}$
f. 45 mol $L^{-1}min^{-1}$

2. The decomposition of N_2O_5 is shown in the following balanced equation:

$$2\,N_2O_5 \rightarrow 4\,NO_2 + O_2$$

In one experiment, the initial concentration of N_2O_5 was 3.15 M and the initial rate of reaction was 5.45×10^{-5} M s^{-1}. In another experiment, the initial concentration of N_2O_5 was 0.78 M and the initial rate of reaction was 1.35×10^{-5} M s^{-1}. What is the overall order of this decomposition reaction?

a. −1
b. 0
c. 0.5
d. 1
e. 2
f. 3
g. 4
h. 5

3. At high temperatures ethylene oxide $(CH_2)_2O$ decomposes by a first-order mechanism and has a half-life of 56.3 minutes. How many grams of a 5.0 gram sample of ethylene oxide remains after 4.0 hours?

a. 0.012 g
b. 0.052 g
c. 0.26 g
d. 0.39 g
e. 0.78 g
f. 0.95 g
g. 2.9 g
h. 4.8 g

4. Butadiene C_4H_6 can dimerize to form C_8H_{12}. This dimerization reaction obeys the rate law:

$$\text{Rate} = k[C_4H_6]^2$$

It takes 1630 seconds for a 0.010 M C_4H_6 sample to decrease to a concentration of 0.005 M. How long does it take the 0.005 M sample to decrease to 0.00125 M?

a. 1630 s
b. 2850 s
c. 3260 s
d. 4890 s
e. 6520 s
f. 7770 s
g. 8000 s
h. 9780 s

5. Which of the following would be expected to increase the rate of a reaction?

A. increase the concentration of the reactants
B. lower the activation energy
C. lower the reaction temperature
D. add a catalyst

a. A only
b. B only
c. C only
d. D only
e. A and B
f. A and D
g. B and D
h. A, B, and D
i. B and C
j. A, B, C, and D

6. The colorless gas N_2O_4 decomposes to the brown gas NO_2 in a first-order reaction:

$$N_2O_4(g) \rightleftharpoons 2\ NO_2(g)$$

At 274 K the rate constant k = $4.5 \times 10^3\ s^{-1}$ and at 283K the rate constant k = $1.00 \times 10^4\ s^{-1}$. What is the activation energy for this reaction?

a. 0.56 kJ mol^{-1}
b. 0.74 kJ mol^{-1}
c. 4.3 kJ mol^{-1}
d. 21 kJ mol^{-1}
e. 36 kJ mol^{-1}
f. 57 kJ mol^{-1}
g. 62 kJ mol^{-1}
h. 74 kJ mol^{-1}

7. Urea $(NH_2)_2CO$ can be prepared by heating ammonium cyanate NH_4OCN. A proposed mechanism for this reaction is:

$NH_4OCN \rightleftharpoons NH_3 + HOCN$ *fast*

$NH_3 + HOCN \rightarrow (NH_2)_2CO$ *slow*

What is the rate law predicted by this mechanism?

a. Rate = $k[NH_4OCN]$
b. Rate = $k[(NH_2)_2CO]$
c. Rate = $k[NH_3]^2$
d. Rate = $k[NH_3][HOCN]$
e. Rate = $k[NH_3][HOCN][NH_4OCN]^{-1}$
f. Rate = $k[NH_3][NH_4OCN]$

8. Consider the following reaction mechanism:

$NO_2Cl(g) \rightleftharpoons NO_2(g) + Cl(g)$

$Cl(g) + NO_2Cl(g) \rightarrow NO_2(g) + Cl_2(g)$

What is the overall chemical equation?

a. $Cl(g) + NO_2Cl(g) \rightarrow NO_2(g) + Cl_2(g)$
b. $2\ Cl(g) + 2\ NO_2Cl(g) \rightarrow 2\ NO_2(g) + Cl_2(g)$
c. $2\ NO_2Cl(g) \rightarrow 2\ NO_2(g) + Cl_2(g)$
d. $Cl(g) + 2\ NO_2Cl(g) \rightarrow 2\ NO_2(g) + 3/2\ Cl_2(g)$
e. $NO_2Cl(g) \rightarrow NO_2(g) + Cl(g)$
f. $2\ Cl(g) + NO_2Cl(g) \rightarrow 2\ NO_2(g) + 3/2\ Cl_2(g)$

9. Calculate the equilibrium constant K for the following equilibrium from the concentrations of the various components at equilibrium shown:

$$N_2(g) + 3\ H_2(g) \rightleftharpoons 2\ NH_3(g)$$

concentrations at equilibrium:
$[N_2(g)]$ = 0.10 mol L^{-1}
$[H_2(g)]$ = 0.10 mol L^{-1}
$[NH_3(g)]$ = 0.002 mol L^{-1}

a. 0.02
b. 0.04
c. 0.2
d. 0.5
e. 2
f. 5
g. 20
h. 25

10. Calculate the equilibrium constant for: $1/2\ H_2(g) + 1/2\ I_2(g) \rightleftharpoons HI(g)$ K = ???
given the following information: $2\ HI(g) \rightleftharpoons H_2(g) + I_2(g)$ K = 1.97×10^{-2}

a. -9.85×10^{-3}
b. 1.97×10^{-2}
c. 0.14
d. 3.86
e. 7.12
f. 26
g. 51
h. 2580

11. Consider the following disturbances on the equilibrium system shown. In which, if any, direction will the system shift to restore equilibrium in each case?

$$3\ Fe(s) + 4\ H_2O(g) \rightleftharpoons Fe_3O_4(s) + 4\ H_2(g) \qquad \Delta H = -149\ kJ$$

	increase temperature	*decrease volume*	*remove Fe(s)*	*add a catalyst*
a.	right	no change	left	no change
b.	left	no change	no change	right
c.	left	left	left	no change
d.	right	right	right	left
e.	left	no change	left	no change
f.	left	no change	no change	no change

12. The following system has an equilibrium constant $K_c = 2.18 \times 10^6$. If a 12.0 L flask initially contains 3.2 moles of HBr(*g*) and no $H_2(g)$ or $Br_2(g)$, What is the concentration of $H_2(g)$ once equilibrium is established?

$$H_2(g) + Br_2(g) \rightleftharpoons 2\ HBr(g) \qquad K_c = 2.18 \times 10^6$$

a. 1.8×10^{-4} M
b. 1.4×10^{-3} M
c. 0.27 M
d. 0.34 M
e. 0.99 M
f. 15 M

13. Which of the following statements is/are correct?

I. The pH of water is always 7.0.
II. The lower the pH, the more acidic the solution.
III. The strongest acid that can exist in aqueous solution is the hydronium ion.
IV. The pH of a 1.0×10^{-8} M solution of HNO_3 is 8.

a. I only
b. II only
c. III only
d. IV only
e. II and III
f. II and IV
g. III and IV
h. I, II, and III
i. II, III, and IV
j. I, II, III, and IV

14. What is the pH of a 0.013 M aqueous LiOH solution?

a. 1.9
b. 3.6
c. 5.5
d. 7.0
e. 9.7
f. 10.2
g. 12.1
h. 14.0

15. How many of the following substances can act as a Lewis base?

F^- $\quad NH_3 \quad$ CO $\quad BF_3 \quad Co^{3+} \quad H_2O$

a. 0
b. 1
c. 2
d. 3
e. 4
f. 5
g. 6

16. The K_{a1} for boric acid H_3BO_3 is 5.8×10^{-10} at 25°C. What is the K_b for $H_2BO_3^-$?

a. 5.8×10^{-24}
b. 1×10^{-14}
c. 1.7×10^{-5}
d. 2.4×10^{-5}
e. 3.6×10^2
f. 5.8×10^4

17. Decide if aqueous solutions of the following compounds will be acidic, basic, or neutral. Choose the row in which all answers are correct.

	NH_4NO_3	KCH_3CO_2	HClO	LiCl
a.	basic	basic	acidic	neutral
b.	acidic	basic	acidic	neutral
c.	neutral	acidic	acidic	basic
d.	acidic	acidic	neutral	neutral
e.	basic	neutral	basic	neutral
f.	acidic	acidic	acidic	neutral

18. What is the pH of a 0.3 M aqueous solution of KCN? The K_a for HCN is 6.2×10^{-10}.

a. 2.2 c. 6.2 e. 9.4 g. 12.6
b. 2.7 d. 7.0 f. 11.3 h. 13.9

19. A 0.2 M solution of carbonic acid H_2CO_3 was titrated with 0.2 M NaOH.
For carbonic acid, $K_{a1} = 4.3 \times 10^{-7}$ and $K_{a2} = 5.6 \times 10^{-11}$.
What is the pH of the solution when the concentration of HCO_3^- is equal to 0.2 M?

a. 0.70 c. 6.37 e. 10.25
b. 3.53 d. 8.31 f. 13.30

20. In the titration described in question 19, what is the pH of the solution when $[HCO_3^-] = [CO_3^{2-}]$?

a. 0.70 c. 6.37 e. 10.25
b. 3.53 d. 8.31 f. 13.30

21. A solution was prepared by combining equal volumes of 0.85 M formic acid HCOOH and 1.4 M sodium formate HCOONa. The K_a for formic acid is 1.8×10^{-4}. What is the pH of the solution?

a. 3.14 c. 3.46 e. 3.67 g. 3.89
b. 3.33 d. 3.52 f. 3.74 h. 3.96

22. K_a values for several weak acids are shown below. Which acid is the strongest acid? Which of the acids will have the strongest conjugate base? Choose the row in which both questions are answered correctly.

benzoic acid	$K_a = 6.5 \times 10^{-5}$	hypoiodous acid	$K_a = 2.0 \times 10^{-11}$
hydroazoic acid	$K_a = 2.2 \times 10^{-5}$	thiocyanic acid	$K_a = 1.3 \times 10^{-1}$
hydrocyanic acid	$K_a = 6.2 \times 10^{-10}$		

	strongest acid	*acid with strongest conjugate base*
a.	thiocyanic acid	hypoiodous acid
b.	benzoic acid	hydroazoic acid
c.	hydrocyanic acid	thiocyanic acid
d.	thiocyanic acid	hydrocyanic acid
e.	hypoiodous acid	thiocyanic acid
f.	hypoiodous acid	hypoiodous acid
g.	hydroazoic acid	benzoic acid

EXAMINATION 1

CHEMISTRY 142 SPRING 2010

Thursday February 18th 2010

1. Propane burns in air to form carbon dioxide and water according to the equation:

$$C_3H_8(g) \ + \ 5\,O_2(g) \ \rightarrow \ 3\,CO_2(g) \ + \ 4\,H_2O(g)$$

After 2.0 moles of C_3H_8 and 10.0 moles of oxygen were placed in a 1.0 L flask, the concentrations of the reactants and products were monitored as a function of time. The graph shown on the right was obtained.
Identify the components A, B, C, and D:

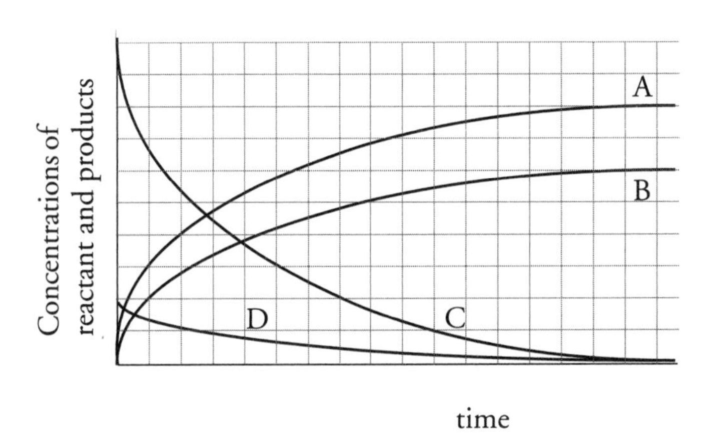

	A	B	C	D
a.	O_2	CO_2	H_2O	C_3H_8
b.	O_2	C_3H_8	H_2O	CO_2
c.	CO_2	O_2	H_2O	C_3H_8
d.	CO_2	H_2O	O_2	C_3H_8
e.	H_2O	CO_2	O_2	C_3H_8
f.	H_2O	C_3H_8	O_2	CO_2
g.	C_3H_8	O_2	H_2O	CO_2
h.	C_3H_8	H_2O	O_2	CO_2

2. The following decomposition of the hypochlorite ion is observed in aqueous solution:

$$3\,ClO^-(aq) \ \rightarrow \ ClO_3^-(aq) \ + \ 2\,Cl^-(aq)$$

The rate law for the reaction is: Rate = $k[ClO^-]^2$

What is the overall order for the reaction and what would happen to the rate if the concentration of the hypochlorite ion was tripled?

	order	rate
a.	1	× 1
b.	1	× 3
c.	1	× 6
d.	2	× 1
e.	2	× 3
f.	2	× 8
g.	2	× 9
h.	3	× 9
i.	3	× 27

3. The concentration of a reactant in a reaction changed as follows as the reaction proceeded:

At time zero (initial concentration)	144 mol L^{-1}
After 12 minutes	36 mol L^{-1}
After *another* 16 minutes	18 mol L^{-1}
One hour after the start of the reaction	9 mol L^{-1}

What is the order of the reaction with respect to this reactant?

a. 0 b. 1 c. 2 d. 3 e. 4 f. 5

4. The reaction of a substance R with another substance Q to form P was investigated in a series of experiments with different initial concentrations of the two reactants. The following initial rates were observed. What is the overall reaction order?

	Initial [R]	*Initial [Q]*	*Initial rate of formation of P*
Experiment 1	0.10 M	0.20 M	3.0×10^{-2} mol L^{-1} s^{-1}
Experiment 2	0.20 M	0.20 M	6.0×10^{-2} mol L^{-1} s^{-1}
Experiment 3	0.30 M	0.40 M	3.6×10^{-1} mol L^{-1} s^{-1}
Experiment 4	0.20 M	0.40 M	2.4×10^{-1} mol L^{-1} s^{-1}

a. 4 c. 2 e. 0 g. −2 i. −4
b. 3 d. 1 f. −1 h. −3 j. −5

5. Calculate the activation energy, E_A, of a first-order reaction if the rate constant k doubles when the temperature is increased from 350°C to 372°C.

a. 1.05 kJ mol^{-1} c. 3.41 kJ mol^{-1} e. 7.12 kJ mol^{-1}
b. 34.1 kJ mol^{-1} d. 71.2 kJ mol^{-1} f. 105 kJ mol^{-1}

6. A radioactive isotope decomposes via a first-order reaction. What is the half-life (in minutes) of the isotope if 75% of the isotope decomposed in 1.3 hours?

a. 3.1 min c. 28 min e. 78 min
b. 18 min d. 39 min f. 188 min

7. The principal reason why the rate of a reaction increases when the temperature is increased is that

a. the rate of collisions increases
b. the molecules collide with greater energy
c. the molecules move faster
d. the activation energy for the forward reaction decreases
e. the equilibrium moves toward the product
f. the activation energy for the reverse reaction increases

8. Consider the following mechanism for the reduction of nitric oxide by hydrogen to form nitrogen and water:

$2\,NO(g) \rightleftharpoons N_2O_2(g)$ *Fast*

$N_2O_2(g) + H_2(g) \rightarrow N_2O(g) + H_2O(g)$ *Slow*

$N_2O(g) + H_2(g) \rightarrow N_2(g) + H_2O(g)$ *Fast*

What is(are) reaction intermediate(s) in this reaction mechanism?

a. NO *and* N_2O
b. N_2O *and* N_2O_2
c. N_2O_2
d. N_2O *and* N_2
e. N_2O *and* H_2
f. N_2, N_2O, *and* N_2O_2

9. Calculate the equilibrium constant K_c for the following equilibrium if the concentrations of the various components at equilibrium are as shown:

$$3\,F_2(g) + Cl_2(g) \rightleftharpoons 2\,ClF_3(g)$$

Concentrations at equilibrium:
$[F_2(g)] = 2.0$ mole/L
$[Cl_2(g)] = 3.0$ mole/L
$[ClF_3(g)] = 6.0$ mole/L

a. 1.0
b. 1.33
c. 1.5
d. 2.0
e. 2.4
f. 2.5
g. 3.0
h. 4.0
i. 4.5
j. 6.0

10. For the reaction illustrated in the previous question, what is the relation between K_p and K_c at 25°C?

a. K_p is larger than K_c
b. K_p equals K_c
c. K_p is smaller than K_c

11. The following system has an equilibrium constant $K_c = 0.0081$. If a 5.0 L flask initially contains 0.20 mole of $HBr(g)$ and no $H_2(g)$ or $Br_2(g)$, what is the concentration of $H_2(g)$ when equilibrium is established?

$$2\,HBr(g) \rightleftharpoons H_2(g) + Br_2(g)$$

a. 0.003 M
b. 0.012 M
c. 0.015 M
d. 0.025 M
e. 0.032 M
f. 0.091 M
g. 0.121 M
h. 0.137 M

12. Consider the following disturbances on the equilibrium system shown. In which, if any, direction will the system shift to restore equilibrium in each case?

$$CO(g) + 2H_2(g) \rightleftharpoons 2\,CH_3OH(g) \qquad \Delta H° = -91\text{ kJ}$$

	decrease temperature	*increase volume*	*add CO(g)*	*add a catalyst*
a.	right	right	right	no change
b.	left	no change	left	right
c.	right	left	right	right
d.	right	right	right	left
e.	left	right	left	right
f.	left	left	right	no change
g.	right	left	right	no change

13. According to the Brønsted-Lowry definitions of acids and bases, the conjugate base of the dihydrogen phosphate ion $H_2PO_4^-$ is

a. H_3PO_4 b. PO_4^{3-} c. HPO_4^{2-} d. $H_2PO_4^-$ e. OH^-

14. Which statement is correct?

a. When salts dissolve in water, the solution is always basic.
b. A strong acid in aqueous solution can always be made stronger by heating the solution.
c. The higher the pH, the more acidic the solution.
d. Lewis acids donate pairs of electrons.
e. The strongest base that can exist in aqueous solution is the hydride ion H^-.
f. If the concentration of acid is very high, the acid is said to be strong.
g. A neutral aqueous solution is defined as one in which $[H_3O^+] = [OH^-]$.
h. The pH of a 1.0×10^{-8} M solution of HNO_3 is 8 (i.e. basic).

15. At 40°C, the K_w for pure water is 3.8×10^{-14}. Calculate the pH of pure water at 40°C, and determine whether the water is acidic, basic, or neutral.

	pH	acidity		pH	acidity
a.	6.7	acidic	d.	7.3	neutral
b.	6.7	neutral	e.	7.3	basic
c.	7.0	neutral	f.	13.4	basic

16. The weak base pyridine (C_5H_5N) has an ionization constant K_b equal to 1.7×10^{-9}. Calculate the pH of a 0.050 M solution of pyridine.

a. 1.30 c. 5.23 e. 8.96
b. 5.04 d. 8.77 f. 12.7

17. How many of the following electrolytes are classified as strong acids or strong bases?

H_2SO_4	HNO_3	$(NH_4)_2S$	KNO_2
$Mg(OH)_2$	HClO	HF	NH_3
HI	CsF	NaOH	HNO_2

a. 1 b. 2 c. 3 d. 4 e. 5 f. 6 g. 7 h. 8

18. The net ionic equation representing the equilibrium established when lithium hydrogen sulfide dissolves in water is shown. Label the species present as either acid or base according to Brønsted-Lowry theory.

	HS^-	+	H_2O	$\rightleftharpoons$	H_2S	+	OH^-
a.	acid		base		acid		base
b.	acid		acid		base		base
c.	acid		base		base		acid
d.	base		acid		base		acid
e.	base		acid		acid		base
f.	base		base		acid		acid

19. The acid ionization constants for the diprotic H_2SO_3 are shown below:

$H_2SO_3 + H_2O \rightleftharpoons H_3O^+ + HSO_3^-$ $K_{a1} = 1.3 \times 10^{-2}$ $pK_{a1} = 1.89$

$HSO_3^- + H_2O \rightleftharpoons H_3O^+ + SO_3^{2-}$ $K_{a2} = 6.3 \times 10^{-8}$ $pK_{a2} = 7.20$

What is the equilibrium constant for the following system that illustrates what happens when potassium hydrogen sulfite dissolves in water?

$HSO_3^- + HSO_3^- \rightleftharpoons H_2SO_3 + SO_3^{2-}$ K = ?

a. 8.2×10^{-10}
b. 6.3×10^{-8}
c. 4.8×10^{-6}
d. 9.2×10^{-4}
e. 3.3×10^{-2}
f. 8.2×10^{4}
g. 4.8×10^{-20}
h. 2.1×10^{5}

20. Based upon the data provided in the previous question, what is the pH of a 0.10 M solution of potassium hydrogen sulfite?

a. 1.89
b. 3.23
c. 4.10
d. 4.55
e. 6.51
f. 7.20
g. 9.09
h. 10.23

21. Which one of the following equimolar solutions in water will act as a buffer solution?

a. KOH and KCl
b. Na_2CO_3 and K_2CO_3
c. LiOH and LiH
d. NaF and NaCl
e. KCN and HCN
f. H_2SO_4 and $NaHSO_4$
g. NaF and NaCl

22. What is the value of the neutralization constant K_{neut} in a titration of hydrocyanic acid (HCN) with ammonia (NH_3)?

$K_a(HCN) = 4.9 \times 10^{-10}$
$K_b(NH_3) = 1.8 \times 10^{-5}$

a. 8.82×10^{-15}
b. 8.82×10^{-1}
c. 2.04×10^{-5}
d. 4.9×10^{4}
e. 5.56×10^{-10}
f. 1.80×10^{9}
g. 2.72×10^{-5}
h. 3.67×10^{4}

EXAMINATION 1

CHEMISTRY 142 FALL 2010

Tuesday October 12th 2010

Kinetics

1. In the reaction

   $3\ ClO^-(aq) \rightarrow 2\ Cl^-(aq) + ClO_3^-(aq)$

   the product Cl^- forms at a rate of 3.6 mol L^{-1} min^{-1}. What are the corresponding rates for consumption of ClO^- and production of ClO_3^-?

	$\Delta[ClO^-]/\Delta t$	$\Delta[ClO_3^-]/\Delta t$
a.	−5.4 mol L^{-1} min^{-1}	1.8 mol L^{-1} min^{-1}
b.	−2.4 mol L^{-1} min^{-1}	7.2 mol L^{-1} min^{-1}
c.	−2.4 mol L^{-1} min^{-1}	1.8 mol L^{-1} min^{-1}
d.	−3.6 mol L^{-1} min^{-1}	3.6 mol L^{-1} min^{-1}
e.	−5.4 mol L^{-1} min^{-1}	7.2 mol L^{-1} min^{-1}
f.	−3.6 mol L^{-1} min^{-1}	7.2 mol L^{-1} min^{-1}

2. Initial rates for the reaction

   $2\ NO(g) + O_2(g) \rightarrow 2\ NO_2(g)$

   were measured using several different starting concentrations for the reactants, and the data collected are summarized below.

Experiment	[NO], *M*	$[O_2]$, *M*	*Initial Rate, M s*$^{-1}$
1	0.0126	0.0125	1.41×10^{-2}
2	0.0252	0.0250	1.13×10^{-1}
3	0.0252	0.0125	5.64×10^{-2}

   The rate law equation for this reaction is

   a. Rate = $k[NO][O_2]$
   b. Rate = $k[O_2]$
   c. Rate = $k[NO][O_2]^2$
   d. Rate = $k[NO]^2[O_2]$
   e. Rate = $k[NO]^2[O_2]^{-1}$
   f. Rate = $k[NO][O_2]^{-2}$
   g. Rate = $k[NO]^2[O_2]^{-2}$
   h. Rate = $k[NO]$

3. The decomposition of hydrogen peroxide H_2O_2 to form water and oxygen follows first-order kinetics. At 20°C the concentration of H_2O_2 will decay to half of its original concentration of 2.00 M in 3.92×10^4 s. What is the rate constant for this reaction?

   a. $2.6 \times 10^{-5}\ s^{-1}$
   b. $3.6 \times 10^{-5}\ s^{-1}$
   c. $1.8 \times 10^{-5}\ s^{-1}$
   d. $7.8 \times 10^{4}\ s^{-1}$
   e. $5.6 \times 10^{4}\ s^{-1}$
   f. $1.3 \times 10^{-5}\ s^{-1}$

4. For the decomposition of hydrogen peroxide described in the previous problem, how many hours will it take for the concentration to reach 0.600 M?

   a. 8.21 hours
   b. 11.4 hours
   c. 18.9 hours
   d. 21.3 hours
   e. 26.2 hours
   f. 32.6 hours

5. The decomposition of hydroxylamine (NH_2OH) in the presence of oxygen obeys the rate law expression

$$Rate = k[NH_2OH][O_2].$$

The rate constant k for this reaction is 2.64×10^{-4} L mol^{-1}s^{-1} at 25 °C; the rate constant increases to 2.02×10^{-3} L mol^{-1}s^{-1} at 50 °C. Find the activation energy for this reaction.

a. –65.2 kJ mol^{-1} c. –0.846 kJ mol^{-1} e. 0.644 kJ mol^{-1} g. 48.9 kJ mol^{-1}
b. –48.9 kJ mol^{-1} d. –0.644 kJ mol^{-1} f. 0.846 kJ mol^{-1} h. 65.2 kJ mol^{-1}

6. Consider the following reaction mechanism:

$$NO_2(g) + NO_2(g) \xrightarrow{k_1} NO_3(g) + NO(g) \quad \textit{slow}$$
$$NO_3(g) + CO(g) \xrightarrow{k_2} NO_2(g) + CO_2(g) \quad \textit{fast}$$

Identify the reactants, products, and intermediates in this reaction.

	Reactants	*Products*	*Intermediates*
a.	NO_2, CO	NO, CO_2	NO_3, NO_2
b.	NO_2, NO_3, CO	NO_3, NO, NO_2, CO_2	none
c.	NO_2	NO, CO_2	NO_3, CO
d.	NO_2	CO_2	NO_3, CO, NO
e.	NO_2, CO	NO, CO_2	NO_3
f.	NO_2, NO_3, CO	NO, CO_2	none
g.	NO_2, CO	NO, NO_2, CO_2	NO_3

7. The rate law corresponding to the mechanism given in the previous problem is

a. Rate = $k[NO_2]$
b. Rate = $k[NO_2]^2$
c. Rate = $k[NO_2]^2[NO_3][CO]$
d. Rate = $k[NO_3][CO]$
e. Rate = $k[NO_2]^2[CO][NO]^{-1}$
f. Rate = $k[NO_2]^2[CO]$

8. Consider the reaction profile corresponding to a two-step reaction mechanism shown below, and complete the following statements.

___ is the activated complex corresponding to the rate-determining step in this mechanism.
___ represents the intermediates in the reaction.
The overall reaction is ______.

a.	B	C	exothermic
b.	C	B	exothermic
c.	D	C	exothermic
d.	B	D	exothermic
e.	B	C	endothermic
f.	C	D	endothermic
g.	D	C	endothermic
h.	D	B	endothermic

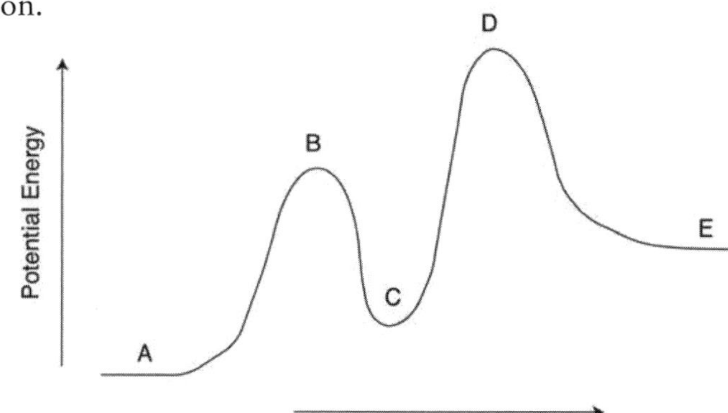

Equilibrium

9. Consider the decomposition reaction of H_2O_2.

$$2\,H_2O_2(g) \rightleftharpoons 2\,H_2O(g) + O_2(g) \qquad K_c = 6.25 \times 10^{-3}$$

What is the concentration of O_2 if $[H_2O_2] = 0.0500$ M and $[H_2O] = 0.100$M at equilibrium?

a. 1.25×10^{-2} M
b. 2.50×10^{-2} M
c. 3.13×10^{-2} M
d. 1.56×10^{-3} M
e. 3.13×10^{-3} M
f. 1.56×10^{-4} M

10. Using the equilibrium constants for the reactions

$$SnO_2(s) + 2\,CO(g) \rightleftharpoons Sn(s) + 2\,CO_2(g) \qquad K_1$$

$$CO(g) + H_2O(g) \rightleftharpoons CO_2(g) + H_2(g) \qquad K_2$$

find an expression for the equilibrium constant for the reaction

$$SnO_2(s) + 2\,H_2(g) \rightleftharpoons Sn(s) + 2\,H_2O(g) \qquad K_3$$

a. $K_3 = K_1/K_2$
b. $K_3 = K_1 - K_2$
c. $K_3 = K_1K_2^2$
d. $K_3 = K_1/K_2^2$
e. $K_3 = K_1 - 2\,K_2$
f. $K_3 = K_1/\,2\,K_2$

11. A reaction mixture at 800 K was found to contain 0.417 M N_2, 0.524 M H_2, and 0.122 M NH_3. The reaction has an equilibrium constant $K_c = 0.278$ at 800 K.

$$N_2(g) + 3\,H_2(g) \rightleftharpoons 2\,NH_3(g)$$

Find the reaction quotient Q, and select the correct response below.

a. Q = 0.248; to reach equilibrium, reactants must be converted to products.
b. Q = 0.248; to reach equilibrium, products must be converted to reactants.
c. Q = 4.03; to reach equilibrium, reactants must be converted to products.
d. Q = 4.03; to reach equilibrium, products must be converted to reactants.
e. Q = 0.558; to reach equilibrium, reactants must be converted to products.
f. Q = 0.558; to reach equilibrium, products must be converted to reactants.

12. Consider the reaction

$$2\,H_2S(g) + 3\,O_2(g) \rightleftharpoons 2\,SO_2(g) + 2\,H_2O(g)$$

How does K_p for this reaction compare to K_c at 25°C?

a. K_p is smaller than K_c.
b. K_p is equal to K_c.
c. K_p is larger than K_c.

13. When pure solid NH_4HSe is placed in an evacuated container the following reaction occurs

$$NH_4HSe(s) \rightleftharpoons NH_3(g) + H_2Se(g) \qquad \Delta H° = 115 \text{ kJ mol}^{-1}$$

and equilibrium is eventually established between reactants and products. Suppose that equilibrium is disturbed by applying the following stresses to the system. In which direction, if any, will the system shift in order to restore equilibrium? Note: "to right" is equivalent to "toward products" and "to left" is equivalent to "toward reactants."

	add NH_4HSe	*remove NH_3*	*decrease volume*	*increase temperature*
a.	to right	to right	to left	to right
b.	to right	to left	to right	to right
c.	to right	to right	to left	no effect
d.	no effect	to right	to left	to left
e.	no effect	to left	to right	no effect
f.	no effect	to right	to left	to right
g.	to left	to left	to right	to left
h.	to left	to right	to right	to left

Acid-Base

14. The net ionic equation for equilibrium established when methylamine CH_3NH_2 is added to water is given below. Classify the participants in the reaction based on the Brønsted-Lowry definitions for acids and bases.

	$CH_3NH_2(aq)$ +	$H_2O(l)$ ⇌	$CH_3NH_3^+(aq)$ +	$OH^-(aq)$
a.	base	base	acid	acid
b.	base	acid	acid	base
c.	acid	base	acid	base
d.	acid	acid	base	base
e.	base	acid	base	acid
f.	acid	base	base	acid

15. Several definitions for acids and bases were discussed in lecture. Which of the following statements correctly describe the attributes of acids?

I. Acids will give up an anion leading to formation of the solvent anion.
II. Acids can donate hydrogen ions resulting formation of the solvent cation.
III. Acids are generally, but not always, neutral or positively charged.
IV. According to one of the acid definitions, acids are electron pair acceptors.

a. I
b. I and III
c. II and IV
d. II and III
e. II, III, IV
f. All of the statements are correct

16. For the reaction given below K < 1.

$$HBrO(aq) + SCN^-(aq) \rightleftharpoons BrO^-(aq) + HSCN(aq)$$

Identify the weaker and stronger Brønsted-Lowry acids.

	weaker acid	*stronger acid*		*weaker acid*	*stronger acid*
a.	BrO^-	SCN^-	d.	BrO^-	HSCN
b.	SCN^-	BrO^-	e.	HBrO	SCN^-
c.	HSCN	HBrO	f.	HBrO	HSCN

17. Find the pH of the solution that results when 500 mL of 0.020 M HCl is combined with 500 mL of 0.040 M NaOH.

a. 1.4 b. 1.7 c. 2.0 d. 7.0 e. 12.0 f. 12.3 g. 12.6

18. Benzoic acid ($C_6H_5CO_2H$) has an acid ionization constant $K_a = 6.5 \times 10^{-5}$. Calculate the pH of a solution that contains 0.25 M benzoic acid.

a. 0.6 b. 2.4 c. 4.2 d. 4.8 e. 7.0 f. 9.2 g. 11.6

19. Classify the solutions produced when the following compounds are dissolved in water as acidic, basic, or neutral.
Choose the row where all the responses are correct.

	Na_2SO_3	$HClO_4$	KNO_3	LiF
a.	basic	acidic	neutral	neutral
b.	basic	basic	basic	neutral
c.	neutral	acidic	neutral	basic
d.	basic	acidic	basic	basic
e.	basic	acidic	neutral	basic
f.	basic	basic	basic	acidic
g.	neutral	acidic	basic	neutral

20. Consider the polyprotic acid carbonic acid H_2CO_3.

$$H_2CO_3(aq) + H_2O(l) \rightleftharpoons HCO_3^-(aq) + H_3O^+(aq) \qquad K_{a1} = 4.3 \times 10^{-7}$$

$$HCO_3^-(aq) + H_2O(l) \rightleftharpoons CO_3^{2-}(aq) + H_3O^+(aq) \qquad K_{a2} = 5.6 \times 10^{-11}$$

Suppose that a 50 mL of 0.100 M solution of H_2CO_3 is titrated with 0.100 M NaOH. What is the pH at the first equivalence point, which corresponds to addition of 50 mL of NaOH?

a. 3.68 b. 6.37 c. 7.00 d. 8.31 e. 10.25 f. 12.70

21. How many of the following aqueous solutions containing equimolar quantitites of the two components will behave as a buffer solution?

HI and KI
$NaCH_3CO_2$ and CH_3CO_2H
KCN and HCN
H_3PO_4 and Na_3PO_4
HNO_3 and $NaNO_3$
LiOH and HCl

a. none b. 1 c. 2 d. 3 e. 4 f. 5 g. 6

22. A buffer solution was prepared by dissolving 0.50 moles of NH_3 ($K_b = 1.8 \times 10^{-5}$) and 0.40 moles of NH_4Br in 1.0 L of water. Find the pH of the resulting solution.

a. 4.64 b. 4.74 c. 4.83 d. 7.00 e. 9.15 f. 9.25 g. 9.35

EXAMINATION 1

CHEMISTRY 142 SPRING 2011

Thursday February 17th 2011

Kinetics

1. In the reaction

$$2\ N_2O_5(g) \rightarrow 4\ NO(g) + O_2(g)$$

the product nitric oxide $NO(g)$ forms at a rate of 8.0 mol L^{-1} min^{-1}. What are the corresponding rates for consumption of $N_2O_5(g)$ and production of $O_2(g)$?

	consumption of $N_2O_5(g)$	production of $O_2(g)$
a.	−2.0 mol L^{-1} min^{-1}	2.0 mol L^{-1} min^{-1}
b.	−4.0 mol L^{-1} min^{-1}	4.0 mol L^{-1} min^{-1}
c.	−8.0 mol L^{-1} min^{-1}	8.0 mol L^{-1} min^{-1}
d.	−16.0 mol L^{-1} min^{-1}	4.0 mol L^{-1} min^{-1}
e.	−2.0 mol L^{-1} min^{-1}	8.0 mol L^{-1} min^{-1}
f.	−4.0 mol L^{-1} min^{-1}	2.0 mol L^{-1} min^{-1}
g.	−1.0 mol L^{-1} min^{-1}	2.0 mol L^{-1} min^{-1}
h.	−2.0 mol L^{-1} min^{-1}	4.0 mol L^{-1} min^{-1}
i.	−4.0 mol L^{-1} min^{-1}	1.0 mol L^{-1} min^{-1}

2. The rate of the reaction between sulfur dioxide and oxygen was studied and the reaction rate was discovered to depend upon the concentration of the product sulfur trioxide SO_3. The rate was discovered to be independent of the concentration of oxygen. The following data were collected for three experiments in which the concentration of sulfur dioxide SO_2 was kept the same.

$$2\ SO_2(g) + O_2(g) \rightarrow 2\ SO_3(g)$$

	Concentration of SO_3	*Rate of reaction*
Expt 1	2.0×10^{-3} M	3.0×10^{-3} mol $L^{-1}s^{-1}$
Expt 2	2.2×10^{-4} M	9.0×10^{-3} mol $L^{-1}s^{-1}$
Expt 3	2.5×10^{-5} M	2.7×10^{-2} mol $L^{-1}s^{-1}$

What is the order of the reaction with respect to SO_3?

a. ½ c. 1.0 e. 1½ g. 2.0 i. 3.0
b. −½ d. −1.0 f. −1½ h. −2.0 j. −3.0

3. In one experiment, the concentration of a reactant changed as follows as the reaction proceeded:

Initial concentration	128 mmol L^{-1}
After 18 minutes	32 mmol L^{-1}
1.50 hours after the start of the reaction	8.0 mmol L^{-1}

How long did it take, from the beginning of the reaction, for the concentration of the reactant to decrease to 2.0 mmol L^{-1}?

a. 1.0 hr 48 min d. 3.0 hr 6 min g. 5.0 hr 24 min
b. 2.0 hr 6 min e. 3.0 hr 48 min h. 6.0 hr 18 min
c. 2.0 hr 54 min f. 4.0 hr 30 min i. 12 hr 42 min

4. The following mechanism has been proposed for the reaction between nitric oxide and bromine:

Step 1: $NO + Br_2 \rightarrow NOBr_2$ *slow*
Step 2: $NOBr_2 + NO \rightarrow NOBr$ *fast*

What is the rate law for the formation of the product NOBr implied by this mechanism?

a. Rate = k [NO]
b. Rate = k $[NO]^2$
c. Rate = k $[NO]^3$
d. Rate = k $[NO][Br_2]$
e. Rate = k $[NO]^2[Br_2]$
f. Rate = k $[NO]^2[Br_2]^2$
g. Rate = k $[NO][Br_2]^2$
h. Rate = k $[NOBr_2]$
i. Rate = k $[NOBr_2][NO]$

5. Calculate the activation energy, E_A, for the following reaction, given that the rate constant, k, equals 2.4×10^{-6} L mol^{-1}s^{-1} at 575K, and 6.0×10^{-5} L mol^{-1}s^{-1} at 630K.

$$2\,HI(g) \rightarrow H_2(g) + I_2(g)$$

a. 44 kJ
b. 66 kJ
c. 88 kJ
d. 100 kJ
e. 128 kJ
f. 176 kJ
g. 212 kJ
h. 244 kJ
i. 272 kJ
j. 302 kJ

6. In the energy profile shown for an exothermic reaction, which labels correspond to the following?

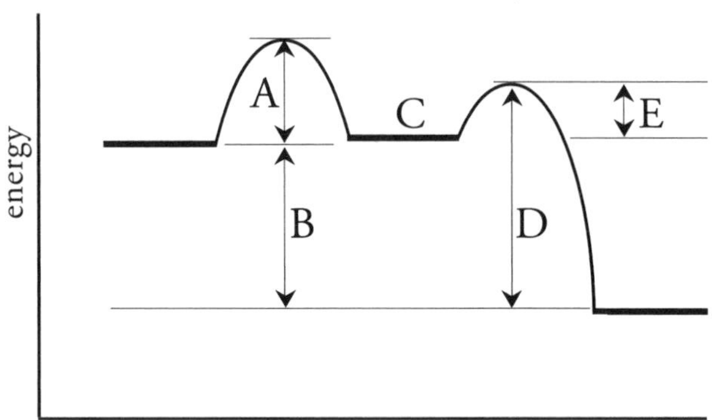

	Activation energy for the first step of the reverse reaction	*The reaction intermediate*	*The energy of reaction ΔE*	*Activation energy for the second step of the forward reaction*
a.	E	C	B	D
b.	D	A	A	E
c.	E	A	D	B
d.	D	C	B	A
e.	E	D	B	A
f.	D	C	B	E
g.	A	C	B	D
h.	D	C	A	E

7. Which of the following is a *correct* statement about reaction mechanisms?

a. The rate constant, k, is always constant, regardless of the conditions.
b. The concentration of a catalyst *cannot* appear in the rate law.
c. The activation energy is the minimum energy the molecules need for a successful reaction.
d. The rate determining step can never be the last step in a reaction mechanism.
e. A catalyst for a reaction must be in the same phase as the reactants.
f. Collisions between molecules that exceed the activation energy always result in product formation.

Equilibria

8. Calculate the equilibrium constant K_c for the following equilibrium if the concentration of the various components at equilibrium are as shown:

$$3\,ClO^-(aq) \rightleftharpoons 2\,Cl^-(aq) + ClO_3^-(aq)$$

Concentrations at equilibrium: $[ClO^-(aq)] = 3.0$ M
$[Cl^-(aq)] = 2.0$ M
$[ClO_3^-(aq)] = 4.0$ M

a. 0.30	c. 0.89	e. 1.5	g. 1.8	i. 3.6
b. 0.59	d. 1.2	f. 1.7	h. 2.0	j. 4.5

9. Using the equilibrium constants K_1 and K_2 for the reactions:

$$TiO_2(s) + 2\,CO(g) \rightleftharpoons Ti(s) + 2\,CO_2(g) \qquad K_1$$

$$CO(g) + H_2O(g) \rightleftharpoons CO_2(g) + H_2(g) \qquad K_2$$

find an expression for the equilibrium constant K_3 for the reaction

$$TiO_2(s) + 2\,H_2(g) \rightleftharpoons Ti(s) + 2\,H_2O(g) \qquad K_3$$

a. $K_3 = K_1 \times K_2$	c. $K_3 = K_1/K_2$	e. $K_3 = K_1K_2^2$	g. $K_3 = K_1 - 2\,K_2$
b. $K_3 = K_1 - K_2$	d. $K_3 = K_1 + K_2$	f. $K_3 = K_1/K_2^2$	h. $K_3 = K_1/\,(2 \times K_2)$

10. Consider the reaction

$$C_3H_8(g) + 5\,O_2(g) \rightleftharpoons 3\,CO_2(g) + 4\,H_2O(g)$$

How does K_p for this reaction compare numerically to K_c at 25°C?

a. K_p is smaller than K_c
b. K_p is equal to K_c
c. K_p is larger than K_c

11. When pure solid NH_4Cl is placed in an evacuated container the following endothermic reaction occurs:

$$NH_4Cl(s) \rightleftharpoons NH_3(g) + HCl(g)$$

Equilibrium is eventually established between reactants and products. Suppose that equilibrium is disturbed by doing the following. In which direction, if any, will the system shift in order to restore equilibrium?

	add $NH_4Cl(s)$	*remove $NH_3\,(g)$*	*increase volume*	*decrease temperature*
a.	to right	to right	to left	to right
b.	to right	to left	to right	to right
c.	to right	no effect	to left	no effect
d.	no effect	to right	to right	to left
e.	no effect	to left	to right	no effect
f.	no effect	to right	to right	to right
g.	no effect	no effect	to left	to right
h.	to left	to right	to right	to left
i.	to left	to left	to right	to left
j.	to left	no effect	to left	to left

Acids and Bases

12. Which of the following electrolytes is the strongest acid?

a. NaH_2PO_4	d. NH_4NO_3	g. $NaHCO_3$
b. HF	e. H_2SO_3	h. HCN
c. K_2SO_4	f. HI	i. CH_3CO_2H

13. When pure water is heated, the hydronium ion concentration increases. This means that...

    1. the pH decreases
    2. the pH increases
    3. the water becomes acidic
    4. the water becomes basic

    Which statements are true?

a. only 1	c. only 3	e. 1 and 3	g. 2 and 3	i. none are true
b. only 2	d. only 4	f. 2 and 4	h. 1 and 4	

14. Which statement is applicable to the Lewis theory of acids and bases but not to the definitions developed by Brønsted and Lowry?

    a. Acids dissolved in any protonic solvent produce the cation of that solvent
    b. The hydronium ion H_3O^+ is the strongest acid than can exist in water
    c. Acids are molecules or ions that accept electron pairs from bases
    d. The stronger an acid, the weaker its conjugate base
    e. Acid-base reactions involve the transfer of a hydrogen ion from the acid to the base
    f. Any acid weaker than hydronium ion H_3O^+ in aqueous solution is classified as weak

15. Consider the following Brønsted-Lowry equilibrium:

$$HSO_4^- + H_2O \rightleftharpoons H_3O^+ + \underline{???????}$$

Identify the missing product, and decide whether that product is acting as a base or an acid in the equilibrium.

a. H_2SO_4	*acting as*	an acid
b. OH^-	*acting as*	a base
c. SO_4^{2-}	*acting as*	a base
d. H_2SO_4	*acting as*	a base
e. SO_4^{2-}	*acting as*	an acid
f. HSO_4^-	*acting as*	a base

16. A 0.10 M solution of hypobromous acid (HBrO) has a pH = 4.85. What is the value of pK_a for hypobromous acid?

a. 4.85	c. 6.22	e. 8.70	g. 10.01
b. 5.17	d. 7.71	f. 9.11	h. 11.34

17. The net ionic equation for equilibrium established when pyridine C_5H_5N is added to water is shown below. Classify the participants in the reaction based on the Brønsted-Lowry definitions for acids and bases.

	$C_5H_5N(aq)$ +	$H_2O(l)$ ⇌	$C_5H_5NH^+(aq)$ +	$OH^-(aq)$
a.	base	base	acid	acid
b.	base	acid	acid	base
c.	base	acid	base	acid
d.	acid	base	acid	base
e.	acid	acid	base	base
f.	acid	base	base	acid

18. Which one of the following salts produces a *neutral* solution when dissolved in water?

a. NH_4Cl
b. $MgCO_3$
c. LiF
d. NaH_2PO_4
e. KNO_3
f. $(NH_4)_2SO_4$
g. NaClO
h. $Ca(CH_3CO_2)_2$
i. NH_4CN

19. What is the value of the neutralization constant K_{neut} in a titration of hydrofluoric acid (HF) with potassium hydroxide (KOH)?

$K_a(HF) = 3.5 \times 10^{-4}$

a. 3.5×10^{-4}
b. 2.9×10^{-14}
c. 2.9×10^{-11}
d. 6.3×10^{11}
e. 5.5×10^{-9}
f. 3.5×10^{10}
g. 2.9×10^{3}
h. 7.0×10^{4}

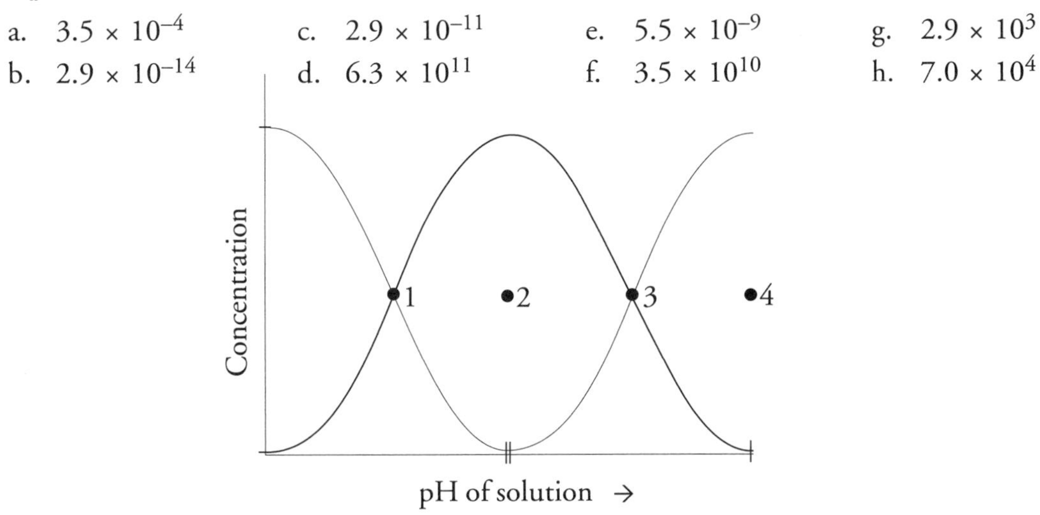

20. The two acid ionization constants for the diprotic acid phosphorous acid, H_3PO_3, are shown below:

$$H_3PO_3 + H_2O \rightleftharpoons H_3O^+ + H_2PO_3^- \qquad K_{a1} = 1.0 \times 10^{-2} \qquad pK_{a1} = 2.00$$
$$H_2PO_3^- + H_2O \rightleftharpoons H_3O^+ + HPO_3^{2-} \qquad K_{a2} = 2.6 \times 10^{-7} \qquad pK_{a2} = 6.59$$

What is the equilibrium constant for the following system that illustrates what happens when sodium dihydrogen phosphite dissolves in water? The distribution diagram is illustrated above for your convenience.

$$H_2PO_3^- + H_2PO_3^- \rightleftharpoons H_3PO_3 + HPO_3^{2-} \qquad K = ?$$

a. 3.8×10^{4}
b. 7.6×10^{-6}
c. 3.8×10^{8}
d. 2.6×10^{-5}
e. 3.8×10^{-4}
f. 5.6×10^{-5}
g. 2.6×10^{-9}
h. 2.6×10^{5}

21. Referring to question 20, what is the pH at point 2?

a. 1.27	c. 1.81	e. 3.96	g. 5.21	i. 7.33
b. 1.44	d. 2.26	f. 4.30	h. 6.22	j. 8.59

22. How many of the following equimolar aqueous solutions will act as a buffer solutions?

NaOH and NaCl

Na_2CO_3 and $KHCO_3$

LiOH and LiH

NaF and HF

KCN and HCN

NaH_2PO_4 and Na_2HPO_4

$NaNO_3$ and HNO_3

a. 1 b. 2 c. 3 d. 4 e. 5 f. 6 g. all of them

EXAMINATION 2

CHEMISTRY 142 SPRING 2006

Wednesday March 29th 2006

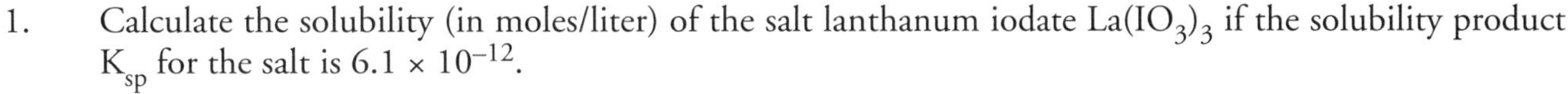

1. Calculate the solubility (in moles/liter) of the salt lanthanum iodate $La(IO_3)_3$ if the solubility product K_{sp} for the salt is 6.1×10^{-12}.

   a. 1.2×10^{-4} M
   b. 1.8×10^{-4} M
   c. 6.9×10^{-4} M
   d. 2.5×10^{-6} M
   e. 1.6×10^{-3} M
   f. 7.9×10^{-5} M

2. *(3 points)* If some potassium iodate KIO_3 was added to the solution of lanthanum iodate, would you expect the solubility of the lanthanum iodate to increase, decrease, or stay the same?

   a. increase
   b. decrease
   c. stay the same

3. Calculate the solubility (in moles/liter) of the same salt lanthanum iodate $La(IO_3)_3$ in the presence of 2.0×10^{-2} M potassium iodate KIO_3.

   a. 3.1×10^{-10} M
   b. 6.2×10^{-3} M
   c. 3.4×10^{-2} M
   d. 1.5×10^{-8} M
   e. 7.6×10^{-7} M
   f. 5.2×10^{-4} M

4. For the salt barium sulfate, $BaSO_4$, the solubility product $K_{sp} = 1.10 \times 10^{-10}$.
   If you were to mix 500 mL of 2.0×10^{-5} M $Ba(NO_3)_2$ and 500 mL of 1.5×10^{-5} M Na_2SO_4, what would you observe in the resulting 1.0 L of solution?

   a. Precipitation of $BaSO_4$*(s)* because $Q > K_{sp}$
   b. Precipitation of $BaSO_4$*(s)* because $Q < K_{sp}$
   c. Precipitation of $BaSO_4$*(s)* because $Q = K_{sp}$
   d. No precipitation of $BaSO_4$*(s)* because $Q = K_{sp}$
   e. No precipitation of $BaSO_4$*(s)* because $Q < K_{sp}$
   f. No precipitation of $BaSO_4$*(s)* because $Q > K_{sp}$

5. The solubility product K_{sp} for lead(II) chloride $PbCl_2$ is 1.7×10^{-5}.
   The solubility product K_{sp} for lead(II) fluoride PbF_2 is 3.6×10^{-8}.

   What is the ratio of the concentration of chloride ions to the concentration of fluoride ions ($[Cl^-]/[F^-]$) in a solution containing Pb^{2+}, Cl^-, and F^- ions in equilibrium with solid $PbCl_2$ and solid PbF_2? Consider the equilibrium:

   $$PbCl_2(s) \; + \; 2\,F^-(aq) \; \rightleftharpoons \; PbF_2(s) \; + \; 2\,Cl^-(aq)$$

   a. 4.7
   b. 2.1×10^{-3}
   c. 21.7
   d. 4.6×10^{-2}
   e. 472
   f. 22,300
   g. 688
   h. 2390

Questions 6 through 11 refer to the following system. Each question is worth 3 points.

The pressure on a system containing 1.5 moles of an ideal monatomic gas decreases from 8.0 atm to 2.0 atm as the system moves from state A to state B at a constant volume of 5.0 L. The system then expands at a constant pressure of 2.0 atm to reach state C at which point the temperature is again what it was at state A.

It is recommended that you draw these changes on the diagram to the right.

$R = 0.08206\ L\ atm\ K^{-1}mol^{-1} = 8.3145\ J\ K^{-1}mol^{-1}$

P

V

6. What is the volume in liters at point C?

   a. 1.25 L b. 2.5 L c. 5.0 L d. 7.5 L e. 10 L f. 15 L g. 20 L h. 30 L

7. What is the temperature at point A?

   a. zero°C b. 3.2°C c. 27°C d. 52°C e. 88°C f. 128°C g. 273°C h. 325°C

8. How much work is done by the system in moving at constant pressure from B to C?

   a. w = –30 J b. w = –40 J c. w = –304 J d. w = –405 J e. w = –1010 J f. w = –2530 J g. w = –3040 J h. w = –4050 J

9. How much heat is absorbed by the system, or liberated by the system, in moving at constant pressure from B to C?

   a. q = –75 J b. q = +75 J c. q = +2250 J d. q = –4560 J e. q = +4560 J f. q = +7600 J g. q = +10,130 J h. q = –10,130 J

10. What is the change in the internal energy of the system ΔE as it moves from B to C?

   a. ΔE = zero J b. ΔE = +45 J c. ΔE = +4560 J d. ΔE = –4560 J e. ΔE = +5070 J f. ΔE = +7600 J g. ΔE = +10,640 J h. ΔE = –10,640 J

11. What is the change in the internal energy of the system ΔE as it moves from A to B?

   a. ΔE = zero J b. ΔE = +45 J c. ΔE = +4560 J d. ΔE = –4560 J e. ΔE = +5070 J f. ΔE = +7600 J g. ΔE = +10,640 J h. ΔE = –10,640 J

12. Suppose a vessel contained just 5 molecules of a gas. What is the probability that all five molecules would be in the left-hand half of the vessel?

   a. 1/2 b. 1/4 c. 1/6 d. 1/8 e. 1/16 f. 1/32 g. 1/64 h. 1/128 i. 1/256 j. 1/512

13. Suppose that these five molecules were constrained to just one-half of the vessel and then allowed to occupy the entire vessel (keeping the temperature constant). What would the entropy change be for this isothermal expansion of the gas?

a. 5.32×10^{-16} JK^{-1}
b. 5.00×10^{25} JK^{-1}
c. 2.84×10^{-1} JK^{-1}
d. 6.11×10^{-11} JK^{-1}
e. 4.79×10^{-23} JK^{-1}
f. 28.8 JK^{-1}

14. Which one of the following statements about entropy is *not true*?

a. All elements at standard temperature and pressure (1 atm and 0°C) have an entropy equal to zero.
b. Entropy is a state function.
c. A positive value for a change in entropy, ΔS, indicates a change toward greater 'disorder'.
d. Entropy values are *never* negative.
e. *All* spontaneous processes ultimately involve an increase in the entropy of the universe.
f. The entropy of any chemical substance increases if the temperature is increased.

15. The enthalpy of fusion $\Delta H°$ of ammonia is 5.65 kJ mol^{-1} and the entropy of fusion $\Delta S°$ of ammonia is 28.9 J K^{-1} mol^{-1}. What is the approximate melting point of ammonia?

a. 25°C
b. 196°C
c. –106°C
d. 77°C
e. 298 K
f. 196 K
g. –35°C
h. 5 K

16. What is the meaning of the Gibbs Free Energy change $\Delta G°$ for a reaction?

a. It is equal to the heat given off or absorbed in the reaction.
b. It is a measure of the change in randomness or disorder caused by the reaction.
c. It is a measure of the temperature dependence of the reaction.
d. It has meaning only when the reaction has reached equilibrium.
e. It is the energy available to do work, since some energy may be required or released by changes in the entropy of the system or its surroundings.
f. It is a fictitious energy invented to make sure that the energy balance adds up correctly—so that the first law of thermodynamics is obeyed.

17. If $\Delta G°$ for the autoionization of water were to decrease (i.e. become more negative or less positive), what would the pH of a neutral aqueous solution be?

a. It would remain equal to 7.
b. It would be less than 7.
c. It would be greater than 7.
d. The hydroxide ion concentration [OH^-] would exceed the hydronium ion concentration [H_3O^+].
e. The hydroxide ion concentration [H_3O^+] would exceed the hydronium ion concentration [OH^-].
f. Water would become more acidic.

18. Calculate $\Delta G°$ for the reaction

$3\,Cl_2(g) + 2\,CH_4(g) \rightarrow CH_3Cl(g) + CH_2Cl_2(g) + 3\,HCl(g)$ if
$\Delta G_f°(CH_4) = -51$ kJ mol^{-1}
$\Delta G_f°(CH_3Cl) = -57$ kJ mol^{-1}
$\Delta G_f°(CH_2Cl_2) = -69$ kJ mol^{-1}
$\Delta G_f°(HCl) = -95$ kJ mol^{-1}

a. –27 kJ
b. +27 kJ
c. –75 kJ
d. +75 kJ
e. –170 kJ
f. +170 kJ
g. –309 kJ
h. +309 kJ
i. –462 kJ
j. +462 kJ

19. What is the value of the equilibrium constant K at 25°C for the reaction described in the previous question?

a. 1.4×10^{54}
b. 7.3×10^{-55}
c. 4.4
d. 2.7×10^{28}
e. 3.1×10^{5}
f. 1.7×10^{41}
g. 7.8×10^{35}
h. 5.4×10^{17}

20. Aluminum metal, Al, reacts with water *in basic conditions* to produce the ion $[Al(OH)_4]^-$ and hydrogen gas, H_2. Write the balanced equation for this reaction. What is the coefficient for water in the balanced equation?

a. 1
b. 2
c. 3
d. 4
e. 5
f. 6
g. 7
h. 8
i. 9
j. 10

21. *(3 points)* A cell is written in shorthand notation as: Ni*(s)* | Ni^{2+}*(aq, 1M)* || Tl^{3+}*(aq, 1M)* | Tl*(s)*

The standard half-cell reduction potentials are:

E° (Ni^{2+}|Ni) = –0.257 V
E° (Tl^{3+}|Tl) = 0.741 V

What is the standard cell potential for this cell?

a. 0.484 V
b. 0.998 V
c. –0.484 V
d. –0.998 V

22. *(3 points)* Is the reaction illustrated by the cell notation in the previous question a spontaneous reaction or a nonspontaneous reaction?

a. spontaneous
b. nonspontaneous

23. *(3 points)* In the cell described in the previous two questions, what is the reducing agent?

a. nickel metal Ni
b. nickel ion Ni^{2+}
c. thallium metal Tl
d. thallium ion Tl^{3+}

24. The standard cell potential E° for the copper/zinc cell: Zn*(s)* | Zn^{2+}*(aq)* || Cu^{2+}*(aq)* | Cu*(s)* is 1.104 V at 25°C. Suppose that this cell has run for some time and the [Cu^{2+}] concentration has decreased to 0.50 M. What is the cell potential now?

a. 1.104 V
b. 0.0141 V
c. 1.118 V
d. 0.965 V
e. 1.090 V
f. 0.886 V

EXAMINATION 2

CHEMISTRY 142 FALL 2006

Thursday November 9th 2006

1. Calculate the concentration of strontium ions $[Sr^{2+}]$ in a saturated solution of strontium(II) phosphate ($Sr_3(PO_4)_2$). K_{sp} for $Sr_3(PO_4)_2 = 1.0 \times 10^{-31}$

   a. 7.5×10^{-13} c. 9.6×10^{-2} e. 7.4×10^{-7} g. 8.3×10^{-20}
   b. 4.6×10^{-11} d. 5.6×10^{-25} f. 1.1×10^{-1} h. 8.1×10^{-15}

2. Calculate the concentration of $[Sr^{2+}]$ ions in a saturated solution of $Sr_3(PO_4)_2$ if the solution also contains 3.0×10^{-2} M phosphate (PO_4^{3-}).

   a. 7.6×10^{-20} c. 4.8×10^{-10} e. 9.3×10^{-22} g. 5.8×10^{-24}
   b. 1.3×10^{-2} d. 7.1×10^{-5} f. 6.2×10^{-25} h. 3.6×10^{-27}

3. Which of the following solutions will result in the precipitation of the greatest amount of silver(I) phosphate (Ag_3PO_4)? K_{sp} for $Ag_3PO_4 = 1.8 \times 10^{-18}$

	$[Ag^+]$	$[PO_4^{3-}]$
a.	0.5 M	0.5 M
b.	0.2 M	0.1 M
c.	0.6 M	0.2 M
d.	0.3 M	0.8 M
e.	1.0 M	0.5 M
f.	0.6 M	0.4 M
g.	0.5 M	1.0 M
h.	2.0 M	0.1 M

Questions 4 to 10 refer to the following system. Each question is worth 4 points.

The volume of a system containing 3 moles of an ideal monotomic gas at 298 K increases from 2.5 L to 11.5 L as the system moves from state A to state B at constant pressure. The system is then cooled to 210 K at constant volume to state C, where the new pressure is 4.5 atm. Finally, the system is cooled at constant pressure until the volume is 2.5 L. This is state D.

It is recommended that you graph these changes on a pressure versus volume plot.

$R = 0.08206$ L atm mol^{-1} K^{-1} = 8.3145 J mol^{-1} K^{-1}

4. What is the pressure at point B?

   a. 1.75 atm c. 0.56 atm e. 29.3 atm g. 7.90 atm
   b. 20.2 atm d. 15.3 atm f. 40.7 atm h. 2.25 atm

5. What is the temperature at point B?

   a. 1298 K c. 518.6 K e. 816.6 K g. 1371 K
   b. 111.7 K d. 6811 K f. 205.3 K h. 186 K

6. How much work is done by the system or by the surroundings as the system moves from state C to state D (C → D)?

   a. +4100 kJ b. +230 kJ c. +55.6 kJ d. +4.10 kJ e. –4.10 kJ f. –55.6 kJ g. –230 kJ h. –4100 kJ

7. How much work is done by the system or by the surroundings as the system moves from A → B → C → D?

   a. –58.2 kJ b. –50.7 kJ c. –22.6 kJ d. –3.87 kJ e. +58.2 kJ f. +50.7 kJ g. +22.6 kJ h. +3.87 kJ

8. How much heat is absorbed or released by the system as it moves from C → D?

   a. –10.2 kJ b. –3.6 kJ c. –45 kJ d. –315 kJ e. +10.2 kJ f. +3.6 kJ g. +45 kJ h. +315 kJ

9. Suppose the after reaching state D, the system returns to state A. What is the change in the internal energy of the system as it moves from A → B → C → D → A?

   a. 210 kJ b. 170 kJ c. –35 k d. 0 kJ e. –42 kJ f. –104 kJ g. –127 kJ h. –196 kJ

10. Which of the following statements about the changes in the system is true?

   a. In moving from A → B, the system expands isothermally.
   b. In moving from A → B, the system expands reversibly.
   c. The total amount of work done by the system or by the surroundings as the system moves from A → B → C → D is zero.
   d. The system does ***not*** undergo any isothermal changes as it moves from A → B → C → D.
   e. No work is done when the system moves from A → B → C.
   f. No work is done when the system moves from B → C → D.

Use the following diagram to answer questions 11 and 12.

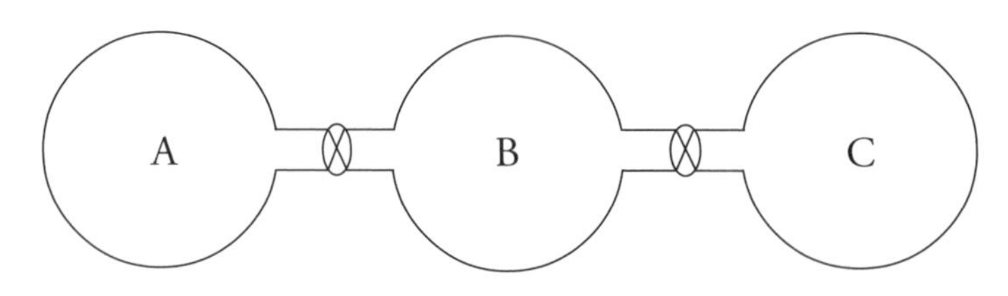

The vessel shown above has three compartments labeled A, B, and C. Compartments A and B are connected by a valve which can be opened and closed. Compartments B and C are also connected by a valve which can be opened and closed. All three compartments are equal in volume.

11. Suppose that two molecules are initially contained within bulb A, with both valves closed. Consider this to be a single microstate of the system. If the valve between A and B is opened, what is the *change* in the number of microstates available to the system?

   a. 0 b. 1 c. 4 d. 3 e. 8 f. 7 g. 15 h. 19

12. Suppose now that the valve between B and C is opened. What is the ***total*** change in the entropy of the system (in J K^{-1}) as the molecules move from occupying only A to occupying A, B, and C?

a. 7.6×10^{2}
b. 1.1×10^{-3}
c. 9.2×10^{-8}
d. 9.2×10^{-10}
e. 9.2×10^{-12}
f. 8.5×10^{-17}
g. 3.0×10^{-23}
h. 4.4×10^{-30}

13. The entropy of vaporization ΔH_{vap} of methylamine (CH_3NH_2) is 25.8 kJ mol^{-1} and the boiling point of methylamine is 266.8 K. What is the entropy of vaporization of methylamine (in J K^{-1} mol^{-1})?

a. 53.1
b. 65.7
c. 80.0
d. 86.2
e. 96.7
f. 103.8
g. 113.9
h. 129.0

14. The entropy of vaporization of most liquids is about 80 J K^{-1} mol^{-1}. Which of the following statements best explains why the entropy of vaporization of methylamine is *greater* than 80 J K^{-1} mol^{-1}?

a. Dispersion forces cause methylamine to be more ordered than most liquids.
b. Hydrogen bonding causes methylamine to be more ordered than most liquids.
c. Hydrogen bonding causes methylamine to be less ordered than most liquids.
d. Methylamine is more ordered than most liquids because it can act as a weak base.
e. Dipole-dipole forces cause methylamine to be more ordered than most liquids.
f. Methylamine is less ordered than most liquids because of its lack of intermolecular forces.

15. Iron reacts with carbon dioxide to produce iron(III) oxide and carbon monoxide according to the following balanced equation:

$$2\,Fe(s) + 3\,CO_2(g) \rightarrow Fe_2O_3(s) + 3\,CO(g)$$

The equilibrium constant for this reaction increases by a factor of two when the temperature is increased from 298 K to some final temperature, T_2. What is the value of T_2? Assume ΔH is independent of temperature.

$\Delta H^\circ_f(CO_2(g)) = -393.5$ kJ mol^{-1}
$\Delta H^\circ_f(Fe_2O_3(s)) = -826$ kJ mol^{-1}
$\Delta H^\circ_f(CO(g)) = -110.5$ kJ mol^{-1}

a. 301 K
b. 322 K
c. 378 K
d. 398 K
e. 426 K
f. 452 K
g. 833 K
h. 1002 K

16. Referring to the previous question, what is ΔG for the reaction at 100 K? Assume ΔH and ΔS are independent of temperature.

$S^\circ(CO_2(g)) = 214$ J K^{-1} mol^{-1}
$S^\circ(Fe_2O_3(s)) = 90$ J K^{-1} mol^{-1}
$S^\circ(CO(g)) = 198$ J K^{-1} mol^{-1}
$S^\circ(Fe(s)) = 27$ J K^{-1} mol^{-1}

a. −208 kJ
b. −188 kJ
c. −121 kJ
d. −24.2 kJ
e. +24.2 kJ
f. +121 kJ
g. +188 kJ
h. +208 kJ

17. *(2 points)* Referring to the previous two questions, is this reaction spontaneous or nonspontaneous at 100 K?

a. spontaneous
b. nonspontaneous

18. Methanol (CH_3OH) reacts with the dichromate ion ($Cr_2O_7^{2-}$) to produce formaldehyde (CH_2O) and chromium(III) (Cr^{3+}) in acidic solution. Write the balanced equation for this reaction. What is the coefficient of water in the balanced equation?

a. 0 b. 1 c. 3 d. 5 e. 7 f. 9 g. 11 h. 13

19. Given the following reduction potentials, calculate the potential (E°) for the following cell at standard conditions:

$$Cr(s)|Cr^{3+}(aq)||La^{3+}(aq)|La(s)$$

E° (Cr^{3+}|Cr) = –0.73 V
E° (La^{3+}|La) = +2.37 V

a. –3.100 V b. –2.370 V c. –0.730 V d. 0 V e. +0.730 V f. +2.370 V g. +3.100 V

20. Suppose that the cell described in the previous question has been running for some time and that the Cr^{3+} concentration has increased to 1.5 M while the La^{3+} concentration has decreased to 0.5 M. What is the cell potential (E) at this point? Assume that the temperature is 298 K.

a. –3.100 V b. –3.091 V c. –1.990 V d. 0 V e. +1.990 V f. +3.091 V g. +3.100 V

21. What will be the potential (E) of the cell described in the previous two questions at equilibrium? Assume that the temperature is 298 K.

a. –3.100 V b. –3.091 V c. –1.990 V d. 0 V e. +1.990 V f. +3.091 V g. +3.100 V

22. Which of the following statements is one of Faraday's laws of electrolysis?

a. In any redox reaction, the number of electrons lost in oxidation must equal the number of electrons gained in reduction.
b. Primary cells are voltaic cells that cannot be recharged.
c. A battery is a collection of cells in series.
d. The amount of substance liberated or deposited at an electrode is directly proportional to the quantity of electric charge passed through the cell.
e. Fuel cells never reach equilibrium.
f. Secondary cells are voltaic cells that can be recharged.

EXAMINATION 2

CHEMISTRY 142 SPRING 2007

Wednesday March 28th 2007

1. Which of the following is the correct solubility product constant expression for magnesium phosphate $Mg_3(PO_4)_2$?

a. $K_{sp} = [Mg^{2+}][PO_4^{3-}]$
b. $K_{sp} = [Mg^{2+}]^2[PO_4^{3-}]^3$
c. $K_{sp} = [Mg^{2+}]^3[PO_4^{3-}]^2$
d. $K_{sp} = [Mg_3][(PO_4)_2]$
e. $K_{sp} = [Mg^{2+}]^3[PO_4^{3-}]^2 / [Mg_3(PO_4)_2]$
f. $K_{sp} = [Mg^{2+}]^3 / [PO_4^{3-}]^2$

2. In a saturated solution of magnesium phosphate in water, what is the relationship between the concentrations of the two ions present?

a. $[Mg^{2+}] = [PO_4^{3-}]$
b. $[Mg^{2+}]^2 = [PO_4^{3-}]^3$
c. $[Mg^{2+}]^3 = [PO_4^{3-}]^2$
d. $3 \times [Mg^{2+}] = 2 \times [PO_4^{3-}]$
e. $[Mg^{2+}] = 2/3 \times [PO_4^{3-}]$
f. $[Mg^{2+}] = 1.5 \times [PO_4^{3-}]$

3. If the solubility product for magnesium phosphate $K_{sp} = 1.0 \times 10^{-24}$ at 25°C, what is the solubility of magnesium phosphate at 25°C in moles of $Mg_3(PO_4)_2$ per liter?

a. 1.2×10^{-5} M
b. 1.9×10^{-5} M
c. 2.5×10^{-5} M
d. 5.6×10^{-5} M
e. 6.2×10^{-6} M
f. 1.2×10^{-7} M

4. Suppose sufficient potassium phosphate K_3PO_4 is added to a saturated solution of magnesium phosphate to make the concentration of phosphate ions in the solution equal to 0.010 M. What would the concentration of magnesium ions in the solution be?

a. 1.0×10^{-20} M
b. 1.0×10^{-10} M
c. 2.2×10^{-7} M
d. 2.2×10^{-9} M
e. 4.6×10^{-10} M
f. 4.6×10^{-5} M

5. Suppose 50 mL of a 1.0×10^{-5} M solution of magnesium nitrate is added to 50 mL of a 4.0×10^{-5} M solution of potassium phosphate K_3PO_4. Would precipitation of magnesium phosphate occur?

a. yes
b. no

6. The solubility product K_{sp} for copper(I) chloride CuCl is 1.7×10^{-7}.
The solubility product K_{sp} for copper(I) cyanide CuCN is 3.5×10^{-20}.

What is the ratio of the concentrations of chloride ions and cyanide ions ($[Cl^-]/[CN^-]$) in a solution containing Cu^+, Cl^-, and CN^- ions in equilibrium with the solids CuCl and CuCN? In other words, consider the two simultaneous equilibria in solution and the overall system:

$$CuCl(s) + CN^-(aq) \rightleftharpoons CuCN(s) + Cl^-(aq)$$ and then determine K

a. 4.5×10^{-7}
b. 2.1×10^{-13}
c. 2.2×10^{6}
d. 4.6×10^{-2}
e. 4.9×10^{12}
f. 7.2×10^{14}
g. 3.1×10^{4}
h. 2.9×10^{19}

There are four fundamental laws of thermodynamics—the zeroth(0), first(1), second(2), and the third(3). Phrases commonly associated with these laws are listed below. Which row has all four laws assigned correctly?

	zero entropy at 0K	*conservation of energy*	*thermal equilibrium*	*spontaneous reaction when ΔS_{univ} > zero*
a.	0	1	2	3
b.	1	2	3	0
c.	2	3	0	1
d.	3	1	0	2
e.	2	1	3	0
f.	1	3	0	2
g.	3	1	2	0
h.	2	1	0	3

8. If a system absorbs 1670 J of heat from its surroundings and expands in volume by 4.6 L against an external pressure P of 0.880 atm, what is the change in its internal energy (ΔE)?

a. −1260 J
b. +1260 J
c. −2080 J
d. +2080 J
e. −1674 J
f. +1674 J

Questions 9, 10, 11, 12, and 13 refer to the following:

By decreasing the temperature, the volume of a system containing 2.0 moles of an ideal monatomic gas is decreased from 12.0 L to 3.0 L as the system moves from state A to state B at a constant pressure of 5.0 atm. The pressure on the system is then increased at constant volume to reach state C at which point the temperature is again what it was at state A.

It is recommended that you draw these changes on the diagram to the right.

P

V

$R = 0.08206 \text{ L atm K}^{-1}\text{mol}^{-1} = 8.3145 \text{ J K}^{-1}\text{mol}^{-1}$

9. What is the pressure in atm at point C?

a. 1.25 atm
b. 2.5 atm
c. 5.0 atm
d. 10 atm
e. 14 atm
f. 15 atm
g. 20 atm
h. 24 atm

10. What is the temperature at point C?

a. zero°C
b. 32°C
c. 42°C
d. 63°C
e. 76°C
f. 92°C
g. 177°C
h. 330°C

11. How much heat is absorbed by the system, or liberated by the system, in moving at constant volume from B to C?

a. q = −1780 J
b. q = +1780 J
c. q = +3650 J
d. q = −3650 J
e. q = +4210 J
f. q = +4210 J
g. q = +6840 J
h. q = −6840 J

12. What work would be done by the system if it were to move back from C to A isothermally and reversibly?

a. w = zero J
b. w = –3310 J
c. w = +4120 J
d. w = –4310 J
e. w = +6170 J
f. w = –8430 J
g. w = –9,230 J
h. w = +10,720 J

13. What is the change in the entropy of the system ΔS as it moves back from C to A via B?

a. ΔS = zero JK^{-1}
b. $\Delta S = -13\ JK^{-1}$
c. $\Delta S = +15\ JK^{-1}$
d. $\Delta S = -19\ JK^{-1}$
e. $\Delta S = +23\ JK^{-1}$
f. $\Delta S = -27\ JK^{-1}$
g. $\Delta S = +32\ JK^{-1}$
h. $\Delta S = +39\ JK^{-1}$

14. The standard entropy S° of $N_2O_4(g)$ is 304 $JK^{-1}\ mol^{-1}$ and the standard entropy S° of $NO_2(g)$ is 240 $JK^{-1}\ mol^{-1}$. Use this information and the enthalpy change given below to calculate $\Delta G°$ for the following reaction at 25°C.

$$N_2O_4(g) \rightleftharpoons 2\ NO_2(g) \qquad \Delta H° = +57.1\ kJ$$

a. +7.22 kJ
b. +52.7 kJ
c. –52.7 kJ
d. +21.7 kJ
e. +4.63 kJ
f. +61.5 kJ

15. What is the value of $\Delta G°$ for the autoionization of water at 25°C represented by the equation:

$$2\ H_2O \rightleftharpoons H_3O^+ + OH^-$$

a. +6.7 kJ
b. $+7.4 \times 10^{-3}$ kJ
c. +788 J
d. $+8.0 \times 10^{6}$ kJ
e. +8.1 kJ
f. +80 kJ

16. What is the oxidation number of S in thiosulfuric acid, $H_2S_2O_3$?

a. +6
b. +4
c. +2
d. 0
e. –2
f. –1

17. Balance the following equation:

$$\_Cl_2(g) \rightarrow \_HClO(aq) + \_Cl^-(aq)$$

The solution is acidic. What is the coefficient for the $H^+(aq)$ in the balanced equation?

a. 1
b. 3
c. 4
d. 5
e. 6
f. 8
g. 10
h. 15

18. What is the value of E° for a cell based on the reaction represented by the following equation:

$$16H^+(aq) + 2MnO_4^-(aq) + 10I^-(aq) \rightarrow 5I_2(aq) + 2Mn^{2+}(aq) + 8H_2O(l)$$

The standard half-cell reduction potentials E° are:

E° ($MnO_4^-|Mn^{2+}$) = +1.51 V
E° ($I_2|I^-$) = +0.54 V

a. +0.40 V
b. –0.40 V
c. +0.97 V
d. –0.97 V
e. +1.10 V
f. –1.10 V
g. +2.05 V
h. –2.05 V

19. What is the value of n, in the expression $\Delta G° = -nFE°$, for the reaction described by the equation listed in Question 18?

a. 1
b. 2
c. 3
d. 4
e. 5
f. 6
g. 7
h. 8
i. 9
j. 10

20. Is the reaction, written in Question 18, spontaneous or not spontaneous?

    a. spontaneous
    b. not spontaneous

21. What is the value of $\Delta G°$ for the reaction as represented by the equation in Question 18?

    a. +468 kJ
    b. −336 kJ
    c. −936 kJ
    d. +187 kJ
    e. −1978 kJ
    f. +98 kJ
    g. −57 kJ
    h. +108 kJ

22. As an electrochemical cell, such as the Daniell cell, is used and "runs down", which of the following processes occurs?

    a. The anode becomes heavier.
    b. The metal ions in solution on the reduction side of the cell increase in concentration.
    c. The metal cathode dissolves.
    d. The concentration of the solution in the anode half of the cell decreases.
    e. Metal ions from the solution are reduced at the cathode.
    f. Charge builds up in the anode half of the cell.

23. How long will it take to plate out 5.00 mol Ag metal from an aqueous solution of Ag^+, if you were to use a current of 100 amp?

    a. 1.08 hr
    b. 1.34 s
    c. 1.34 hr
    d. 13.2 hr
    e. 2.68 hr
    f. 20.5 s

The following two questions (24 and 25) *are written in pairs* (24a & 24b and 25a & 25b). *You may choose which of each pair you answer. The first question of each pair is written specifically for students attending the 2:40 pm lectures by Prof. McHarris. You may, however, answer either one of each pair regardless of the lecture you attend.*

Answer either a or b:

24a. Who was the narrator in the movie shown in the 2:40 pm class before Spring Break?

    a. Dan Rather
    b. Wolf Blitzer
    c. Barbara Walters
    d. Dick Cheney
    e. Richard Nixon
    f. Al Gore

24b. What is the value of $\Delta G°$ at equilibrium?

    a. The slope of the free energy curve as the value of Q (the composition) changes.
    b. zero.
    c. Its maximum value for the system.
    d. The change in the enthalpy of the system when pure reactants are completely changed to pure products.
    e. The standard entropy change for the reaction.
    f. The same value that it has at any point during the reaction.

Answer either a or b:

25a. In the ACIA (Arctic Climate Impact Assessment), what might cause Europe to become much colder as a result of global warming?

a. Europe's heat becoming transferred toward Africa.
b. Melting of Greenland's glaciers.
c. Formation of a new land bridge between Alaska and Asia.
d. Icebergs drifting northward from Antarctica.
e. An increase in the ozone hole over Antarctica.
f. None of the above.

25b. What must be true of a redox reaction?

a. Only one element may be reduced and only one element may be oxidized.
b. The number of electrons lost in the oxidization must equal the number gained in the reduction.
c. The reaction can only occur in aqueous solution.
d. All metals are equally easily oxidized.
e. Metals always lose two electrons when they are oxidized.
f. Copper is more easily oxidized than zinc.

The following question (26) is a bonus question. You do not have to answer this question to achieve a 100% score on the exam. It is provided for those of you who have time on your hands — you should make sure that you have answered all other questions properly before trying this one.

26. Balance the equation for the reaction of chloric acid with arsenic(III) sulfide in water:

$$As_2S_3(s) + HClO_3(aq) \rightarrow H_3AsO_4(aq) + H_2SO_4(aq) + HCl(aq)$$

Note that $As_2S_3(s)$ is insoluble and should be written as this formula unit in the balanced equation. What stoichiometric coefficient is necessary for the HCl in the balanced equation?

a. 3
b. 4
c. 5
d. 6
e. 8
f. 9
g. 12
h. 14
i. 16
j. 20

EXAMINATION 2

CHEMISTRY 142 FALL 2007

Thursday November 8th 2007

1. Calculate the solubility (in moles/liter) of the salt silver phosphate Ag_3PO_4 if the solubility product K_{sp} for the salt is 8.9×10^{-17}.

   a. 2.8×10^{-6} M
   b. 7.8×10^{-4} M
   c. 4.3×10^{-5} M
   d. 2.2×10^{-4} M
   e. 5.4×10^{-5} M
   f. 7.2×10^{-6} M

2. If some potassium phosphate K_3PO_4 was added to the solution of silver phosphate, would you expect the solubility of the silver phosphate to increase, decrease, or stay the same?

   a. decrease
   b. increase
   c. stay the same

3. Calculate the solubility (in moles/liter) of the same salt silver phosphate Ag_3PO_4 in the presence of 5.0×10^{-2} M potassium phosphate K_3PO_4.

   a. 1.2×10^{-5} M
   b. 4.0×10^{-6} M
   c. 2.4×10^{-6} M
   d. 1.9×10^{-7} M
   e. 6.6×10^{-6} M
   f. 4.3×10^{-4} M

4. For this same salt, silver phosphate Ag_3PO_4, consider mixing 500 mL of a 2.0×10^{-5} M potassium phosphate K_3PO_4 solution and 1000 mL of a 1.5×10^{-4} M silver nitrate $AgNO_3$ solution. What would you observe in the resulting 1.5 L solution?

   a. No precipitation of $Ag_3PO_4(s)$ because $Q_{sp} > K_{sp}$
   b. No precipitation of $Ag_3PO_4(s)$ because $Q_{sp} < K_{sp}$
   c. No precipitation of $Ag_3PO_4(s)$ because $Q_{sp} = K_{sp}$
   d. Precipitation of $Ag_3PO_4(s)$ because $Q_{sp} > K_{sp}$
   e. Precipitation of $Ag_3PO_4(s)$ because $Q_{sp} < K_{sp}$
   f. Precipitation of $Ag_3PO_4(s)$ because $Q_{sp} = K_{sp}$

5. The solubility product K_{sp} for silver(I) iodide AgI is 8.3×10^{-17}.
   The solubility product K_{sp} for silver(I) chloride AgCl is 1.8×10^{-10}.

   What is the ratio of the concentration of chloride ions to the concentration of iodide ions ($[Cl^-]/[I^-]$) in a solution containing Ag^+, Cl^-, and I^- ions in equilibrium with solid AgI and solid AgCl? Consider the equilibrium:

   $$AgCl(s) + I^-(aq) \rightleftharpoons AgI(s) + Cl^-(aq)$$

   a. 7.3×10^{8}
   b. 5.0×10^{6}
   c. 4.1×10^{7}
   d. 1.5×10^{3}
   e. 9.2×10^{9}
   f. 4.6×10^{-7}
   g. 6.8×10^{2}
   h. 2.2×10^{6}

6. Magnesium carbonate is a sparingly soluble salt ($K_{sp} = 6.8 \times 10^{-6}$). How would the addition of acid affect the solubility of magnesium carbonate?

   a. The solubility would decrease
   b. The solubility would increase
   c. The addition of acid would have no effect

Questions 7 through 14 refer to the following system. Some questions are worth only 6 points each.

The pressure on a system containing 1.5 moles of an ideal monatomic gas decreases from 12.0 atm to 4.0 atm as the system moves from state A to state B at a constant volume of 3.0 L. The system then expands at a constant pressure of 4.0 atm to reach state C at which point the temperature is again what it was at state A.

It is recommended that you draw these changes on the diagram to the right.

$R = 0.08206 \text{ L atm K}^{-1}\text{mol}^{-1} = 8.3145 \text{ J K}^{-1}\text{mol}^{-1}$

P

V

7. *(6 points)* What is the volume in liters at point C?

a. 1.0 L b. 2.0 L c. 4.5 L d. 6.0 L e. 7.5 L f. 9.0 L g. 12.0 L h. 15.0 L

8. *(6 points)* What is the temperature at point A?

a. zero°C b. 6.7°C c. 7.3°C d. 11.2°C e. 13.9°C f. 19.3°C g. 25.8°C h. 45.7°C

9. *(6 points)* What is the temperature at point B?

a. 23.8K b. 44.7K c. 52.9K d. 61.2K e. 66.1K f. 73.1K g. 82.1K h. 97.5K

10. How much work is done by the system in moving at constant pressure from B to C?

a. w = –12 J b. w = –24 J c. w = –36 J d. w = –556 J e. w = –1860 J f. w = –2430 J g. w = –3650 J h. w = –4450 J

11. How much heat is absorbed by the system, or liberated by the system, in moving at constant pressure from B to C?

a. q = –60 J b. q = +60 J c. q = +487 J d. q = –487 J e. q = +731 J f. q = –731 J g. q = +6080 J h. q = –6080 J

12. What is the change in the internal energy of the system ΔE as it moves from B to C?

a. ΔE = zero J b. ΔE = +36 J c. ΔE = –2370 J d. ΔE = –3650 J e. ΔE = +3650 J f. ΔE = +8510 J g. ΔE = –8510 J h. ΔE = –12,330 J

13. *(6 points)* What is the change in the internal energy of the system ΔE as it moves from A to B?

a. ΔE = zero J b. ΔE = +36 J c. ΔE = –2370 J d. ΔE = –3650 J e. ΔE = +3650 J f. ΔE = +8510 J g. ΔE = –8510 J h. ΔE = –12,330 J

14. How much work must be done on the system to compress it reversibly and isothermally from C to A?

a. 40 J b. 265 J c. 2500 J d. 3000 J e. 3500 J f. 4000 J g. 4500 J h. 5000 J

15. Which of the following properties is not a state function?

a. enthalpy	c. energy	e. temperature	g. pressure
b. volume	d. heat	f. free energy	h. entropy

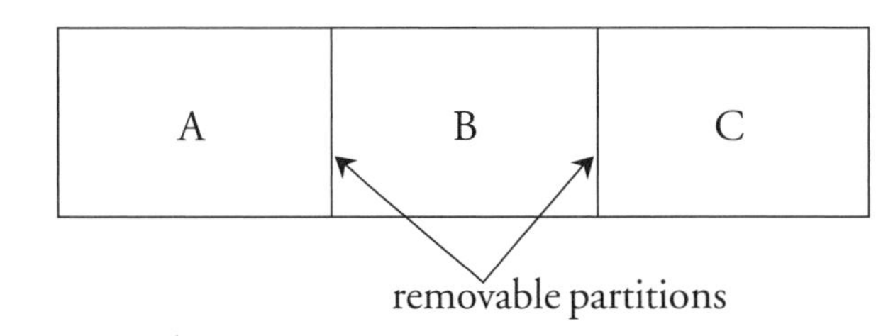

16. The vessel shown above has three compartments labeled A, B, and C separated by removable partitions. All three compartments are equal in volume. Suppose that three molecules are initially contained within compartment A. Consider this to be a single microstate of the system. If the partition between A and B is removed, what is the *change* in the number of microstates available to the system?

a. 0	c. 4	e. 8	g. 15
b. 1	d. 3	f. 7	h. 19

17. Refer again to the diagram for the previous question. Suppose now that the partition between compartments B and C is removed. What is the change in the entropy of the system (in JK^{-1}) as the molecules move from occupying *only* A to occupying A, B, and C?

a. 7.6×10^{-22}	c. 2.3×10^{-23}	e. 4.7×10^{-21}	g. 7.1×10^{-24}
b. 1.1×10^{-19}	d. 4.6×10^{-23}	f. 5.9×10^{-22}	h. 3.7×10^{-21}

18. A system consisting of 2.5 moles of an ideal gas is heated. As a result the temperature rises from 298K to 545K and the volume increases from 3.5 liters to 6.4 liters. The pressure at the end is the same as it was at the beginning. What is the change in the entropy of the system?

a. 12.6 JK^{-1}	c. 23.6 JK^{-1}	e. 31.4 JK^{-1}	g. 51.2 JK^{-1}
b. 18.8 JK^{-1}	d. 28.7 JK^{-1}	f. 42.7 JK^{-1}	h. 61.9 JK^{-1}

19. The Second Law of Thermodynamics can be expressed in many ways—including some of the statements below. Which statement is not an expression of the Second Law?

a. All systems spontaneously approach equilibrium.
b. For a spontaneous change, the entropy of the universe must increase.
c. The change in the free energy of a system, ΔG, equals $\Delta H - T\Delta S$.
d. In an isolated system, any spontaneous change results in an increase in the entropy of the system.
e. For a spontaneous process, ΔG must be less than zero.

20. What is the oxidation number of chromium in rubidium dichromate $Rb_2Cr_2O_7$?

a. +1	c. +3	e. +5	g. +8	i. +14
b. +2	d. +4	f. +6	h. +12	j. +16

21. Consider the following redox reaction. What is the total number of electrons lost by the element being oxidized, and the number of electrons being gained by the element being reduced, in this equation for the reaction?

$$3\ PbO + 2\ NH_3 \rightarrow 3\ Pb + N_2 + 3\ H_2O$$

a. 1 b. 2 c. 3 d. 4 e. 5 f. 6 g. 7 h. 8 i. 9 j. 10

22. Write and balance the equation for the oxidation of phosphine PH_3 by chromate CrO_4^{2-} to produce phosphorus P_4 and the chromium(III) hydroxide $Cr(OH)_4^-$ in basic solution. How many water molecules are on which side of the balanced equation?

a. 2 on the left
b. 3 on the left
c. 4 on the left
d. 6 on the left
e. 8 on the left
f. 2 on the right
g. 3 on the right
h. 4 on the right
i. 6 on the right
j. 8 on the right

23. The value of E° for a cell based upon the reaction represented by the following equation

$$2\ Ag(s) + Zn^{2+}(aq) + 2\ OH^-(aq) \rightarrow Ag_2O(s) + Zn(s) + H_2O(l)$$

is –1.104 V. The fact that this value is negative indicates that...

a. zinc is not oxidized in the reaction
b. the reaction only occurs in basic conditions
c. the reaction is not spontaneous in the direction in which it is written
d. silver(I) oxide is insoluble in water
e. silver loses only one electron in the oxidation process

24. If the standard half-cell reduction potential for zinc, $E^\circ_{red}(Zn^{2+}|Zn) = -0.762$ V, calculate the standard half-cell reduction potential for silver, $E^\circ_{red}(Ag^+|Ag)$ in basic conditions, based upon the data provided in the previous question.

a. –0.342 V b. +0.342 V c. –0.762 V d. +0.762 V e. –1.104 V f. +1.104 V g. –1.866 V h. +1.866 V

25. What is the value of ΔG° for the reaction represented by the equation in Question 23?

a. –107 kJ b. +107 kJ c. –147 kJ d. +147 kJ e. –213 kJ f. +213 kJ g. –426 kJ h. +426 kJ

26. How long will it take to plate out 50 grams of copper metal from an aqueous solution of Cu^{2+} if you were to use a current of 15 amp?

a. 42 min b. 1.04 hr c. 1.92 hr d. 2.81 hr e. 3.56 hr f. 3.77 hr g. 4.21 hr h. 23 days

EXAMINATION 2

CHEMISTRY 142 SPRING 2008

Wednesday March 26th 2008

For the first four questions, consider the following data for various hydroxides at 25°C:

AgOH	$K_{sp} = 1.5 \times 10^{-8}$	$Fe(OH)_3$	$K_{sp} = 1.1 \times 10^{-36}$	$Zn(OH)_2$	$K_{sp} = 4.5 \times 10^{-17}$
$Fe(OH)_2$	$K_{sp} = 1.6 \times 10^{-14}$	$Mg(OH)_2$	$K_{sp} = 1.2 \times 10^{-11}$		

1. *(3 pts.)* Which of the following is the correct solubility product constant expression for iron(III) hydroxide $Fe(OH)_3$?

   a. $K_{sp} = [Fe^{3+}][OH^-]$
   b. $K_{sp} = [Fe^{3+}]^2[OH^-]^3$
   c. $K_{sp} = [Fe^{3+}]^3[OH^-]^2$
   d. $K_{sp} = [Fe^{3+}][OH^-]^3$
   e. $K_{sp} = [Fe^{3+}]^3[OH^-]$
   f. $K_{sp} = [Fe^{3+}][OH^-]^3/[Fe(OH)_3]$

2. Which hydroxide has the highest molar solubility in water at 25°C?

   a. AgOH
   b. $Fe(OH)_2$
   c. $Fe(OH)_3$
   d. $Mg(OH)_2$
   e. $Zn(OH)_2$

3. What is the hydroxide ion concentration in a solution saturated in $Fe(OH)_2$ and containing 0.025 M $FeCl_2$ at 25°C?

   a. 1.6×10^{-5} M
   b. 2.0×10^{-5} M
   c. 1.3×10^{-7} M
   d. 4.0×10^{-7} M
   e. 8.0×10^{-7} M
   f. 6.3×10^{-8} M
   g. 3.2×10^{-13} M
   h. 6.4×10^{-13} M

4. Which reagent would you add to a system consisting of solid $Mg(OH)_2$ in equilibrium with a saturated solution of $Mg(OH)_2$ to increase the solubility of Mg^{2+} ions in the solution?

   a. sodium hydroxide
   b. nitric acid
   c. more water
   d. any weak base
   e. more magnesium hydroxide
   f. ethanol

The diagram on the following page illustrates the change in the free energy of a system as the reactants form products. In this particular case the system is isolated from its surroundings.
Questions 5 through 9 refer to this diagram (3 pts.each).

5. How is the free energy of the system changing in the direction of the arrow at point S?

   a. ΔG is constant throughout the reaction
   b. ΔG is negative (the free energy is decreasing)
   c. ΔG is positive (the free energy is increasing)

6. What is the difference X called?

   a. ΔH°
   b. ΔG
   c. ΔS°
   d. ΔH
   e. ΔG°
   f. ΔE

7. How does X change as the reaction proceeds?

a. It becomes less negative until it reaches zero.
b. It decreases in value until it reaches zero; the reaction then stops.
c. It doesn't change.

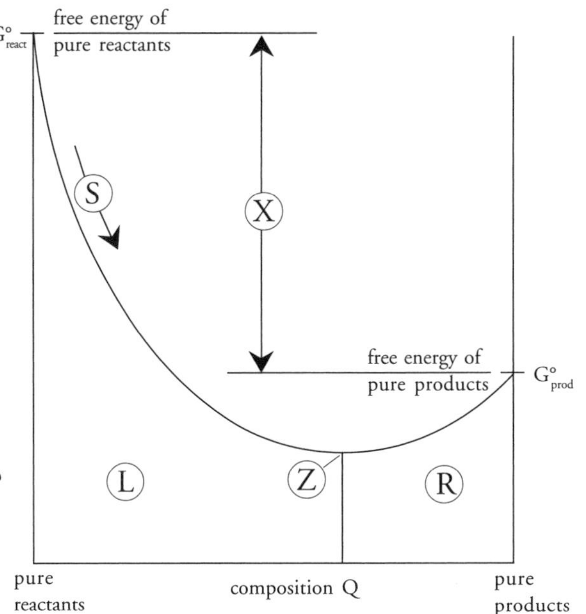

8. What can you say about the entropy of this isolated system at point Z?

a. It's equal to zero
b. It's at a maximum
c. It's at a minimum

9. What is the relationship between K and Q at point L?

a. Q = K
b. Q < K
c. Q > K

10. Which of the following statements is(are) true?

I. The entropy of a liquid increases when it is vaporized.
II. An exothermic reaction will always be product–favored.
III. If $\Delta G^\circ_{rxn} < 0$ (i.e., a negative value), then the reaction will have an equilibrium constant greater than 1.
IV. Reactions with a positive ΔH°_{rxn} and a positive ΔS°_{rxn} can never be product–favored.

a. All are true
b. I and II
c. I and III
d. II and III
e. III and IV
f. II and IV
g. I, II, and III
h. II, III, and IV

11. The Second Law of Thermodynamics can be expressed in many ways—including some of the statements below. Which statement is *not* an expression of the Second Law?

a. All systems spontaneously approach equilibrium.
b. For spontaneous change, the entropy of the universe must increase.
c. $\Delta G^\circ_{rxn} = \Sigma[\Delta G^\circ_f(\text{products}) - \Delta G^\circ_f(\text{reactants})]$.
d. Not all the energy released in a chemical reaction is necessarily available to do work.
e. For spontaneous change, ΔG must be less than zero.

12. A system consisting of 4.0 moles of an ideal gas is heated. As a result the temperature rises from 298K to 447K and the volume increases from 5.0 liters to 7.5 liters. The pressure at the end is the same as it was at the beginning. What is the change in the entropy of the system?

a. 9.6 JK^{-1}
b. 13.5 JK^{-1}
c. 20.2 JK^{-1}
d. 22.7 JK^{-1}
e. 29.1 JK^{-1}
f. 33.7 JK^{-1}
g. 41.8 JK^{-1}
h. 53.9 JK^{-1}

13. Consider a box containing 6 Ar atoms; what is the entropy change associated with doubling the volume available to the 6 Ar atoms at 25°C?

a. 5.74×10^{-23} J K^{-1}
b. 9.57×10^{-24} J K^{-1}
c. 2.47×10^{-23} J K^{-1}
d. -3.43×10^{-23} J K^{-1}
e. 4.95×10^{-23} J K^{-1}
f. -4.95×10^{-23} J K^{-1}
g. 4.79×10^{-23} J K^{-1}
h. 3.43×10^{-23} J K^{-1}
i. -5.74×10^{-23} J K^{-1}

14. *(3 pts.)* What is the probability that all six Ar atoms would remain in the original one-half of the vessel and not spread out to fill the entire volume ?

a. 1/2	c. 1/6	e. 1/16	g. 1/36	i. 1/64
b. 1/4	d. 1/8	f. 1/24	h. 1/48	j. 1/128

For the following questions (15–23), consider the following system (3 pts.each):

A 2.5 mol sample of an ideal monatomic gas at a pressure of 3.0 atm occupies a volume of 18.0 L. Call this state of the system A. Suppose that the gas is then heated at constant volume to reach state B, which has a pressure of 9.0 atm. Subsequently, the temperature of the gas is reduced at constant pressure to reach state C, where the temperature is the same as at state A.

P

V

15. What is the temperature of the gas in state A?

a. –270°C	c. 0°C	e. 263°C
b. –10°C	d. 10°C	f. 2.6°C

16. How much work is done on the system as it moves *isothermally* and *reversibly* from state A to state C?

a. 0.0 J	c. 59 J	e. 3.6 kJ	g. 6.0 kJ
b. –23 J	d. –59 J	f. –3.6 kJ	h. –6.0 kJ

17. How much work is done on the system as it moves from state A to state B?

a. 0.0 J	c. 54 J	e. 162 J	g. 10.9 kJ
b. 23 J	d. 108 J	f. 5.5 kJ	h. –10.9 kJ

18. How much work is done on the system as it moves from state B to state C?

a. 0.0 J	c. –108 J	e. –6.0 kJ	g. 10.9 kJ
b. 23 J	d. 108 J	f. 6.0 kJ	h. –10.9 kJ

19. How much heat is absorbed or released when the system moves *reversibly* from state A to state C?

a. 0.0 J	c. 54 J	e. 3.6 kJ	g. 6.0 kJ
b. 23 J	d. –54 J	f. –3.6 kJ	h. –6.0 kJ

20. How much heat does the system absorb or liberate as it moves from state A to state C along the path ABC?

a. 0.0 J	c. –5.0 kJ	e. –10.9 kJ	g. –16.4 kJ	i. –27.4 kJ
b. 5.0 kJ	d. 10.9 kJ	f. 16.4 kJ	h. 27.4 kJ	j. –43.8 kJ

21. What is the change in the internal energy of the system in moving from state A to state B?

a. 0.0 J	c. –6.6 kJ	e. –10.9 kJ	g. –16.4 kJ	i. –27.4 kJ
b. 6.6 kJ	d. 10.9 kJ	f. 16.4 kJ	h. 27.4 kJ	j. –43.8 kJ

22. What is the change in the internal energy of the system in moving from state B to state C?

a. 0.0 J b. 6.6 kJ c. −6.6 kJ d. 10.9 kJ e. −10.9 kJ f. 16.4 kJ g. −16.4 kJ h. 27.4 kJ i. −27.4 kJ j. −43.8 kJ

23. *(4 pts.)* What are the signs of ΔG, ΔH, and ΔS in the condensation of water at 1.0 atm pressure and at 50°C and at 140°C?

	50°C			140°C		
	ΔG	ΔH	ΔS	ΔG	ΔH	ΔS
a.	−	+	+	−	+	−
b.	+	+	+	−	+	+
c.	−	−	+	−	−	−
d.	+	−	+	−	−	+
e.	−	+	−	+	+	−
f.	+	+	−	+	+	+
g.	−	−	−	+	−	−
h.	+	−	−	+	−	+

24. *(3 pts.)* Determine the oxidation numbers of nitrogen in the following compounds. Then add them up using the correct signs. What is the sum?

$NaNH_2$ N_2H_4 HNO_2 N_2O_4

a. −6 b. −4 c. −3 d. −2 e. −1 f. 0 g. +1 h. +2 i. +3 j. +4

25. Balance the following redox reaction under basic conditions

$$H_2O_2(aq) + Cl_2O_7(aq) \rightarrow ClO_2^-(aq) + O_2(g)$$

What is the stoichiometric coefficient of water in the balanced equation? Indicate also whether water appears as a reactant or product in the balanced equation.

a. 1 reactant
b. 2 reactant
c. 3 reactant
d. 4 reactant
e. 6 reactant
f. 1 product
g. 2 product
h. 3 product
i. 5 product
j. 6 product

26. *(3 pts.)* Find the standard cell potential E° at 25°C for a cell in which Zn is oxidized to Zn^{2+} at the anode and Fe^{3+} is reduced to Fe^{2+} at the cathode.

$E^\circ_{red}(Zn^{2+}|Zn) = -0.763$ V and $E^\circ_{red}(Fe^{3+}|Fe^{2+}) = +0.770$ V

a. +0.007 V b. −0.007 V c. +0.770 V d. −0.770 V e. +0.763 V f. −0.763 V g. +1.533 V h. −1.533 V

27. Now determine the cell potential E at 25°C for the following cell. Note that the platinum electrode is inert and it may help if you write the balanced equation for the reaction that occurs in the cell.

$$Zn(s) \mid Zn^{2+}(aq,\ 0.726M) \parallel Fe^{3+}(aq,\ 0.274M) \mid Fe^{2+}(aq,\ 0.736M) \mid Pt(s)$$

a. +1.554 V b. +1.524 V c. +1.490 V d. +1.512 V e. −0.014 V f. +0.028 V g. +0.024 V h. −0.010 V

EXAMINATION 2

CHEMISTRY 142 FALL 2008

Thursday November 6th 2008

1. What is the solubility of copper(II) hydroxide if the K_{sp} for copper(II) hydroxide is 2.2×10^{-20}?

a.	5.5×10^{-21} M	c.	1.5×10^{-10} M	e.	2.2×10^{-7} M	g.	4.4×10^{-5} M
b.	6.3×10^{-15} M	d.	2.8×10^{-7} M	f.	1.8×10^{-7} M	h.	3.7×10^{-4} M

2. What is the solubility of copper(II) hydroxide in a solution with pH equal to 12.0? The K_{sp} for copper(II) hydroxide is 2.2×10^{-20}.

a.	2.2×10^{-16} M	c.	2.8×10^{-7} M	e.	3.6×10^{-6} M	g.	7.2×10^{3} M
b.	1.5×10^{-10} M	d.	1.8×10^{-7} M	f.	4.4×10^{-5} M	h.	2.2×10^{4} M

3. How will the addition of each of the following solutes affect the solubility of copper(I) bromide? The K_{sp} for copper(I) bromide is 4.2×10^{-8}.
Choose the row in which all responses are correct.

	HBr	HCl	NH_3	KBr
a.	decrease	increase	increase	decrease
b.	decrease	no change	increase	decrease
c.	increase	no change	decrease	increase
d.	increase	increase	decrease	decrease
e.	no change	no change	increase	decrease
f.	no change	increase	increase	increase

4. 100 mL of a 5.0×10^{-4} M Ag^+ solution is added to 75 mL of a 1.0×10^{-4} M CrO_4^{2-} solution. The K_{sp} for silver chromate is 9.0×10^{-12}. Which of the following statements is true regarding this solution?

a. $Q_{sp} = K_{sp}$; the system is at equilibrium
b. $Q_{sp} > K_{sp}$; precipitation occurs
c. $Q_{sp} > K_{sp}$; no precipitation occurs
d. $Q_{sp} < K_{sp}$; precipitation occurs
e. $Q_{sp} < K_{sp}$; no precipitation
f. $Q_{sp} < K_{sp}$; the system is at equilibrium

5. A system consisting of a cylinder with a movable piston absorbs 280 J of energy as heat. As a result, the piston moves up in the cylinder and does 130 J of work against a constant external pressure of 1.0 atmosphere. What is the change in the internal energy ΔE of the system?

a.	+130 J	c.	+150 J	e.	+280 J	g.	+410 J
b.	−130 J	d.	−150 J	f.	−280 J	h.	−410 J

6. Match each one of the following processes with the correct statement regarding the change in internal energy of the system ΔE. *(Choose the row in which all responses are correct.)*

	adiabatic	*isothermal*	*constant volume*	*constant pressure*
a.	ΔE = zero	ΔE = w	ΔE = zero	ΔE = ΔH
b.	ΔE = w	ΔE = zero	ΔE = q_v	ΔE = q_p – PΔV
c.	ΔE = w	ΔE = zero	ΔE = q_v	ΔE = q_p
d.	ΔE = q + w	ΔE = zero	ΔE = w	ΔE = q_v
e.	ΔE = q + w	ΔE = –q	ΔE = PΔV	ΔE = w
f.	ΔE = w	ΔE = –w	ΔE = q_v	ΔE = ΔH

Questions 7 - 11 refer to the following system.

The pressure on a system containing 2.0 moles of an ideal monatomic gas decreases from 15.0 atm to 3.0 atm as the system moves from state A to state B at a constant volume of 4.0 L. The system then expands at a constant pressure of 3.0 atm to reach state C at which point the temperature is the same as it was at state A.

7. What is the temperature at state A?

a. zero °C
b. 19.3 °C
c. 25.8 °C
d. 45.7 °C
e. 73.1 °C
f. 92.5 °C
g. 200 °C
h. 366 °C

8. How much work must be done on the system as it moves reversibly and isothermally from C to A?

a. 0.72 kJ
b. 1.9 kJ
c. 4.0 kJ
d. 9.8 kJ
e. 15 kJ
f. 23 kJ
g. 29 kJ
h. 34 kJ

9. How much work is done by the system moving at constant pressure from B to C?

a. zero J
b. –48 J
c. –556 J
d. –1520 J
e. –2640 J
f. –3180 J
g. –4860 J
h. –6850 J

10. How much heat is absorbed by the system, or liberated by the system, in moving at constant pressure from B to C?

a. +6080 J
b. –6080 J
c. +7300 J
d. –7300 J
e. +12160 J
f. –12160 J
g. +22370 J
h. –22370 J

11. What is the change in the internal energy of the system ΔE as it moves from A to B?

a. zero J
b. –7300 J
c. –3650 J
d. –6850 J
e. +7300 J
f. +6850 J
g. +3650 J
h. –12160 J

12. Suppose a vessel contains 6 molecules of gas that are initially constrained to one-fourth of the vessel and then are allowed to occupy the entire vessel. The expansion occurs at a constant temperature. What is the entropy change for the isothermal expansion of the gas?

a. 4.61×10^{-23} JK^{-1}
b. 69.2 JK^{-1}
c. 3.72×10^{-18} JK^{-1}
d. 1.15×10^{-22} JK^{-1}
e. 9.90×10^{-23} JK^{-1}
f. 59.6 J K^{-1}
g. 2.20×10^{-17} JK^{-1}
h. 1.07×10^{-23} JK^{-1}

13. Calculate $\Delta S°$ for the reaction $H_2(g) + Cl_2(g) \rightarrow 2\,HCl(g)$

given $S°\ H_2(g) = 131.0\ JK^{-1}mol^{-1}$
$S°\ Cl_2(g) = 223.0\ JK^{-1}mol^{-1}$
$S°\ HCl(g) = 187.0\ JK^{-1}mol^{-1}$

a. $+541\ JK^{-1}$
b. $-541\ JK^{-1}$
c. $+167\ JK^{-1}$
d. $-167\ JK^{-1}$
e. $+95\ JK^{-1}$
f. $-95\ JK^{-1}$
g. $+20\ JK^{-1}$
h. $-20\ JK^{-1}$

14. What are the expected signs for ΔG, ΔH, and ΔS for the exothermic reaction shown below?

$$2\,NH_4NO_3(s) \rightarrow 2\,N_2(g) + 4\,H_2O(g) + O_2(g)$$

	ΔG	ΔH	ΔS		ΔG	ΔH	ΔS
a.	–	–	–	e.	+	–	–
b.	–	–	+	f.	+	–	+
c.	–	+	–	g.	+	+	–
d.	–	+	+	h.	+	+	+

15. Consider this equilibrium: $N_2O_4(g) \rightleftharpoons 2\,NO_2(g)$

At 298 K, the equilibrium constant K_p is 0.113 and standard free-energy change $\Delta G°$ for this reaction is 5.4 kJ mol^{-1}. In one experiment, the initial pressure of N_2O_4 is 0.453 atm and the initial pressure of NO_2 is 0.122 atm. What is the value of ΔG under these conditions and in which direction must the reaction spontaneously shift in order to establish equilibrium?

a. $\Delta G = -8.5$ kJ; shift left
b. $\Delta G = -8.5$ kJ; shift right
c. $\Delta G = -8.5$ kJ; no shift
d. $\Delta G = -3.1$ kJ; shift left
e. $\Delta G = -3.1$ kJ; shift right
f. $\Delta G = -3.1$ kJ; no shift
g. $\Delta G = +2.1$ kJ; shift left
h. $\Delta G = +2.1$ kJ; shift right
i. $\Delta G = +2.1$ kJ; no shift

16. What are the oxidation numbers for H, S, and O in the molecule $H_2S_2O_6$?
Choose the row in which all responses are correct.

	H	S	O		H	S	O
a.	+1	+5	–12	e.	+2	+10	–12
b.	+1	+10	–12	f.	+2	+5	–2
c.	+1	+5	–2	g.	+2	+5	–12
d.	+1	+10	–2	h.	+2	+10	–2

17. Balance the following redox reaction in basic solution: $CN^- + MnO_4^- \rightarrow CNO^- + MnO_2$

What is the coefficient for the water in the balanced equation? Also indicate if the water is on the reactant or product side of the equation.

a. 1 reactant
b. 2 reactant
c. 3 reactant
d. 4 reactant
e. 5 reactant
f. 1 product
g. 2 product
h. 3 product
i. 4 product
j. 5 product

18. Calculate the standard cell potential for the cell $Cl^-(aq) \mid Cl_2(g) \parallel Ce^{4+}(aq) \mid Ce^{3+}(aq)$

Given the half-cell standard reduction potentials: $E°(Ce^{4+} \mid Ce^{3+}) = +1.61$ V
$E°(Cl_2 \mid Cl^-) = +1.36$ V

a. +0.25 V c. +1.11 V e. +2.97 V g. +4.59 V
b. –0.25 V d. –1.11 V f. –2.97 V h. –4.59 V

19. What is the value of $\Delta G°$ for the reaction described in question 18?
(Hint: Write the balanced equation for the cell!)

a. –7.3 kJ c. –24 kJ e. –48 kJ g. –287 kJ
b. –15 kJ d. –36 kJ f. –136 kJ h. –573 kJ

20. Calculate the cell potential at 25°C based on the reaction:

$$Sn(s) + Pb^{2+}(aq) \rightarrow Pb(s) + Sn^{2+}(aq) \quad E° = 0.012 \text{ V}$$

when $[Pb^{2+}] = 0.0010$ M and $[Sn^{2+}] = 0.50$ M

a. +0.092 V c. +0.055 V e. zero V g. –0.068 V
b. +0.078 V d. +0.031 V f. –0.160 V h. –1.12 V

21. A constant current of 0.912 A is passed through an electrolytic cell containing molten $MgCl_2$ for 18 hours. What mass of magnesium is produced?

a. 0.912 g c. 2.64 g e. 5.72 g g. 10.8 g
b. 1.84 g d. 3.31 g f. 7.45 g h. 14.9 g

22. Estimate the melting point of an unknown substance in °C if the heat of fusion of the substance is 10.9 kJ mol^{-1} and the entropy change for the change in state of the substance from solid to liquid at its melting point is 39.1 $JK^{-1}mol^{-1}$.

a. 0.28 °C c. 5.6 °C e. 75.2 °C g. 157 °C
b. 3.6 °C d. 16.4 °C f. 126 °C h. 279 °C

EXAMINATION 2

CHEMISTRY 142 Spring 2009

Thursday April 2nd 2009

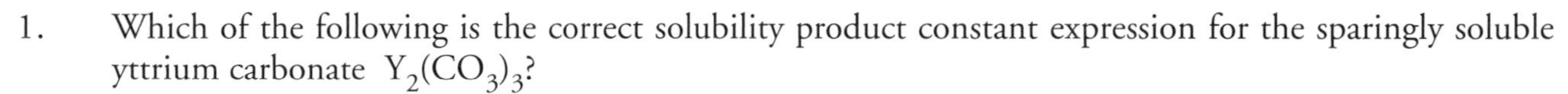

1. Which of the following is the correct solubility product constant expression for the sparingly soluble yttrium carbonate $Y_2(CO_3)_3$?

a. $K_{sp} = [Y^{3+}][CO_3^{2-}]$
b. $K_{sp} = [Y_2][(CO_3)_3]$
c. $K_{sp} = [Y^{3+}]^3[CO_3^{2-}]^2$
d. $K_{sp} = [Y^{3+}]^2[CO_3^{2-}]^3/[Y_2(CO_3)_3]$
e. $K_{sp} = [Y^{3+}]^2[CO_3^{2-}]^3$
f. $K_{sp} = [Y^{3+}]^3[CO_3^{2-}]^2/[Y_2(CO_3)_3]$
g. $K_{sp} = (2[Y^{3+}])^2(3[CO_3^{2-}])^3$

2. In a saturated solution of yttrium carbonate, what is the relationship between the concentrations of the yttrium ions and the carbonate ions?

a. $[Y^{3+}] = [CO_3^{2-}]$
b. $2 \times [Y^{3+}] = 3 \times [CO_3^{2-}]$
c. $3 \times [Y^{3+}] = 2 \times [CO_3^{2-}]$
d. $[Y^{3+}]^2 = [CO_3^{2-}]^3$
e. $[Y^{3+}]^3 = [CO_3^{2-}]^2$
f. $2 \times [Y^{3+}]^3 = 3 \times [CO_3^{2-}]^2$

3. Calculate the concentration of thallium ions Tl^+ in a saturated solution of thallium sulfide that also contains 0.10 M sodium sulfide Na_2S.

K_{sp} for thallium sulfide $Tl_2S = 6.0 \times 10^{-22}$

a. 3.0×10^{-21} M
b. 6.0×10^{-21} M
c. 1.9×10^{-11} M
d. 3.9×10^{-11} M
e. 7.7×10^{-11} M
f. 5.3×10^{-8} M
g. 6.0×10^{-20} M
h. 9.2×10^{-21} M

4. The solubility product for BaF_2 is $K_{sp} = 2.4 \times 10^{-5}$.
The acid ionization constant for HF is $K_a = 7.2 \times 10^{-4}$.
K_w for water $= 1.0 \times 10^{-14}$.
Calculate the equilibrium constant for the reaction at 25°C:

$$BaF_2(s) + 2\,H_2O(l) \rightleftharpoons Ba^{2+}(aq) + 2\,HF(aq) + 2\,OH^-(aq)$$

a. 4.6×10^{1}
b. 3.3×10^{-2}
c. 1.7×10^{-8}
d. 1.2×10^{-11}
e. 3.3×10^{-16}
f. 2.4×10^{-19}
g. 5.3×10^{-20}
h. 4.6×10^{-27}
i. 1.7×10^{-32}
j. 3.3×10^{-33}

5. An excess amount of barium hydroxide ($Ba(OH)_2$) is added to water to make a saturated solution. The pH of the solution was determined to be 13.33 at 25°C. Estimate the value of the solubility product K_{sp} for barium hydroxide.

a. 2.3×10^{-2}
b. 4.9×10^{-3}
c. 4.7×10^{-4}
d. 5.8×10^{-5}
e. 1.3×10^{-6}
f. 1.8×10^{-8}
g. 4.4×10^{-11}
h. 7.2×10^{-14}
i. 2.4×10^{-17}
j. 5.1×10^{-21}

Questions 6 through 11 refer to the following system.

Consider a sample of an ideal monatomic gas at a pressure of 8.0 atm, a temperature of 50.0°C, occupying a volume of 15.0. Call this state of the system A.

Suppose the gas is then cooled at constant volume to reach state B, where the pressure is 3.0 atm.

Subsequently, the gas is permitted to expand at constant pressure to reach state C, which has the same temperature as state A.

It is recommended that you draw these changes on the diagram to the right.

$R = 0.08206 \text{ L atm K}^{-1}\text{mol}^{-1} = 8.3145 \text{ J K}^{-1}\text{mol}^{-1}$

6. How many moles of gas are present in this system?

a. 0.034 b. 0.045 c. 0.22 d. 4.5 e. 11 f. 22 g. 29 h. 34 i. 45 j. 57

7. How much work is done as the system moves isothermally and reversibly from State A to State C?

a. w = –12 kJ b. w = –2.6 kJ c. w = –1.8 kJ d. w = –180 J e. w = +180 J f. w = +1.8 kJ g. w = +2.6 kJ h. w = +12 kJ

8. How much work is done as the system moves from State A to State B?

a. w = 0.0 J b. w = –12 kJ c. w = –4.5 kJ d. w = –75 J e. w = –7.6 kJ f. w = +4.5 kJ g. w = +7.6 kJ h. w = +12 kJ

9. How much work is done as the system moves from State B to State C?

a. w = 0.0 J b. w = –12 kJ c. w = –4.5 kJ d. w = –75 J e. w = –7.6 kJ f. w = +4.5 kJ g. w = +7.6 kJ h. w = +12 kJ

10. What is the change in the internal energy of the system ΔE as it moves from A to B?

a. ΔE = 4.2 kJ b. ΔE = –2.5 kJ c. ΔE = +11 kJ d. ΔE = +2.5 kJ e. ΔE = –4.2 kJ f. ΔE = –11 kJ g. ΔE = +19 kJ h. ΔE = –19 J

11. How much heat does the system liberate or absorb as it moves from State A to State C along the path ABC?

a. q = zero J b. q = +2.6 kJ c. q = –2.6 kJ d. q = +7.6 kJ e. q = –7.6 kJ f. q = +12 kJ g. q = –12 kJ h. q = –19 kJ

12. Determine whether the following statements are true or false.

I At equilibrium, total entropy is minimized.
II When $\Delta G°$ is negative (< 0), the system favors the reactants at equilibrium.
III If ΔG is negative (< 0), then the reaction quotient Q > K.
IV At equilibrium, $\Delta G°$ = zero.

	I	II	III	IV
a.	true	true	true	true
b.	false	true	true	true
c.	true	false	false	true
d.	false	false	true	false
e.	false	false	false	true
f.	true	false	false	false
g.	false	true	false	false
h.	true	true	false	false
i.	true	false	true	false
j.	false	false	false	false

13. Some changes that occur in a system are path-independent. In other words, it doesn't matter what route is taken between the initial and final states of the system. Other changes depend upon the path taken. These are referred to as path-dependent. Which one of the following changes is path-dependent?

a. an enthalpy change
b. an increase in the volume of the system
c. the work done on or by the system
d. a change in free energy
e. a change in internal energy of the system
f. a change in temperature
g. an increase in the entropy of the system
h. a decrease in pressure with no change in volume

14. Imagine a 5 liter section of a larger 25 liter vessel. Now imagine 20 molecules in the 25 liter vessel. What is the probability that all 20 molecules will occupy only the smaller 5 liter section within the larger vessel, rather than spreading out to occupy the entire 25 liters?

a. 1.0×10^{-3}
b. 1.0×10^{-6}
c. 1.0×10^{-9}
d. 1.0×10^{-12}
e. 1.0×10^{-14}
f. 1.0×10^{-17}
g. 1.0×10^{-21}
h. 1.0×10^{-29}

15. Refer again to the previous question. Suppose that the 20 molecules were confined to the smaller 5 liter section of the larger vessel. The gas is then allowed to expand so that the 20 molecules are able to occupy the entire 25 liter space. What is the entropy change (in JK^{-1}) for this process?

a. 7.6×10^{-8}
b. 2.7×10^{-10}
c. 2.2×10^{-11}
d. 3.3×10^{-15}
e. 4.4×10^{-22}
f. 7.6×10^{-25}
g. 3.1×10^{-27}
h. 3.4×10^{-32}

16. What is the oxidation number of carbon in potassium oxalate $K_2C_2O_4$?

a. +6
b. +5
c. +4
d. +3
e. +2
f. +1
g. 0
h. −1
i. −2
j. −4

17. Some amount of ice at 0°C is added to an insulated vessel containing 100 grams of hot water at 80°C. As a result, the hot water cools down to 0°C and 15 grams of ice remain unmelted. What mass of ice was initially added?

Latent heat of fusion of ice = 333 Jg^{-1}
Specific heat of water = 4.184 $JK^{-1}g^{-1}$

a. 45 g
b. 55 g
c. 65 g
d. 75 g
e. 85 g
f. 95 g
g. 105 g
h. 115 g
i. 125 g
j. 135 g

18. Consider again the system described in the previous question. What is the total change in the entropy of the system when the ice is added to the hot water and the system reaches equilibrium at 0°C?

a. +15 JK^{-1}
b. −15 JK^{-1}
c. +107.5 JK^{-1}
d. −107.5 JK^{-1}
e. +122.5 JK^{-1}
f. −122.5 JK^{-1}
g. +230 JK^{-1}
h. −230 JK^{-1}

19. Write and balance the equation for the oxidation of thiosulfate $S_2O_3^{2-}$ by triiodide I_3^- to produce tetrathionate $S_4O_6^{2-}$ and iodide I^- in acidic solution. What is the sum of *all* the coefficients in the balanced equation?

a. 6
b. 7
c. 8
d. 11
e. 12
f. 13
g. 14
h. 15
i. 17
j. 21

20. Calculate the standard cell potential E° at 25°C for the following electrochemical cell. Note that the platinum electrode is inert:

$$Ag(s) \mid Ag^+(aq,\ 1M) \parallel Cl^-(aq,\ 1M) \mid Cl_2(g,\ 1\ atm) \mid Pt(s)$$

$E^\circ_{red}(Ag^+|Ag)$ = +0.800 V
$E^\circ_{red}(Cl_2|Cl^-)$ = +1.358 V

a. +0.242 V
b. −0.242 V
c. +0.558 V
d. −0.558 V
e. +2.158 V
f. −2.158 V
g. +2.958 V
h. −2.958 V

21. What is the equilibrium constant K for the following reaction at 25°C?

$$Ni^{2+}(aq) + Cd(s) \rightleftharpoons Ni(s) + Cd^{2+}(aq) \qquad E^\circ_{cell} = 0.15\ V$$

a. 2.1×10^{-12}
b. 1.2×10^{5}
c. 8.5×10^{-6}
d. 3.6×10^{7}
e. 1.0×10^{-1}
f. 1.3×10^{15}
g. 1.2×10^{1}
h. 3.5×10^{1}

22. Calculate the number of grams of aluminum produced in 1.00 hour by the electrolysis of molten $AlCl_3$ using a current of 20.0 A and a voltage of 6.5 V.

a. 1.86 mg
b. 112 mg
c. 336 mg
d. 6.71 g
e. 20.1 g
f. 60.4 g
g. 648 kg
h. 1940 kg

EXAMINATION 2

CHEMISTRY 142 FALL 2009

Thursday November 17th 2008

1. Methylamine CH_3NH_2 is a weak base like ammonia. Its K_b at 25°C is equal to 4.4×10^{-4}. What is the value of the neutralization constant K_{neut} in a titration of methylamine with nitric acid HNO_3?

   a. 1.0×10^{-14} c. 4.4×10^{-4} e. 2.3×10^{3} g. 6.5×10^{7}
   b. 2.3×10^{-11} d. 1.6×10^{-2} f. 4.7×10^{4} h. 4.4×10^{10}

2. What is the solubility of calcium fluoride CaF_2 if the solubility product K_{sp} for the salt is 5.3×10^{-9}.

   a. 1.1×10^{-3} M c. 3.6×10^{-5} M e. 7.3×10^{-5} M
   b. 1.7×10^{-3} M d. 5.1×10^{-5} M f. 2.6×10^{-9} M

3. What is the solubility of calcium fluoride CaF_2 (K_{sp} is 5.3×10^{-9}) in the presence of 0.010 M sodium fluoride NaF?

   a. 1.3×10^{-7} M c. 5.3×10^{-7} M e. 2.7×10^{-7} M
   b. 1.3×10^{-5} M d. 5.3×10^{-5} M f. 2.7×10^{-5} M

4. Suppose 25 mL of a 1.0×10^{-5} M solution of calcium chloride $CaCl_2$ is added to 25 mL of 1.5×10^{-5} M sodium fluoride NaF. Will precipitation of calcium fluoride CaF_2 occur and what is the relationship between Q_{sp} and K_{sp}?

   a. yes, $Q_{sp} < K_{sp}$ d. no, $Q_{sp} < K_{sp}$
   b. yes, $Q_{sp} > K_{sp}$ e. no, $Q_{sp} > K_{sp}$
   c. yes, $Q_{sp} = K_{sp}$ f. no, $Q_{sp} = K_{sp}$

5. Silver bromide is a sparingly soluble salt. The K_{sp} for silver bromide is 5×10^{-13} and the K_f for the formation of $[Ag(NH_3)_2]^+$ from Ag^+ and ammonia in solution is 1.6×10^{7}. Which of the following solutes could be added to increase the solubility of silver bromide?

   I. NH_3 II. HNO_3 III. HBr

   a. I only c. III only e. I and III g. I, II, and III
   b. II only d. I and II f. II and III

6. When a mixture of methane and oxygen burns, the internal energy of the system decreases by 180 J. The process causes the surroundings to be heated by 40 J. How much work was done by or on this system?

   a. −40 J c. −140 J e. −180 J g. −220 J
   b. +40 J d. +140 J f. +180 J h. +220 J

Questions 7 - 10 refer to the following system.

The pressure on a system containing 0.50 mole of an ideal monatomic gas increases from 2.0 atm to 8.0 atm as the system moves from state A to state B at a constant volume of 12.0 L. The system then decreases in volume at a constant pressure of 8.0 atm to reach state C. The temperature at state C is the same as the temperature at state A.

P

V

7. What is the volume at state C?

a. 1.0 L c. 3.0 L e. 5.0 L g. 18.0 L
b. 2.0 L d. 4.0 L f. 6.0 L h. 48.0 L

8. How much work is done on the system moving at constant pressure from state B to state C?

a. 9.0 J c. 36 J e. 96 J g. 7.3 kJ
b. 18 J d. 72 J f. 2.4 kJ h. 9.7 kJ

9. How much work is done when the system moves reversibly from state A to state C?

a. 1.3 kJ c. 3.4 kJ e. 5.7 kJ g. 8.8 kJ
b. 2.5 kJ d. 4.6 kJ f. 6.2 kJ h. 11 kJ

10. What is the change in the internal energy of the system in moving from state A to state B?

a. 7.3 kJ c. 12.1 kJ e. 18.2 kJ g. 23.8 kJ
b. 10.9 kJ d. 17.5 kJ f. 20.4 kJ h. 25.0 kJ

11. 10.0 g of steam at 100 °C is bubbled into a mixture of 250 g of water and 100 g of ice at 0 °C. All of the steam condenses to water. How much of the ice melts?

Specific heat of water = 4.184 J K^{-1} g^{-1}
Latent heat of fusion of water = 333 J g^{-1}
Latent heat of vaporization of water = 2260 J g^{-1}

a. 18 g c. 36 g e. 50 g g. 80 g
b. 20 g d. 40 g f. 68 g h. 95 g

12. What is the change in entropy for the process described in Question 11?

a. +6 J K^{-1} c. +12 J K^{-1} e. +24 J K^{-1} g. +48 J K^{-1}
b. –6 J K^{-1} d. –12 J K^{-1} f. –24 J K^{-1} h. –48 J K^{-1}

13. Which of the following are state functions?

I. entropy II. heat III. enthalpy

a. I only c. III only e. II and III g. I, II, and III
b. II only d. I and II f. I and III

14. Calculate the standard molar entropy change for this reaction:

$N_2(g) + 3H_2(g) \rightarrow 2\,NH_3(g)$ Given: $\Delta S°\ H_2(g) = 130.7\ J\ mol^{-1}\ K^{-1}$
$\Delta S°\ N_2(g) = 191.6\ J\ mol^{-1}\ K^{-1}$
$\Delta S°\ NH_3(g) = 192.5\ J\ mol^{-1}\ K^{-1}$

a. $-129.8\ J\ K^{-1}$ c. $-198.7\ J\ K^{-1}$ e. $-253.4\ J\ K^{-1}$ g. $-514.8\ J\ K^{-1}$
b. $+129.8\ J\ K^{-1}$ d. $+198.7\ J\ K^{-1}$ f. $+253.4\ J\ K^{-1}$ h. $+514.8\ J\ K^{-1}$

15. Consider the following reaction at 25°C: $N_2O_4(g) \rightarrow 2\,NO_2(g)$
Which one of the following statements is false?

a. The change in entropy for this reaction is positive.
b. The change in enthalpy for this reaction is positive.
c. If the temperature is increased, the equilibrium constant K will also increase.
d. If the temperature is increased, the value of DG° will become more positive.
e. As the temperature increases, the reaction becomes more product favored.
f. The equilibrium constant expression for this reaction is: $K = [NO_2]^2 / [N_2O_4]$

16. Which of the following statements is/are true about a system at equilibrium?

I. The value of $\Delta G°$ at equilibrium is the same as the value of $\Delta G°$ at any other point in the reaction.
II. At equilibrium, $\Delta G°$ is equal to zero.
III. At equilibrium, entropy is at a minimum.
IV. If $\Delta G° < 0$ (a negative value), then the reaction will have an equilibrium constant greater than 1.

a. I only c. III only e. I and III g. II and III i. I, III, and IV
b. II only d. IV only f. I and IV h. II and IV j. II, III, and IV

17. What is the oxidation number for chlorine in each of the following molecules / ion?
(*Choose the row in which all answers are correct.*)

	$HClO_3$	ClO_4^-	Cl_2	HCl
a.	+5	+7	0	−1
b.	+5	+1	−1	−1
c.	+1	+1	0	−1
d.	+5	+7	0	+1
e.	+1	+7	−1	0
f.	+1	+3	0	−1
g.	+5	+7	−1	0
h.	+1	+3	+1	−1

18. Balance the following redox reaction under basic conditions:

$Br_2(l) \rightarrow BrO_3^-(aq) + Br^-(aq)$

How many water molecules are on which side of the balanced equation?

a. 1, reactant c. 2, reactant e. 3, reactant g. 4, reactant
b. 1, product d. 2, product f. 3, product h. 4, product

(19.) A galvanic cell is made from a Cd electrode in a 1.0 M $Cd(NO_3)_2$ solution and a Pb electrode in a 1.0M $Pb(NO_3)_2$ solution. What half-reaction would occur at the anode to produce the most product favored reaction? What is the standard potential for this cell?
(*Choose the row in which both answers are correct.*)

E°_{red} (Pb^{2+}|Pb) = –0.13 V
E°_{red} (Cd^{2+}|Cd) = –0.40 V

	anode half-reaction	*standard cell potential*
a.	$Pb^{2+}(aq) + 2e^- \rightarrow Pb(s)$	E° = +0.27 V
b.	$Pb^{2+}(aq) + 2e^- \rightarrow Pb(s)$	E° = –0.27 V
c.	$Pb(s) \rightarrow Pb^{2+}(aq) + 2e^-$	E° = –0.53 V
d.	$Pb(s) \rightarrow Pb^{2+}(aq) + 2e^-$	E° = –0.27 V
e.	$Cd^{2+}(aq) + 2e^- \rightarrow Cd(s)$	E° = +0.27 V
f.	$Cd^{2+}(aq) + 2e^- \rightarrow Cd(s)$	E° = +0.53 V
g.	$Cd(s) \rightarrow Cd^{2+}(aq) + 2e^-$	E° = +0.27 V
h.	$Cd(s) \rightarrow Cd^{2+}(aq) + 2e^-$	E° = +0.53 V

(20) What is the value of ΔG° for the cell described in Question 19?

a. +52 kJ c. +102 kJ e. +26 kJ g. +77 kJ
b. –52 kJ d. –102 kJ f. –26 kJ h. –77 kJ

(21) Calculate the cell potential, at 25°C, based upon the overall reaction:

$$3Zn(s) + 2Cr^{3+}(aq) \rightarrow 3Zn^{2+}(aq) + 2Cr(s)$$ if $[Cr^{3+}] = 0.0100$ M and $[Zn^{2+}] = 0.0085$ M

The standard reduction potentials are: $Cr^{3+}(aq) + 3e^- \rightarrow Cr(s)$ E° = –0.74 V
$Zn^{2+}(aq) + 2e^- \rightarrow Zn(s)$ E° = –0.76 V

a. 0.01 V c. 0.03 V e. 0.05 V g. 0.07 V
b. 0.02 V d. 0.04 V f. 0.06 V h. 0.08 V

22. How long will it take to plate out 60.2 g of aluminum from an aqueous solution of Al^{3+} ions if you were to use a current of 50 amp?

a. 1.2 hr c. 10.8 hr e. 47.1 hr g. 71.8 hr
b. 3.6 hr d. 20.6 hr f. 65.3 hr h. 96.8 hr

EXAMINATION 2

CHEMISTRY 142 Spring 2010

Thursday April 1st 2010

1. The molar concentration of lead nitrate $Pb(NO_3)_2$ in an aqueous solution is 3.6×10^{-4} M. What is the concentration of nitrate ions in the solution?

a. 3.6×10^{-4} M
b. 1.8×10^{-4} M
c. 1.2×10^{-4} M
d. 4.8×10^{-4} M
e. 7.2×10^{-4} M
f. 1.1×10^{-3} M

2. Which of the following is the correct solubility product constant expression for the sparingly soluble lead sulfate?

a. $K_{sp} = [Pb^{2+}][SO_4^{2-}]$
b. $K_{sp} = [Pb^{2+}]^2[SO_4^{2-}]$
c. $K_{sp} = [Pb^{2+}][SO_4^{2-}]^2$
d. $K_{sp} = [Pb^{2+}]^2[SO_4^{2-}]^3$
e. $K_{sp} = [Pb^{2+}][SO_4^{2-}]^3$
f. $K_{sp} = [Pb^{2+}][SO_4^{2-}]/[PbSO_4]$
g. $K_{sp} = [Pb^{2+}]^3[SO_4^{2-}]^2$

3. 500 mL of the lead nitrate solution described in question 1 is added to 500 mL of a 5.4×10^{-2} M sodium sulfate solution. What is the concentration of lead ions in the resulting solution?

K_{sp} for lead sulfate = 1.8×10^{-8}.

a. 3.3×10^{-7} M
b. 2.7×10^{-4} M
c. 1.8×10^{-4} M
d. 1.1×10^{-6} M
e. 6.7×10^{-7} M
f. 3.3×10^{-4} M

4. The solubility product for a sparingly soluble metal salt, MX_2, equals 6.0×10^{-18}. The formation constant, K_f, for the formation of the tetraammine complex of the metal ion $[M(NH_3)_4]^{2+}$ in aqueous solution equals 5.0×10^{10}. What is the value of the equilibrium constant for the overall solution process for this sparingly soluble salt in the presence of ammonia?

$$MX_2 + 4\,NH_3 \rightleftharpoons [M(NH_3)_4]^{2+} + 2\,X^- \qquad K = ?$$

a. 3.3×10^{6}
b. 1.5×10^{4}
c. 1.2×10^{-28}
d. 6.7×10^{-5}
e. 3.0×10^{-7}
f. 8.3×10^{27}

5. Calcium carbonate, $CaCO_3$, is a sparingly soluble salt with a $K_{sp} = 8.7 \times 10^{-9}$. If you mix 200 mL of 1.0×10^{-4} M Na_2CO_3 and 200 mL of 2.0×10^{-4} M $CaCl_2$, what would you observe in the resulting solution?

a. No precipitation of $CaCO_3(s)$, because $Q_{sp} > K_{sp}$
b. No precipitation of $CaCO_3(s)$, because $Q_{sp} = K_{sp}$
c. No precipitation of $CaCO_3(s)$, because $Q_{sp} < K_{sp}$
d. Precipitation of $CaCO_3(s)$, because $Q_{sp} > K_{sp}$
e. Precipitation of $CaCO_3(s)$, because $Q_{sp} = K_{sp}$
f. Precipitation of $CaCO_3(s)$, because $Q_{sp} < K_{sp}$

Questions 6 through 11 refer to the following system. It is recommended that you draw the corresponding PV diagram.

P

V

The initial state A of the system consists of one mole of an ideal gas at a pressure of 10.0 atm and a volume of 2.00 L. The gas expands isothermally and reversibly to a new state of the system B where the volume is 5.00 L.

An alternative adiabatic expansion to the same volume (5.00 L) leads to a state C of the system. Note that in an adiabatic reversible expansion of an ideal gas, no heat is transferred between the system and the surroundings. Any work done must result in a decrease in the temperature of the system. Also, since no heat is transferred in an adiabatic expansion, the entropy change must be zero.

6. What is the temperature at point B?

a. 79 K
b. 102 K
c. 132 K
d. 168 K
e. 221 K
f. 244 K
g. 273 K
h. 312 K
i. 378 K
j. 391 K

7. How much work is done as the system moves isothermally and reversibly from A to B?

a. w = zero J
b. w = –763 J
c. w = –1109 J
d. w = –1256 J
e. w = –1389 J
f. w = –1452 J
g. w = –1529 J
h. w = –1857 J
i. w = –2067 J

8. How much work is done as the system moves from B to C?

a. w = zero J
b. w = –763 J
c. w = –1109 J
d. w = –1256 J
e. w = –1389 J
f. w = –1452 J
g. w = –1529 J
h. w = –1857 J
i. w = –2067 J

9. What is the entropy change ΔS as the system moves from A to B and then from B to C?

a. $+1.41\ JK^{-1}$
b. zero JK^{-1}
c. $+2.70\ JK^{-1}$
d. $+4.83\ JK^{-1}$
e. $+7.61\ JK^{-1}$
f. $+8.39\ JK^{-1}$
g. $+11.2\ JK^{-1}$
h. $+14.7\ JK^{-1}$
i. $+15.1\ JK^{-1}$
j. $-7.61\ JK^{-1}$

10. What is the temperature at point C?

a. 79 K
b. 102 K
c. 132 K
d. 168 K
e. 221 K
f. 244 K
g. 273 K
h. 312 K
i. 378 K
j. 391 K

11. How much work is done as the system moves adiabatically and reversibly from A to C?

a. w = zero J
b. w = –763 J
c. w = –1109 J
d. w = –1256 J
e. w = –1389 J
f. w = –1452 J
g. w = –1529 J
h. w = –1857 J
i. w = –2067 J

12. Suppose a vessel contained just 8 molecules of a gas.
    a) What is the probability that all eight molecules would be in the left-most half of the vessel?
    b) What is the probability that 1 molecule would be in the right half and 7 molecules would be in the left half?

	a) answer	b) answer
a.	1/2	1/8
b.	1/16	1/32
c.	1/64	1/4
d.	1/32	1/64
e.	1/128	1/16
f.	1/8	1/16
g.	1/256	1/32
h.	1/4	1/16
i.	1/512	1/64

13. Match each one of the following processes with the correct statement regarding the change in internal energy of the system ΔE. *(Choose the row in which all responses are correct.)*

	isothermal	*constant volume*	*constant pressure*	*adiabatic*
a.	$\Delta E = w$	$\Delta E =$ zero	$\Delta E = \Delta H$	$\Delta E =$ zero
b.	$\Delta E =$ zero	$\Delta E = q_v$	$\Delta E = q_p - P\Delta V$	$\Delta E = w$
c.	$\Delta E =$ zero	$\Delta E = q_v$	$\Delta E = q_p$	$\Delta E = \Delta H$
d.	$\Delta E =$ zero	$\Delta E = w$	$\Delta E = q_v$	$\Delta E = q + w$
e.	$\Delta E = -q$	$\Delta E = P\Delta V$	$\Delta E = w$	$\Delta E = q + w$
f.	$\Delta E = -w$	$\Delta E = q_v$	$\Delta E = \Delta H$	$\Delta E = w$

14. Which of the following statement is *incorrect*?

    a. For a system changing at constant volume, the work done is equal to zero.
    b. The 3rd law of thermodynamics states that the entropy of a perfectly crystalline solid at 0 K is zero.
    c. Under isothermal conditions, the change in internal energy equals zero.
    d. The 2nd law of thermodynamics states that $\Delta S_{univ} >$ zero for all spontaneous processes.
    e. At equilibrium, $\Delta G°$ equals zero.
    f. In an adiabatic change no energy is transferred between the system and the surroundings as heat.

15. How many of the following properties depend upon the path taken?

enthalpy	entropy	heat	internal energy	free energy
volume	work	temperature	pressure	

a. 0	c. 2	e. 4	g. 6	i. 8
b. 1	d. 3	f. 5	h. 7	j. 9

16. What is the oxidation number of the underlined element in each of the following compounds. Add them up using the correct signs. What is the sum?

$(NH_4)_2\underline{Cr}_2O_7$ $\quad$ $XeO\underline{F}_4$ $\quad$ $\underline{C}H_2Br_2$ $\quad$ $Na_2\underline{S}_2O_3$

a. +1	c. +3	e. +5	g. +7	i. +9
b. +2	d. +4	f. +6	h. +8	j. +10

17. Using the standard half-cell reduction potentials below, identify which metal species would make the best reducing agent?

$E^\circ_{red}(Fe^{3+}|Fe^{2+}) = +0.77$ V
$E^\circ_{red}(Cr^{3+}|Cr) = -0.74$ V
$E^\circ_{red}(Al^{3+}|Al) = -1.66$ V
$E^\circ_{red}(Li^{+}|Li) = -3.05$ V

a. Fe^{3+} c. Cr^{3+} e. Al^{3+} g. Li^{+}
b. Fe^{2+} d. Cr f. Al h. Li

18. In the equation representing the redox reaction:

$$3\ Cl_2(g) + 6\ OH^-(aq) \rightarrow ClO_3^-(aq) + 5\ Cl^-(aq) + 3\ H_2O(l)$$

what is the change in the oxidation number of the element that is oxidized?

a. +1 c. +3 e. +5 g. +7
b. +2 d. +4 f. +6 h. +8

19. Balance the equation:

$$\_MnO_4^- + \_SO_3^{2-} \rightarrow \_MnO_2 + \_SO_4^{2-}$$

in basic aqueous solution. How many water molecules are on which side in the balanced equation?

a. 1, right c. 2, right e. 3, right g. 4, right i. no water
b. 1, left d. 2, left f. 3, left h. 4, left

20. Calculate the standard cell potential for the cell:

$$Pb(s)\ |\ PbSO_4(s)\ |\ H^+(aq)\ ||\ H^+(aq)\ |\ PbO_2(s)\ |\ PbSO_4(s)$$

given the half-cell standard reduction potentials:

E°_{red} $PbSO_4(s)\ |\ Pb(s) = -0.356$ V
E°_{red} $PbO_2(s)\ |\ PbSO_4(s) = +1.685$ V

a. +1.329 V b. +2.041 V c. −1.329 V d. −2.041 V

21. What is the value of ΔG° for the reaction (described in the previous question):

$$Pb(s) + PbO_2(s) + 2SO_4^{2-}(aq) + 4H_3O^+(aq) \rightarrow 2\ PbSO_4(s) + 6H_2O(l)$$

a. −98 kJ c. −197 kJ e. −394 kJ g. −914 kJ
b. −124 kJ d. −286 kJ f. −788 kJ h. −1180 kJ

22. In an electrolytic cell used for the purification of copper metal, a current of 10 amps was passed through the cell for 24 hours. How much copper metal plated out on the cathode?

a. 28 g c. 142 g e. 347 g g. 442 g
b. 71 g d. 285 g f. 403 g h. 498 g

EXAMINATION 2

CHEMISTRY 142 Fall 2010

Tuesday November 16th 2010

This exam consists of 24 questions. Questions are worth 10 points each, unless otherwise indicated.

Solubility Equilibria

1. Which of the following is the correct solubility product expression for lead(II) hydroxide, $Pb(OH)_2$?

a. $K_{sp} = [Pb^{2+}][OH^-]^2/[Pb(OH)_2]$
b. $K_{sp} = [Pb^{2+}][OH^-]$
c. $K_{sp} = [Pb^{2+}][OH^-]^2$
d. $K_{sp} = [Pb^{2+}]^2[OH^-]$
e. $K_{sp} = [Pb][(OH)_2]$
f. $K_{sp} = [Pb^{2+}](2[OH^-])$
g. $K_{sp} = [Pb^{2+}][OH^-]/[Pb(OH)_2]$
h. $K_{sp} = [Pb^{2+}](2[OH^-])^2$

2. Determine the solubility of silver phosphate, Ag_3PO_4 ($K_{sp} = 1.8 \times 10^{-18}$).

a. 1.6×10^{-5} M
b. 2.1×10^{-5} M
c. 2.8×10^{-5} M
d. 3.7×10^{-5} M
e. 4.1×10^{-7} M
f. 8.4×10^{-7} M
g. 1.3×10^{-9} M
h. 7.7×10^{-10} M

3. The insoluble salt MgF_2 has a solubility product constant $K_{sp} = 6.4 \times 10^{-9}$. Find the fluoride ion concentration in a saturated solution that also contains 0.050 M $Mg(NO_3)_2$.

a. 1.8×10^{-4} M
b. 2.5×10^{-4} M
c. 3.6×10^{-4} M
d. 8.0×10^{-5} M
e. 9.0×10^{-5} M
f. 1.3×10^{-7} M
g. 2.6×10^{-7} M
h. 6.4×10^{-8} M

4. Iron(II) fluoride, FeF_2, is a sparingly soluble salt, $K_{sp} = 2.4 \times 10^{-6}$. Suppose that 500 mL of a 1.0×10^{-2} M solution of $Fe(NO_3)_2$ is combined with 500 mL of a 2.4×10^{-2} M solution of NaF. What will be observed in the resulting solution?

a. No precipitation of FeF_2 because $Q < K_{sp}$
b. No precipitation of FeF_2 because $Q = K_{sp}$
c. No precipitation of FeF_2 because $Q > K_{sp}$
d. Precipitation of FeF_2 because $Q < K_{sp}$
e. Precipitation of FeF_2 because $Q = K_{sp}$
f. Precipitation of FeF_2 because $Q > K_{sp}$

5. How will the addition of the three solutes given below affect the solubility of lead(II) bromide, $PbBr_2$, in water ($K_{sp} = 4.6 \times 10^{-6}$)? In the presence of ethylenediaminetetraacetic acid (EDTA), lead(II) forms the complex ion $[Pb(EDTA)]^{2-}$ ($K_f = 2.0 \times 10^{18}$). Select the row in which all of the responses are correct.

	HNO_3	NaBr	EDTA
a.	increases	decreases	increases
b.	no effect	decreases	decreases
c.	decreases	increases	decreases
d.	decreases	decreases	increases
e.	no effect	decreases	increases
f.	increases	increases	no effect
g.	increases	decreases	no effect
h.	no effect	increases	increases

Thermodynamics

6. Suppose that 25 kJ of heat is transferred as a gas is cooled, and 10 kJ of work is done as the same gas is compressed. What is the change in the internal energy of the gas?

   a. –35 kJ
   b. –25 kJ
   c. –15 kJ
   d. –10 kJ
   e. 10 kJ
   f. 15 kJ
   g. 25 kJ
   h. 35 kJ

7. Complete the following statements:
   According to the first law of thermodynamics, ______
   The second law of thermodynamics indicates that for any spontaneous process, _____

   a. $\Delta E_{univ} = 0$ $\Delta S_{univ} < 0$
   b. $\Delta E_{univ} = 0$ $\Delta S_{sys} > 0$
   c. $\Delta E_{univ} > 0$ $\Delta S_{univ} = 0$
   d. $\Delta E_{univ} < 0$ $\Delta S_{sys} < 0$
   e. $\Delta E_{univ} > 0$ $\Delta S_{univ} < 0$
   f. $\Delta E_{univ} = 0$ $\Delta S_{univ} > 0$
   g. $\Delta E_{univ} < 0$ $\Delta S_{univ} > 0$

Questions 8 through 12 refer to the system described below. It is recommended that you draw a PV diagram to illustrate the changes described and/or make a table to keep track of P, V, T, and n for each of the states.

P

V

Consider an ideal monatomic gas that has a pressure of 6.80 atm and occupies a volume of 14.3 L at 45°C in state A. Suppose that this gas is cooled at constant volume to reach a new state B where the pressure is 2.44 atm. Subsequently, the gas is allowed to expand at constant pressure to reach state C which has the same temperature as state A.
$R = 0.08206 \text{ L atm mol}^{-1}\text{K}^{-1} = 8.3145 \text{ J mol}^{-1}\text{K}^{-1}$

8. *(5 points)* How many moles of the ideal monatomic gas are present in the system?

   a. 0.260 mol
   b. 0.269 mol
   c. 0.591 mol
   d. 1.86 mol
   e. 2.40 mol
   f. 3.72 mol
   g. 26.3 mol

9. Calculate the work done in going from state A to state B.

   a. 10.1 kJ
   b. 9.48 kJ
   c. 6.32 kJ
   d. 0.00 kJ
   e. –6.32 kJ
   f. –9.48 kJ
   g. –10.1 kJ

10. How much heat is transferred and in what direction is the heat transferred in a reversible, isothermal expansion from state A to state C?

    a. No heat is transferred between the sample and the surroundings.
    b. 9.48 kJ is transferred from the surroundings to the gas sample.
    c. 10.1 kJ is transferred from the surroundings to the gas sample.
    d. 15.8 kJ is transferred from the surroundings to the gas sample.
    e. 9.48 kJ is transferred from the gas sample to the surroundings.
    f. 10.1 kJ is transferred from the gas sample to the surroundings.
    g. 15.8 kJ is transferred from the gas sample to the surroundings.

11. Determine the internal energy change that occurs in going from state C to state B.

    a. –15.8 kJ c. –6.32 kJ e. 6.32 kJ g. 15.8 kJ
    b. –9.48 kJ d. 0.00 kJ f. 9.48 kJ

12. What is the internal energy change ΔE in going from state A to state C along the path from state A to state B and then from state B to state C?

    a. –15.8 kJ c. –9.48 kJ e. 9.48 kJ g. 15.8 kJ
    b. –10.1 kJ d. 0.00 J f. 10.1 kJ

13. Suppose that 9 gas atoms initially confined to a volume V_i are allowed to expand into volume that is 4 times the initial volume ($4V_i$). What is the entropy change associated with this process?

    a. 1.04×10^{2} J K^{-1} d. 7.31×10^{1} J K^{-1} g. -1.22×10^{-22} J K^{-1}
    b. 1.72×10^{-22} J K^{-1} e. -1.04×10^{2} J K^{-1} h. -7.31×10^{1} J K^{-1}
    c. 1.22×10^{-22} J K^{-1} f. -1.72×10^{-22} J K^{-1}

14. Suppose that two 100g copper blocks are placed in an insulated container and are brought into thermal contact. Block A has an initial temperature of 300 K, and block B has an initial temperature of 400 K. The specific heat of copper is 0.385 J K^{-1}g^{-1}. Which of the following responses correctly describes the entropy changes associated with the system reaching thermal equilibrium?

	The entropy of block A	*The entropy of block B*	*The total entropy of the system*
a.	increases	decreases	no change
b.	decreases	increases	increases
c.	increases	increases	increases
d.	increases	decreases	decreases
e.	decreases	decreases	decreases
f.	increases	decreases	increases
g.	decreases	increases	decreases

15. Calculate the standard free energy change at 25°C for the reaction

$$C(\textit{graphite}) + 2\,H_2(g) \rightarrow CH_4(g)$$

given
$\Delta H_f^\circ[C(\textit{graphite})] = 0$ kJ mol^{-1} $S^\circ[C(\textit{graphite})] = 5.6$ J K^{-1}mol^{-1}
$\Delta H_f^\circ[H_2(g)] = 0$ kJ mol^{-1} $S^\circ[H_2(g)] = 130.7$ J K^{-1}mol^{-1}
$\Delta H_f^\circ[CH_4(g)] = -74.9$ kJ mol^{-1} $S^\circ[CH_4(g)] = 186.3$ J K^{-1}mol^{-1}

    a. –50.8 kJ mol^{-1} c. –76.9 kJ mol^{-1} e. –99.0 kJ mol^{-1} g. -1.50×10^{4} kJ mol^{-1}
    b. –72.9 kJ mol^{-1} d. –89.8 kJ mol^{-1} f. -2.40×10^{4} kJ mol^{-1}

16. Find the equilibrium constant for the reaction

$$Ag^+(aq) + 2\,NH_3(aq) \rightarrow [Ag(NH_3)_2]^+(aq)$$

given the standard free energy change $\Delta G^\circ_{rxn} = -41.1$ kJ mol^{-1} at 25°C.

a. 6.3×10^{-8}
b. 1.7×10^{-2}
c. 0.19
d. 1.0
e. 5.4
f. 1.7×10^{2}
g. 1.6×10^{7}
h. 6.0×10^{18}

Redox Reactions & Electrochemistry

17. What is the oxidation number of the bromine atom in $HBrO_3$?

a. −1
b. 0
c. +1
d. +3
e. +5
f. +7

18. Balance the following redox reaction under acidic conditions

$$S_2O_6^{2-}(aq) + HClO_2(aq) \rightarrow SO_4^{2-}(aq) + Cl_2(g)$$

How many water molecules are present in the balanced equation, and do they appear as reactants or products?

a. 2, reactant
b. 4, reactant
c. 6, reactant
d. 2, product
e. 4, product
f. 6, product

19. *(5 points)* In a voltaic/galvanic cell, ______ occurs at the anode and electrons flow from the ______ in the external circuit.

a. oxidation; anode to cathode
b. oxidation; cathode to anode
c. reduction; anode to cathode
d. reduction; cathode to anode

20. Determine the standard cell potential for the cell

$$Pt(s) \mid Fe^{2+}(aq), Fe^{3+}(aq) \parallel Cl^-(aq) \mid Cl_2(g) \mid C(s, graphite)$$

Note that the Pt*(s)* and C*(s, graphite)* in this cell serve as inert electrodes. The standard reduction potentials are:

$E^\circ_{red}(Fe^{3+}|Fe^{2+}) = 0.77$ V
$E^\circ_{red}(Cl_2|Cl^-) = 1.36$ V

a. 2.90 V
b. 2.13 V
c. 0.59 V
d. 0.18 V
e. −0.18 V
f. −0.59 V
g. −2.13 V
h. −2.90 V

21. *(5 points)* In the cell described above, which species is the strongest oxidizing agent?

a. $Fe^{2+}(aq)$
b. $Fe^{3+}(aq)$
c. $Cl_2(g)$
d. $Cl^-(aq)$

22. Calculate the cell potential at 298K for an electrochemical cell in which the following reaction occurs

$$3\ Pb^{2+}(aq) + 2\ Cr(s) \rightarrow 3\ Pb(s) + 2\ Cr^{3+}(aq)$$

when the Pb^{2+} concentration is 8.9×10^{-1} M and the Cr^{3+} concentration is 6.5×10^{-4} M.

Standard reduction potentials:

$Pb^{2+}(aq) + 2\ e^- \rightarrow Pb(s)$, $E° = -0.13$ V
$Cr^{3+}(aq) + 3\ e^- \rightarrow Cr(s)$, $E° = -0.74$ V

a. 0.43 V
b. 0.49 V
c. 0.55 V
d. 0.67 V
e. 0.73 V
f. 0.79 V
g. 1.15 V

23. *(5 points)* Is the oxidation-reduction reaction for the electrochemical cell given in the preceding problem spontaneous or nonspontaneous?

a. spontaneous
b. nonspontaneous

24. A current of 20.0 A was passed through an electrolytic cell containing $CuSO_4(aq)$ for 6.00 hours. Find the mass of copper plated out on the cathode.

a. 2.37 g
b. 71.1 g
c. 120 g
d. 142 g
e. 285 g
f. 357 g
g. 715 g

EXAMINATION 2

CHEMISTRY 142 Spring 2011

Thursday March 31st 2011

This exam consists of 22 questions worth 10 points each.

1. When calcium phosphate $Ca_3(PO_4)_2$ dissolves in water, what is the relationship between the concentration of calcium ions and the concentration of phosphate ions?

   a. $[Ca^{2+}] = 3 \times [PO_4^{3-}]$
   b. $[Ca^{2+}] = 2 \times [PO_4^{3-}]$
   c. $[Ca^{2+}] = (3/2) \times [PO_4^{3-}]$
   d. $[Ca^{2+}] = (2/3) \times [PO_4^{3-}]$
   e. $[Ca^{2+}] = (1/2) \times [PO_4^{3-}]$
   f. $[Ca^{2+}] = (1/3) \times [PO_4^{3-}]$

2. Which of the following is the correct solubility product expression for calcium phosphate, $Ca_3(PO_4)_2$?

   a. $K_{sp} = [Ca^{2+}]^2[PO_4^{3-}]^3/[Ca_3(PO_4)_2]$
   b. $K_{sp} = [Ca^{2+}]^2[PO_4^{3-}]^3$
   c. $K_{sp} = [Ca^{2+}]^3[PO_4^{3-}]^2$
   d. $K_{sp} = [Ca^{2+}][PO_4^{3-}]$
   e. $K_{sp} = [Ca_3]^2[(PO_4)_2]^3$
   f. $K_{sp} = 3[Ca^{2+}](2[PO_4^{3-}])$
   g. $K_{sp} = [Ca^{2+}]^3[PO_4^{3-}]^2/[Ca_3(PO_4)_2]$
   h. $K_{sp} = [3Ca^{2+}]^3[2PO_4^{3-}]^2$

3. If the solubility product K_{sp} for calcium phosphate is 1.2×10^{-26}, what is the molar solubility of calcium phosphate in water at the same temperature?

   a. 7.1×10^{-6} M
   b. 2.1×10^{-5} M
   c. 2.6×10^{-6} M
   d. 5.4×10^{-6} M
   e. 3.9×10^{-6} M
   f. 8.4×10^{-7} M
   g. 2.4×10^{-27} M
   h. 7.7×10^{-6} M

4. What is the molar solubility of calcium phosphate in a saturated solution that also contains 0.250 M Na_3PO_4?

   a. 3.6×10^{-4} M
   b. 5.6×10^{-5} M
   c. 1.9×10^{-9} M
   d. 8.2×10^{-10} M
   e. 5.8×10^{-9} M
   f. 1.3×10^{-7} M
   g. 7.3×10^{-7} M
   h. 1.9×10^{-25} M

5. A saturated solution is prepared by adding sodium sulfide to a mixture of solid lead(II) carbonate and a saturated solution of lead(II) carbonate. As a result, the solution contains lead(II) ions, carbonate ions, sulfide ions, and sodium ions. What is the ratio $[CO_3^{2-}]/[S^{2-}]$, the ratio of the concentrations of carbonate ions and sulfide ions, in this solution?

   K_{sp} for lead(II) carbonate = 3.3×10^{-14}
   K_{sp} for lead(II) sulfide = 3.4×10^{-28}

   a. 9.4×10^{27}
   b. 9.9×10^{6}
   c. 9.7×10^{13}
   d. 1.3×10^{-14}
   e. 1.1×10^{-28}
   f. 1.0×10^{-7}

6. How many of the following are state functions?

heat	volume	free energy	enthalpy
entropy	temperature	work	pressure

a. 1 c. 3 e. 5 g. 7
b. 2 d. 4 f. 6 h. 8

7. There are four fundamental laws of thermodynamics—the zeroth, first, second, and third. Phrases commonly associated with these laws are listed below. Which row has all four laws assigned correctly?

	spontaneous reaction when ΔG_{sys} is negative	*zero entropy at 0K*	*thermal equilibrium*	*conservation of energy*
a.	0	1	2	3
b.	1	2	3	0
c.	2	3	0	1
d.	3	0	1	2
e.	2	0	3	1
f.	1	3	0	2
g.	3	1	2	0
h.	2	3	1	0

Questions 8 through 12 refer to the system described below. It is recommended that you draw a PV diagram to illustrate the changes described and/or make a table to keep track of P, V, T, and n for each of the states.

Consider an ideal monatomic gas that has a pressure of 8.00 atm and occupies a volume of 3.00 L at 292.5K in state A. The gas expands isothermally and reversibly to reach a new state B where the pressure is 2.00 atm. An adiabatic expansion from state A leads to another state C of the system, where the volume is the same as at state B.

P

V

8. How many moles of the ideal gas are present in the system?

a. 0.250 mol c. 0.750 mol e. 1.50 mol g. 3.00 mol i. 5.00 mol
b. 0.500 mol d. 1.00 mol f. 2.00 mol h. 4.00 mol j. 10.0 mol

9. Calculate the work done in going along the isotherm from state A to state B.

a. −33.3 J c. −1.52 kJ e. −5.66 kJ g. −10.1 kJ
b. −223 kJ d. −3.37 kJ f. −7.21 kJ h. zero

10. What is the entropy change in the system in going from state B to state C?

a. +0.11 JK^{-1} c. +11.5 JK^{-1} e. zero JK^{-1} g. −11.5 JK^{-1}
b. +6.43 JK^{-1} d. +23.0 JK^{-1} f. −6.43 JK^{-1} h. −23.0 JK^{-1}

11. Suppose that the system goes from state A to state B via a different route: a decrease in pressure at constant volume from state A to 2.00 atm and then an increase in volume at constant pressure (2.00 atm) to state B. How much work is done via this route?

a. +5612 J
b. +3371 J
c. +884 J
d. 0.00 J
e. −884 J
f. −1824 J
g. −3371 J
h. −5612 J

12. What is the internal energy change ΔE of the system in going from state A to state B along the path from state A to state C and then from state C to state B?

a. +5612 J
b. +3371 J
c. +884 J
d. 0.00 J
e. −884 J
f. −1824 J
g. −3371 J
h. −5612 J

13. Suppose a vessel contained 8 molecules of an ideal gas. Imagine the vessel divided into two imaginary halves. What is the probability that one molecule would be in the right half and seven molecules would be in the left half? Express the probability as a fraction of the total number of arrangements possible.

a. 1 in 2
b. 1 in 4
c. 1 in 8
d. 1 in 10
e. 1 in 16
f. 1 in 32
g. 1 in 64
h. 1 in 128
i. 1 in 256

14. A 150 gram block of iron at 30°C is placed in an insulated container with an 80 gram block of iron at 150°C. Which of the following statement(s) is/are true about the resulting process?

I. ΔS for the process is positive.
II. ΔS for the surroundings is equal to zero.
III. ΔS for the 80 gram block of iron is negative.

a. I only
b. II only
c. III only
d. I and II
e. I and III
f. II and III
g. I, II, and III
h. none

15. Estimate the melting point of an unknown substance in °C if the heat of fusion of the substance is 6.12 kJ mol^{-1} and the entropy change for the change in state of the substance from solid to liquid at its melting point is 22.4 $JK^{-1}mol^{-1}$.

a. −28°C
b. −10°C
c. 0°C
d. 16°C
e. 24°C
f. 41°C
g. 121°C
h. 189°C
i. 273°C
j. 373°C

16. In a voltaic cell such as the Daniell cell, ______ occurs at the anode and electrons flow from the ______ to the ______ in the external circuit.

a. oxidation anode cathode
b. oxidation cathode anode
c. reduction anode cathode
d. reduction cathode anode

17. What is the oxidation number of the bromine atom in magnesium bromate $Mg(BrO_3)_2$?

a. −3
b. −1
c. 0
d. +1
e. +2
f. +3
g. +4
h. +5
i. +6
j. +7

18. Balance the equation for this redox reaction in basic solution, using whole number coefficients:

$$VO_2^+(aq) \quad + \quad Zn(s) \quad \rightarrow \quad VO^{2+}(aq) \quad + \quad Zn^{2+}(aq)$$

How many water molecules are on which side of the balanced equation?

a. 1, reactant side
b. 1, product side
c. 2, reactant side
d. 2, product side
e. 4, reactant side
f. 4, product side
g. 6, reactant side
h. 6, product side

(19.) Calculate the standard cell potential for the cell $Al(s) \mid Al^{3+}(aq) \parallel Cu^{2+}(aq) \mid Cu(s)$ given the half-cell standard reduction potentials

E°_{red} $Cu^{2+}|Cu$ = +0.337 V
E°_{red} $Al^{3+}|Al$ = −1.66 V

a. +1.997 V
b. −1.997 V
c. +1.323 V
d. −1.323 V

(20.) Write the simplest equation, using whole number coefficients, for the cell reaction described in the previous question. What is the equilibrium constant K for the reaction described by the balanced equation at 25°C?

a. 182
b. 271
c. 466
d. 3.29×10^4
e. 1.16×10^6
f. 6.14×10^9
g. 8.47×10^{25}
h. 3.44×10^{202}

21. Predict the values for K, ΔG°, and E° for a spontaneous, product–favored reaction.

	K	ΔG°	E°
a.	>1	<0	<0
b.	>1	<0	>0
c.	>1	zero	<0
d.	>1	zero	>0
e.	>1	>0	>0
f.	<1	zero	<0
g.	<1	zero	>0
h.	<1	<0	<0
i.	<1	<0	>0
j.	<1	>0	>0

22. How long does it take to produce 86 kg of sodium from an electrolysis cell using a current of 25×10^3 A?

a. 1.0 hour
b. 1.5 hours
c. 2.0 hours
d. 2.5 hours
e. 3.0 hours
f. 3.5 hours
g. 4.0 hours
h. 4.5 hours
i. 5.0 hours
j. 6.0 hours

FINAL EXAMINATION

CHEMISTRY 142 SPRING 2006

Monday May 1st 2006

1. Methanol, CH_3OH, burns in air according to the equation:

$$2\ CH_3OH(g) + 3\ O_2(g) \rightarrow 2\ CO_2(g) + 4\ H_2O(g)$$

The rate of the reaction was investigated, and it was determined at one point that oxygen was consumed at a rate of of 2.4×10^2 mol L^{-1} min^{-1}. At what rate was water produced at this time?

a. 1.2×10^2 mol L^{-1} min^{-1}
b. 1.6×10^2 mol L^{-1} min^{-1}
c. 1.8×10^2 mol L^{-1} min^{-1}
d. 3.2×10^2 mol L^{-1} min^{-1}
e. 3.6×10^2 mol L^{-1} min^{-1}
f. 4.8×10^2 mol L^{-1} min^{-1}

2. An initial concentration of a reactant (0.640 M) was found to drop to one-half of this value (0.320 M) in 5.0 minutes in a *first-order* reaction. How long does the reaction take for the concentration of this reactant to drop from its initial value (0.640 M) to one-eighth of this value (0.080 M)?

a. 1.25 min
b. 2.5 min
c. 10 min
d. 15 min
e. 20 min
f. 25 min
g. 35 min
h. 75 min

3. A suggested mechanism for the reduction of nitric oxide by hydrogen to produce water and nitrogen is:

$$2\ NO \rightleftharpoons N_2O_2 \qquad \textit{fast}$$
$$N_2O_2 + H_2 \rightarrow N_2O + H_2O \qquad \textit{slow}$$
$$N_2O + H_2 \rightarrow N_2 + H_2O \qquad \textit{fast}$$

Label the species involved in this reaction appropriately.

	reactant	*intermediates*	*product*
a.	H_2	N_2O N_2O_2	N_2O
b.	H_2O	N_2O_2 N_2O	N_2
c.	NO	NO H_2	N_2O
d.	NO	N_2O_2 N_2O	H_2O
e.	N_2O_2	H_2 N_2O	H_2O
f.	NO	N_2O N_2O_2	N_2O
g.	H_2	H_2O N_2O_2	N_2

4. Wine contains several organic compounds, which in turn contain C, H, N, and O atoms. Suppose that someone was trying to sell you a bottle of "Vintage 1959" wine—supposedly 47 years old. To confirm the age of the wine, you might try radioactive dating. Which of the following radioactive isotopes should you use in your radioactive dating experiment? Assume that you know what the activities of each of the isotopes was when the wine was bottled.

The half-lives of the appropriate isotopes are:

^{14}C	5730 years	^{15}O	125 seconds
^{13}N	9.96 minutes	^{3}H	12.5 years

a. ^{14}C
b. either ^{14}C or ^{13}N
c. either ^{13}N or ^{15}O
d. ^{13}N
e. ^{15}O
f. ^{3}H
g. ^{14}C, ^{13}N or ^{3}H
h. any of the four

5. A catalyst can do all of the following except one. Which one can it *not* do?

   a. speed up a chemical reaction
   b. change the path of a chemical reaction
   c. change the order of the reaction with respect to a particular reactant
   d. change the mechanism of a chemical reaction
   e. change the equilibrium position of a chemical reaction

6. In a 12 L flask *at equilibrium*, there are 2.0 moles of ammonia, 1.0 mole of chlorine, 5.0 moles of hydrazine N_2H_4, and an unknown concentration of hydrogen chloride HCl. The value of the equilibrium constant for the system at this temperature is 20. How many moles of HCl are in the flask?

$$2\,NH_3(g) + Cl_2(g) \rightleftharpoons N_2H_4(g) + 2\,HCl(g)$$

   a. 0.33 mol   c. 1.0 mol   e. 2.5 mol   g. 5.0 mol
   b. 0.50 mol   d. 2.0 mol   f. 4.0 mol   h. 10 mol

7. As indicated above for the following equilibrium, the equilibrium constant is

$$2\,NH_3(g) + Cl_2(g) \rightleftharpoons N_2H_4(g) + 2\,HCl(g) \qquad K = 20$$

   What is the equilibrium constant for the reaction:

$$\tfrac{1}{2}\,N_2H_4(g) + HCl(g) \rightleftharpoons NH_3(g) + \tfrac{1}{2}\,Cl_2(g) \qquad K = ?$$

   a. 0.050   c. 4.5   e. 10   g. 0.22
   b. 400   d. 0.010   f. 2.5×10^{-3} h. 0.10

8. If one mole of each of the following acids are added to separate 1.0 L samples of water, which will produce the highest concentration of hydronium H_3O^+ ions?

   a. HClO   c. HCN   e. HNO_2 g. CH_3CO_2H
   b. H_2CO_3 d. H_3PO_4 f. HF   h. HBr

9. Label the species in the following aqueous equilibrium as acid or base according to the definitions of Brønsted and Lowry.

	S^{2-} +	H_2O ⇌	SH^- +	OH^-
a.	acid	base	acid	base
b.	acid	acid	base	base
c.	acid	base	base	acid
d.	base	acid	acid	base
e.	base	base	acid	acid
f.	base	acid	base	acid

10. Benzoic acid $C_6H_5CO_2H$ is a relatively weak acid; the pH of a 0.10 M solution of the acid is 2.59. What would the pH be if enough sodium benzoate ($C_6H_5CO_2Na$) was added to make the solution 0.20 M in sodium benzoate? Assume the volume doesn't change when the sodium benzoate is added.

   a. 2.29   c. 4.02   e. 5.18   g. 7.27
   b. 3.87   d. 4.47   f. 6.33   h. 9.52

Use the distribution diagram and acid ionization constants shown below for the diprotic H_2SO_3 to answer the following questions (11 and 12). *It is recommended that you label the curves.*

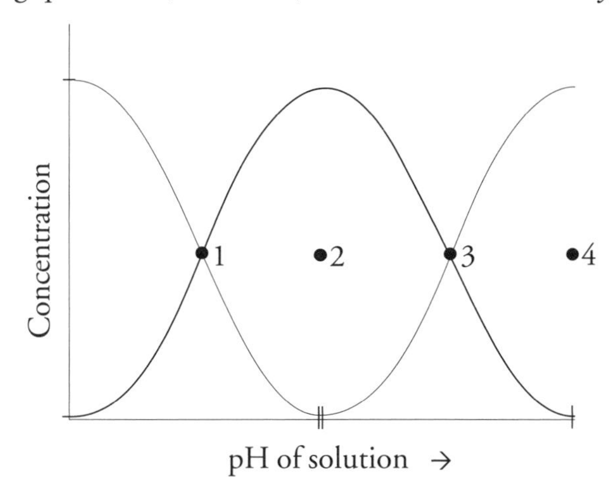

H_2SO_3	+	H_2O	$\rightleftharpoons$	H_3O^+	+	HSO_3^-	$K_{a1} = 1.3 \times 10^{-2}$		$pK_{a1} = 1.89$
HSO_3^-	+	H_2O	$\rightleftharpoons$	H_3O^+	+	SO_3^{2-}	$K_{a2} = 6.3 \times 10^{-8}$		$pK_{a2} = 7.20$

11. What is (are) the predominant sulfur-containing species in solution at point 3?

a. H_2SO_3
b. H_2SO_3 and HSO_3^-
c. HSO_3^-
d. HSO_3^- and SO_3^{2-}
e. SO_3^{2-}

12. What is the pH at point 2?

a. 1.89
b. 2.12
c. 2.87
d. 3.65
e. 4.55
f. 5.32
g. 7.20
h. 8.67

13. Acetylsalicylic acid (aspirin) is a weak acid with an ionization constant $K_a = 3.0 \times 10^{-4}$. What is the approximate pH of a buffered aspirin solution consisting of 0.10 M acetylsalicylic acid and 0.030 M sodium acetylsalicylate?

a. 1.0
b. 2.0
c. 3.0
d. 4.0
e. 5.0
f. 6.0
g. 7.0
h. 8.0
i. 10.0
j. 12.0

14. The pH of gastric juice in a typical person's stomach is approximately 2.1. What is the concentration (in mol L^{-1}) of hydroxide ions $[OH^-]$ in this person's stomach?

a. 2.1
b. 11.9
c. 1.3×10^{-12}
d. 8.0×10^{-3}
e. approximately 7
f. greater than 7

15. Calculate the molar solubility (in moles/liter) of the salt copper(II) iodate $Cu(IO_3)_2$ if the solubility product K_{sp} for the salt is 1.4×10^{-7}.

a. 3.3×10^{-3} M
b. 1.4×10^{-7} M
c. 3.7×10^{-4} M
d. 5.2×10^{-3} M
e. 9.9×10^{-3} M
f. 1.0×10^{-7} M

16. Calculate the solubility (in moles/liter) of the same salt copper(II) iodate $Cu(IO_3)_2$ in the presence of 0.50 M potassium iodate KIO_3.

a. 4.7×10^{-5} M
b. 6.2×10^{-8} M
c. 6.2×10^{-6} M
d. 5.6×10^{-7} M
e. 7.8×10^{-8} M
f. 3.9×10^{-6} M

Questions 17 through 19 refer to the following system.

The pressure on a system containing 0.50 mole of an ideal monatomic gas increases from 2.0 atm to 8.0 atm as the system moves from state A to state B at a constant volume of 12.0 L. The system then decreases in volume at a constant pressure of 8.0 atm to reach state C at which point the temperature is again what it was at state A.

It is recommended that you draw these changes on the diagram to the right.

$R = 0.08206$ L atm $K^{-1}mol^{-1} = 8.3145$ J $K^{-1}mol^{-1}$

17. How much work is done on the system in moving at constant pressure from B to C?

a. w = +2.3 kJ
b. w = +3.3 kJ
c. w = +4.3 kJ
d. w = +5.3 kJ
e. w = +6.3 kJ
f. w = +7.3 kJ
g. w = +8.3 kJ
h. w = +9.3 kJ

18. How much work is done on the system in moving *reversibly* from A to C at constant temperature?

a. w = +2.4 kJ
b. w = +3.4 kJ
c. w = +4.4 kJ
d. w = +5.4 kJ
e. w = +6.4 kJ
f. w = +7.4 kJ
g. w = +8.4 kJ
h. w = +9.4 kJ

19. What is the change in the internal energy of the system ΔE as it moves from A to C?

a. ΔE = zero kJ
b. ΔE = +2.5 kJ
c. ΔE = +4.5 kJ
d. ΔE = +5.5 kJ
e. ΔE = +7.5 kJ
f. ΔE = +8.5 kJ
g. ΔE = +9.5 kJ
h. ΔE = +10.5 kJ

20. Which one of the following has the largest entropy?

a. 1 mole of ice at 0°C
b. 1 mole of liquid water at 0°C
c. 1 mole of ice at 0°F
d. 1 mole of ice at 0 K
e. 1 mole of water vapor at 100°C
f. 1 mole of superheated liquid water at 105°C

21. Which of the following reactions is(are) most likely to progress further toward the product as the temperature is increased?

a. an endothermic reaction
b. an *entropy-driven* reaction
c. a reaction for which the term $T\Delta S$ is positive
d. a reaction in which the products are more disordered than the reactants
e. a reaction that produces a lot of gaseous products
f. all of the above

22. In which one of the following compounds does oxygen have an oxidation number of –1?

a. NaOH
b. H_2SO_3
c. K_2O
d. $H_2S_2O_3$
e. $CsNO_3$
f. H_2O_2
g. $KMnO_4$
h. $K_2Cr_2O_7$

23. The overall reaction for a cell in a lead storage battery is:

$$Pb(s) + PbO_2(s) + 2\,H^+(aq) + 2\,HSO_4^-(aq) \rightarrow 2\,PbSO_4(s) + 2\,H_2O(l)$$

Calculate the cell potential E at 25°C for this cell when the sulfuric acid concentration is 4.5 M, i.e. when the $[H^+] = [HSO_4^-] = 4.5$ M. The standard cell potential E° for this cell = 2.04 V.

a. 0 V
b. 1.96 V
c. 2.04 V
d. 2.12 V
e. 1.88 V
f. 2.18 V
g. 1.72 V
h. 2.36 V

24. Calculate the equilibrium constant for the reaction (at 600 K)

$H_2O(g) + \frac{1}{2}\,O_2(g) \rightleftharpoons H_2O_2(g)$ using the following data (same temperature)

$H_2(g) + O_2(g) \rightleftharpoons H_2O_2(g)$ $K = 2.3 \times 10^6$
$2\,H_2(g) + O_2(g) \rightleftharpoons 2\,H_2O(g)$ $K = 1.8 \times 10^{37}$

a. 2.36×10^{-19}
b. 1.02×10^{-25}
c. 7.89×10^{-16}
d. 5.42×10^{-13}
e. 1.28×10^{-31}
f. 9.76×10^{24}
g. 2.15×10^{-12}
h. 4.24×10^{18}

25. Calculate ΔG° for the reaction (at 600 K) described in the previous question:

$$H_2O(g) + \tfrac{1}{2}\,O_2(g) \rightleftharpoons H_2O_2(g)$$

a. 100 kJ
b. 120 kJ
c. 130 kJ
d. 141 kJ
e. 150 kJ
f. 160 kJ
g. 170 kJ
h. 180 kJ
i. –462 kJ
j. +462 kJ

26. Potassium dichromate (K_2CrO_7) is a very strong oxidizing agent that can be reduced to Cr^{3+}. It will oxidize ethanol as follows:

$$\_\,H^+(aq) + \_\,Cr_2O_7^{2-}(aq) + \_\,C_2H_5OH(l) \rightarrow \_\,Cr^{3+}(aq) + \_\,CO_2(g) + \_\,H_2O(l)$$

When this equation is properly balanced, what is the coefficient for the dichromate ion?

a. 1
b. 2
c. 3
d. 4
e. 5
f. 6
g. 7
h. 8
i. 9
j. 10

27. Which list of elements matches the following descriptions (in order):

A The alkaline earth metal whose chemistry is most similar to that of Al,
B The element in the series Li, Be, B that exhibits the most covalent character in its compounds,
C The noble gas (not including the radioactive Rn) that is most easily oxidized.

	A	B	C
a.	Be	Li	He
b.	Mg	Be	Xe
c.	Ca	B	He
d.	Be	B	Xe
e.	Mg	Be	He
f.	Ca	Li	Xe

28. In the electrolysis of aqueous silver nitrate ($AgNO_3$), 0.67 g of silver was deposited at the cathode. What quantity of electricity was required to deposit this silver?

a. 100 C
b. 200 C
c. 300 C
d. 400 C
e. 500 C
f. 600 C
g. 900 C
h. 1200 C

29. A predominant reason for the great abundance of organic compounds is

a. the presence of carbon dioxide in the atmosphere
b. the ability of carbon to catenate—to form strong covalent bonds with itself
c. the ability of carbon to form covalent bonds with hydrogen
d. the natural abundance of carbon on earth
e. the strong bond that carbon forms with oxygen

30. The reason for the difference between the structures of the elements nitrogen and phosphorus, and between the elements oxygen and sulfur, is

a. the ability of second period elements to form multiple bonds incorporating p_π–p_π bonds
b. the fact that oxygen and nitrogen are both principal constituents of the atmosphere
c. the higher electronegativity of nitrogen and oxygen
d. the fact that sulfur and phosphorus are solids, whereas nitrogen and oxygen are gases
e. the presence of d orbitals in the valence shells of sulfur and phosphorus

31. Cisplatin, the very successful anti-cancer drug developed by Barnett Rosenberg at MSU, is a square-planar complex with the formula *cis*–$Pt(NH_3)_2Cl_2$. This *cis*–isomer is the one that has an effective biological activity. Are there any other *geometrical* isomers of this platinum compound $Pt(NH_3)_2Cl_2$? And, if so, how many?

a. no other isomers
b. 1 other isomer
c. 2 other isomers
d. 3 other isomers
e. 4 other isomers
f. 5 other isomers

32. In the octahedral complex ion $[Fe(H_2O)_x]^{2+}$, what is the value of x?

a. 1
b. 2
c. 3
d. 4
e. 5
f. 6

33. In this octahedral complex $[Fe(H_2O)_x]^{2+}$, what is the LFSE (ligand field stabilization energy)? Ignore the pairing energy, P, if applicable.

a. –4 Dq
b. –6 Dq
c. –8 Dq
d. –12 Dq
e. –16 Dq
f. –20 Dq
g. –24 Dq

34. Meitnerium–266 (element 109) has been prepared by bombarding bismuth–209 (element 83) nuclei with iron–58 nuclei. Write a balanced equation for this nuclear reaction and determine what particle is emitted in the reaction.

a. ${}^{1}_{1}p$
b. ${}^{4}_{2}\alpha$
c. ${}^{1}_{0}n$
d. ${}^{0}_{-1}\beta$
e. ${}^{0}_{+1}\beta$

35. The mass defect of iodine–127 is 1.1555 u. Calculate the nuclear binding energy for this nucleus in J mol^{-1}.

a. 1.04×10^{14} J mol^{-1}
b. 2.57×10^{13} J mol^{-1}
c. 3.88×10^{15} J mol^{-1}
d. 4.56×10^{14} J mol^{-1}
e. 5.72×10^{16} J mol^{-1}
f. 2.83×10^{13} J mol^{-1}

FINAL EXAMINATION

CHEMISTRY 142 FALL 2006

Thursday December 14th 2006

1. Phosphine (PH_3) decomposes to produce phosphorus (P_4) and hydrogen (H_2):

$$4\ PH_3(g) \rightarrow P_4(g) + 6\ H_2(g)$$

If phosphine PH_3 is used at a rate of 12 mol L^{-1} s^{-1} and the initial concentration of phosphorus is zero, what is the concentration of phosphorus after 5 seconds?

a. 1.0 M c. 0.25 M e. 12 M g. 22 M
b. 4.0 M d. 3.0 M f. 15 M h. 25 M

2. Dimethyl ether $(CH_3)_2O$ decomposes to methane (CH_4), hydrogen (H_2), and carbon monoxide (CO) in a first-order process. If the rate constant for this reaction is 3.2×10^{-4} s^{-1}, how long does it take for the concentration of dimethyl ether to decrease from 1.25 M to 0.25 M?

a. 1.0×10^2 s c. 5.0×10^3 s e. 8.5×10^5 s g. 3.0×10^9 s
b. 2.0×10^2 s d. 6.0×10^4 s f. 2.0×10^9 s h. 4.5×10^9 s

3. When the temperature increases from 212 K to 945 K, the rate constant for a reaction increases from 4×10^{-2} s^{-1} to 4.24×10^6 s^{-1}. What is the activation energy E_A for this reaction?

a. 42.0 kJ mol^{-1} c. 108 kJ mol^{-1} e. 188 kJ mol^{-1} g. 267 kJ mol^{-1}
b. 15.0 kJ mol^{-1} d. 124 kJ mol^{-1} f. 231 kJ mol^{-1} h. 287 kJ mol^{-1}

4. Hexachloromolybdate(IV) ($MoCl_6^{2-}$) reacts with the nitrate ion (NO_3^-) to produce pentachlorooxygen-molybdate(IV) ($OMoCl_5^-$), the nitrite ion (NO_2^-), and the chloride ion (Cl^-):

$$MoCl_6^{2-} + NO_3^- \rightarrow OMoCl_5^- + NO_2^- + Cl^-$$

The mechanism for this reaction includes two steps, which are:

$$MoCl_6^{2-} \rightleftharpoons MoCl_5^- + Cl^- \qquad \text{FAST}$$
$$NO_3^- + MoCl_5^- \rightarrow OMoCl_5^- + NO_2^- \qquad \text{SLOW}$$

Identify the intermediate in the above reaction mechanism.

a. $MoCl_6^{2-}$ c. $OMoCl_5^-$ e. NO_2^-
b. $MoCl_5^-$ d. NO_3^- f. Cl^-

5. Which is the rate law for the reaction described in question 4?

a. Rate = k $[MoCl_6^{2-}]$
b. Rate = kK $[MoCl_6^{2-}][NO_3^-]$
c. Rate = k $[NO_3^-][NO_2^-]$
d. Rate = k $[NO_3^-][MoCl_5^-]^2/[Cl^-]$
e. Rate = k $[NO_3^-]^2[MoCl_5^-]$
f. Rate = k $[NO_3^-]^2[MoCl_5^-]^2$
g. Rate = kK $[OMoCl_5^-][NO_2^-]$
h. Rate = kK$[NO_3^-][MoCl_6^{2-}]/[Cl^-]$

6.

HF	HBr	CH_3CH_2OH	HNO_2
HNO_3	HI	HCl	$Mg(OH)_2$

How many of the compounds above are strong acids?

a. 1 c. 3 e. 5 g. 7
b. 2 d. 4 f. 6 h. 8

7. Label the species in the following aqueous equilibrium as acid or base according to the definitions of Brønsted and Lowry.

	HPO_4^{2-} +	H_2O ⇌	$H_2PO_4^-$ +	OH^-
a.	acid	acid	base	base
b.	acid	base	acid	base
c.	acid	acid	base	acid
d.	base	base	acid	acid
e.	base	acid	acid	base
f.	base	acid	base	acid

8. In which direction will the following gas-phase equilibrium shift if the indicated changes are made? *Choose the row in which all answers are correct.*

$$2H_2S(g) + SO_2(g) \rightleftharpoons 3S(s) + 2H_2O(g)$$

	add a catalyst	*add S*	*add H_2O*	*decrease V*
a.	shift left	shift right	shift right	shift right
b.	shift left	shift right	shift left	shift right
c.	no change	shift left	shift right	shift right
d.	no change	shift right	shift right	shift right
e.	no change	no change	shift left	shift right
f.	shift right	no change	shift right	shift right

9. How many of the following atoms, molecules, or ions can act as a Lewis acid?

H_2	F^-	F_2	Ti
Ne	Fe	CH_4	Mn

a. none c. 2 e. 4 g. 6
b. 1 d. 3 f. 5 h. all

10. Calculate the equilibrium constant K_p for the following gas phase equilibrium at 25°C given the equilibrium partial pressures shown:

$$2\,C_2H_2(g) + 5\,O_2(g) \rightleftharpoons 4\,CO_2(g) + 2\,H_2O(l)$$

$$P_{C2H2} = 2 \text{ atm};\ P_{O2} = 5 \text{ atm};\ P_{CO2} = 4 \text{ atm}$$

a. 4.03×10^5 c. 5.15×10^1 e. 2.05×10^{-5} g. 3.79×10^{-10}
b. 8.62×10^4 d. 2.05×10^{-2} f. 6.34×10^{-7} h. 2.15×10^{-12}

11. $BaCl_2$ KBr $Sr(CN)_2$ $CsNO_2$
$Mg(CH_3CO_2)_2$ NH_4NO_3 NH_4Cl $NaNO_3$
How many of the salts above produce an acidic solution when dissolved in water?

a. none	c. 2	e. 5	g. 7
b. 1	d. 4	f. 6	h. all

12. Calculate the pH of a 3.0 M solution of the weak acid phenol C_6H_5OH. $K_a = 1.6 \times 10^{-10}$.

a. 1.22	c. 4.66	e. 7.88	g. 11.2
b. 3.02	d. 6.20	f. 9.70	h. 13.9

13. Calculate the pH of a 3.0 M solution of phenol (see previous question C_6H_5OH) and a 0.75 M $C_6H_5O^-$ solution. $K_a = 1.6 \times 10^{-10}$ for C_6H_5OH.

a. 13.3	c. 11.2	e. 9.20	g. 3.20
b. 12.5	d. 10.7	f. 5.80	h. 1.90

14. Chlorine trifluoride (ClF_3) reacts with ammonia (NH_3) to produce nitrogen (N_2), hydrogen fluoride (HF), and chlorine (Cl_2) according to the following reaction:

$$2\ ClF_3(g) + 2\ NH_3(g) \rightleftharpoons N_2(g) + 6\ HF(g) + Cl_2(g)$$

At equilibrium, the partial pressures of the reactants and products are:

$P_{ClF3} = 0.5$ atm, $P_{NH3} = 0.5$ atm, $P_{N2} = 1.5$ atm, $P_{HF} = 9$ atm, $P_{Cl2} = 1.5$ atm

What is K_c for this reaction? Assume T = 25°C.

a. 8.99×10^{-2}	c. 4.55×10^{2}	e. 53.5	g. 9.10×10^{-1}
b. 67.8	d. 23.6	f. 3.22×10^{2}	h. 71.0

15. What is the pH of a 0.75 M solution of sodium hydrogen oxalate $NaHC_2O_4$? Oxalic acid is a weak diprotic acid with $K_{a1} = 6.5 \times 10^{-2}$ and $K_{a2} = 6.1 \times 10^{-5}$.

a. 0.95	c. 2.70	e. 5.50	g. 8.40
b. 1.41	d. 4.30	f. 6.70	h. 12.6

16. Monochloracetic acid $HC_2H_2ClO_2$ is a weak acid with $K_a = 1.35 \times 10^{-3}$. What is the value of the neutralization constant K_{neut} in a titration of monochloracetic acid with the weak base ethylamine $CH_3CH_2NH_2$.

$K_b = 5.6 \times 10^{-4}$ for $CH_3CH_2NH_2$

a. 8.2×10^{-2}	c. 2.7×10^{4}	e. 7.6×10^{7}	g. 8.8×10^{13}
b. 9.9×10^{2}	d. 3.1×10^{5}	f. 6.9×10^{10}	h. 7.3×10^{15}

17. Calculate the concentration of silver(I) ions $[Ag^+]$ in a saturated solution of silver(I) phosphate Ag_3PO_4. K_{sp} for $Ag_3PO_4 = 1.8 \times 10^{-18}$.

a. 1.2×10^{-1} M	c. 5.5×10^{-7} M	e. 8.7×10^{-10} M	g. 6.1×10^{-18} M
b. 4.8×10^{-5} M	d. 3.4×10^{-8} M	f. 6.1×10^{-13} M	h. 6.1×10^{-20} M

18. Calculate the concentration of [Ag^+] ions in a saturated solution of Ag_3PO_4 if the solution also contains 4.5×10^{-2} M phosphate (PO_4^{3-}).

a. 2.1×10^{-2} M
b. 3.1×10^{-4} M
c. 1.8×10^{-5} M
d. 3.4×10^{-6} M
e. 5.2×10^{-9} M
f. 7.7×10^{-12} M
g. 5.3×10^{-16} M
h. 7.1×10^{-20} M

Questions 19 – 23 refer to the following system.

The volume of a system containing 4 moles of an ideal monatomic gas at 300 K decreases from 13 L to 7.5 L as the system moves from state A to state B at constant pressure. The system is then heated at constant volume to state C, where the new pressure is 12 atm.

R = 0.08206 L atm mol^{-1} K^{-1} = 8.314 J mol^{-1} K^{-1}

Pressure

Volume

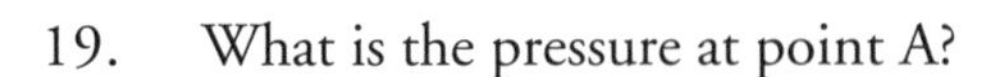

19. What is the pressure at point A?

a. 1.30 atm
b. 10.2 atm
c. 7.60 atm
d. 3.20 atm
e. 2.50 atm
f. 15.3 atm
g. 0.90 atm
h. 13.4 atm

20. What is the temperature at point C?

a. 15.0 K
b. 274 K
c. 50.0 K
d. 162 K
e. 79.0 K
f. 402 K
g. 189 K
h. 26.0 K

21. How much work is done on the system as it moves from state A to state B?

a. 32.5 kJ
b. 25.3 kJ
c. 105 kJ
d. 92.7 kJ
e. 4.24 kJ
f. 8.56 kJ
g. 13.5 kJ
h. 67.9 kJ

22. How much heat is released by the system as it moves from state A to state B?

a. 2.27 kJ
b. 3.96 kJ
c. 7.71 kJ
d. 33.2 kJ
e. 14.3 kJ
f. 10.5 kJ
g. 12.8 kJ
h. 17.1 kJ

23. What is the change in internal energy ΔE as the system moves from A to B?

a. –30.3 kJ
b. –14.5 kJ
c. –6.32 kJ
d. –1.38 kJ
e. 0 kJ
f. +1.38 kJ
g. +6.32 kJ
h. +14.5 kJ

24. Calculate the entropy change ΔS° for the following reaction in J K^{-1}:

$$2\,H_2S(g) + SO_2(g) \rightarrow 3\,S(s) + 2\,H_2O(g)$$

S°($H_2S(g)$) = 206 J mol^{-1} K^{-1}
S°($SO_2(g)$) = 248 J mol^{-1} K^{-1}
S°($S(s)$) = 33 J mol^{-1} K^{-1}
S°($H_2O(g)$) = 189 J mol^{-1} K^{-1}

a. –222 J K^{-1}
b. –454 J K^{-1}
c. –183 J K^{-1}
d. –239 J K^{-1}
e. +183 J K^{-1}
f. +239 J K^{-1}
g. +454 J K^{-1}
h. +222 J K^{-1}

25. Referring to the previous question, what is $\Delta G°$ for this reaction at 150 K? Assume $\Delta H°$ and $\Delta S°$ are independent of temperature.

$\Delta H_f°(H_2S(g)) = -21 \text{ kJ mol}^{-1}$
$\Delta H_f°(SO_2(g)) = -297 \text{ kJ mol}^{-1}$
$\Delta H_f°(H_2O(g)) = -242 \text{ kJ mol}^{-1}$

a. −263 kJ
b. −297 kJ
c. −118 kJ
d. −318 kJ
e. −539 kJ
f. −891 kJ

26. Referring to the previous two questions, what is the value of the equilibrium constant at 770 K?

a. 4.08×10^{-3}
b. 2.72
c. 5.63×10^{-4}
d. 6.58×10^{-5}
e. 7.92×10^{-8}
f. 9.66×10^{15}
g. 3.21×10^{19}
h. 7.92×10^{23}

27. Solid silver (Ag) reacts with the cyanide ion (CN^-) and oxygen (O_2) to produce dicyanoargentate(I) $[Ag(CN)_2]^-$ in basic solution. Write the balanced equation for this reaction. What is the coefficient of the cyanide ion (CN^-) in the balanced equation?

a. 1
b. 2
c. 3
d. 4
e. 5
f. 6
g. 7
h. 8

28. Given the following reduction potentials, calculate the potential (E°) for the following cell at standard conditions:

$K(s)|K^+(aq)||MnO_4^-(aq)|\ MnO_4^{2-}(aq)$

$E°(MnO_4^-|\ MnO_4^{2-}) = +0.56 \text{ V}$
$E°(K^+|K) = -2.92 \text{ V}$

a. −0.56 V
b. +0.56 V
c. −2.36 V
d. +2.36 V
e. −2.92 V
f. +2.92 V
g. −3.48 V
h. +3.48 V

29. In the electrolysis of manganese(IV) sulfate $Mn(SO_4)_2$, 0.44 g of manganese was deposited at the cathode. If the same amount of electricity is passed through a separate electrolysis cell containing silver(I) sulfate Ag_2SO_4, how much silver will be deposited at the cathode?

a. 0.220 g
b. 0.440 g
c. 0.980 g
d. 1.22 g
e. 3.45 g
f. 10.2 g
g. 57.9 g
h. 80.0 g

30. What is the correct formula for the coordination compound pentaamminechlorocobalt(III) chloride?

a. $[Co(NH_3)_2Cl_4]^-$
b. $[Co(NH_3)_3Cl_3]$
c. $[Co(NH_3)_5]^{3+}$
d. $[Co(NH_3)_5]Cl$
e. $[Co(NH_3)_5Cl]Cl_2$
f. $[Co(NH_3)_5Cl_2]Cl$
g. $[Co(NH_3)_5Cl_3]$
h. $[Co(NH_3)_5Cl]^{2+}$

31. How many geometrical isomers can be drawn for the square planar complex $[Pt(NH_3)_2(SCN)_2]$?

a. 1
b. 2
c. 3
d. 4
e. 5
f. 6
g. 7
h. 8

32. In the octahedral complex ion $[Mn(F)_x]^{3-}$, what is the value of x?

a. 1 c. 3 e. 5 g. 7
b. 2 d. 4 f. 6 h. 8

33. In the octahedral complex ion $[Mn(F)_x]^{3-}$, what is the LFSE (ligand field stabilization energy)? Ignore the pairing energy, P, if applicable.

a. –4Dq c. –8Dq e. –12Dq g. –18Dq
b. –6Dq d. –10Dq f. –16Dq h. –24Dq

34. Balance the following nuclear reaction:

$$^{239}_{93}Np \rightarrow ^{239}_{94}Pu + ?$$

What particle is emitted in this nuclear reaction?

a. $^{1}_{1}p$ b. $^{4}_{2}\alpha$ c. $^{1}_{0}n$ d. $^{0}_{-1}\beta$ e. $^{0}_{+1}\beta$

35. According to experiment, the mass of a $^{24}_{12}Mg$ nucleus is 23.9784 u (amu). What is the binding energy of this nucleus in J mol^{-1}?

mass of proton = 1.007825 amu
mass of neutron = 1.008665 amu

a. 3.28×10^{-9} c. 3.28×10^{-19} e. 4.40×10^{5} g. 5.10×10^{20}
b. 1.97×10^{13} d. 7.93×10^{-23} f. 5.63×10^{7} h. 8.75×10^{23}

FINAL EXAMINATION

CHEMISTRY 142 SPRING 2007

Tuesday May 1st 2007

1. The solid fuel of the booster for the space shuttle is a mixture of ammonium perchlorate and aluminum powder. The reaction that takes place when the mixture is ignited is:

$$6\ NH_4ClO_4(s) + 10\ Al(s) \rightarrow 5\ Al_2O_3(s) + 3\ N_2(g) + 6\ HCl(g) + 9\ H_2O(g)$$

At one stage in the reaction, water is produced at a rate of 7200 mol s^{-1}. At what rate is the aluminum being used up ?

a. 2400 mols^{-1}
b. 4800 mols^{-1}
c. 6300 mol s^{-1}
d. 6480 mol s^{-1}
e. 7000 mol s^{-1}
f. 8000 mol s^{-1}

2. A compound undergoing a second-order decomposition decreases in concentration by 180 mol L^{-1} from 192 mol L^{-1} to 12 mol L^{-1} in exactly one hour. How long does it take for the 12 mol L^{-1} to decrease to 3.0 mol L^{-1}?

a. 4 min
b. 16 min
c. 24 min
d. 36 min
e. 48 min
f. 64 min
g. 72 min
h. 128 min
i. 180 min
j. 192 min

3. In the energy profile for a reaction shown, _____ is the energy of the reactants, _____ is the energy of the reaction, _____ is the energy of the activated complex, and _____ is the activation energy for the forward reaction.
(Choose your answers in the correct order.)

a. A E F B
b. A F B G
c. A G B E
d. B E F G
e. B F G E
f. B G F E
g. C E B F
h. C E B F
i. C F E A
j. C G F E

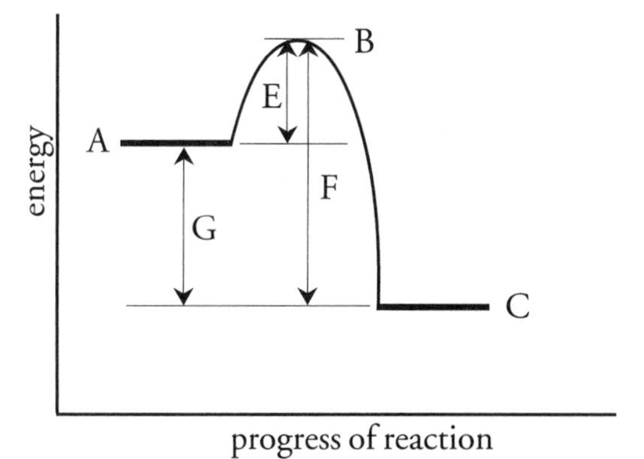

4. The following kinetic data were obtained for the reaction between nitrogen dioxide and ozone:

$$NO_2(g) + O_3(g) \rightarrow NO_3(g) + O_2(g)$$

Concentrations are expressed in *m*mol L^{-1}. What is the reaction order with respect to the NO_2?

	Initial [NO_2]	*Initial* [O_3]	*Initial rate of formation of* O_2
Experiment 1	0.21	0.70	6.3 *m*mol L^{-1} s^{-1}
Experiment 2	0.21	1.39	12.5 *m*mol L^{-1} s^{-1}
Experiment 3	0.38	0.70	11.4 *m*mol L^{-1} s^{-1}
Experiment 4	0.66	0.18	*see the following question*

a. −4
b. −3
c. −2
d. −1
e. 0
f. 1
g. 2
h. 3
i. 4
j. 5

5. What would the initial rate of formation of oxygen O_2 be for the Experiment 4 described in the previous question, where $[NO_2]_{init}$ = 0.66 *m*mol L^{-1} and $[O_3]_{init}$ = 0.18 *m*mol L^{-1} as shown?

a. 1.9 *m*mol L^{-1} s^{-1}	d. 3.5 *m*mol L^{-1} s^{-1}	g. 6.4 *m*mol L^{-1} s^{-1}
b. 2.7 *m*mol L^{-1} s^{-1}	e. 4.7 *m*mol L^{-1} s^{-1}	h. 7.3 *m*mol L^{-1} s^{-1}
c. 3.2 *m*mol L^{-1} s^{-1}	f. 5.1 *m*mol L^{-1} s^{-1}	i. 8.7 *m*mol L^{-1} s^{-1}

6. Raw (unpasteurized) milk goes sour in about 4.0 hours at 28°C but takes about 48 hours to go sour in a refrigerator at 5.0°C. What is the activation energy for the souring of raw milk?

a. 15 kJmol^{-1}	c. 25 kJmol^{-1}	e. 45 kJmol^{-1}	g. 75 kJmol^{-1}	i. 95 kJmol^{-1}
b. 20 kJmol^{-1}	d. 35 kJmol^{-1}	f. 60 kJmol^{-1}	h. 80 kJmol^{-1}	j. 105 kJmol^{-1}

7. Classify the following solutes in aqueous solution.
Choose the row where all responses are correct.

	Weak acid	*Strong base*	*Acidic salt*	*Neutral salt*
a.	NaCN	KOH	NH_4Cl	KNO_2
b.	H_2CO_3	$NaNO_3$	KCH_3CO_2	KBr
c.	NH_3	KF	Na_2HPO_4	$NaHSO_4$
d.	H_2SO_3	NaOH	NH_4Br	NaCl
e.	HF	KOH	$NaHCO_3$	$NaNO_3$
f.	CH_3CO_2H	NaOH	KCN	KCl

8. What is the pH of a solution prepared by diluting 25.0 mL of a 0.020*M* NaOH solution to 500 mL?

a. 2.0	c. 4.0	e. 6.0	g. 9.0	i. 11.0
b. 3.0	d. 5.0	f. 8.0	h. 10.0	j. 12.0

9. In a 2.0 L flask *at equilibrium*, there are 2.0 moles of methane CH_4, 3.0 moles of water, 6.0 moles of carbon monoxide, and an unknown concentration of hydrogen. The value of the equilibrium constant K_c for the system at this particular temperature is 54.0. What is the concentration of hydrogen gas in the 2.0 L container at equilibrium?

$$CH_4(g) + H_2O(g) \rightleftharpoons CO(g) + 3\,H_2(g)$$

a. 0.50 M	c. 1.9 M	e. 3.0 M	g. 6.0 M
b. 1.0 M	d. 2.5 M	f. 3.8 M	h. 8.0 M

10. Label the species in the following aqueous equilibrium as acid or base according to the definitions of Brønsted and Lowry:

$$HS^- + H_2O \rightleftharpoons H_2S + OH^-$$

	HS^-	H_2O	H_2S	OH^-
a.	acid	base	acid	base
b.	base	acid	base	acid
c.	base	acid	acid	base
d.	acid	base	base	acid
e.	base	base	acid	acid
f.	acid	acid	base	base

11. Calculate the equilibrium constant K at 25°C for the reaction between hydrocyanic acid and sodium hydroxide. The acid ionization constant K_a for hydrocyanic acid = 4.0×10^{-10}.

a. 2.0×10^{2}
b. 2.5×10^{-5}
c. 5.0×10^{-3}
d. 4.5×10^{-6}
e. 4.0×10^{4}
f. 3.0×10^{3}
g. 1.0×10^{10}
h. 2.5×10^{6}

12. What is the pH of the solution that results when 100 mL of 0.60 M NaOH is added to 100 mL of 1.20 M acetic acid?
The acid ionization constant K_a for acetic acid = 1.8×10^{-5}.

a. 2.43
b. 3.23
c. 4.13
d. 4.74
e. 5.66
f. 7.32
g. 8.13
h. 9.57
i. 10.21
j. 11.29

13. Carbonic acid H_2CO_3 is a weak diprotic acid:

$H_2CO_3 + H_2O \rightleftharpoons H_3O^+ + HCO_3^-$ $K_{a1} = 4.3 \times 10^{-7}$ $pK_{a1} = 6.37$
$HCO_3^- + H_2O \rightleftharpoons H_3O^+ + CO_3^{2-}$ $K_{a2} = 5.6 \times 10^{-11}$ $pK_{a2} = 10.25$

Determine the acidity or basicity of solutions of the two salts:

	$KHCO_3$	K_2CO_3
a.	acidic	acidic
b.	acidic	basic
c.	basic	acidic
d.	basic	basic

14. Calculate the value of pK for the autoprotolysis reaction in which the bicarbonate ion acts as *both* acid *and* base: (Data are provided in the previous question.)

$HCO_3^- + HCO_3^- \rightleftharpoons H_2CO_3 + CO_3^{2-}$ pK = ??

a. 1.97
b. 2.31
c. 3.89
d. 5.27
e. 6.37
f. 7.22
g. 10.25
h. 16.62

15. Which of the following is the correct solubility product constant expression for the very sparingly soluble salt antimony sulfide Sb_2S_3?

a. $K_{sp} = [Sb^{3+}][S^{2-}]$
b. $K_{sp} = [Sb^{3+}]^2[S^{2-}]^3$
c. $K_{sp} = [Sb^{3+}]^3[S^{2-}]^2$
d. $K_{sp} = [Sb_3][S_2]$
e. $K_{sp} = [Sb^{3+}]^3[S^{2-}]^2 / [Sb_2S_3]$
f. $K_{sp} = [Sb^{3+}]^2 / [S^{2-}]^3$

16. In a saturated solution of antimony sulfide in water, what is the relationship between the concentrations of the two ions present?

a. $[Sb^{3+}] = [S^{2-}]$
b. $[Sb^{3+}]^2 = [S^{2-}]^3$
c. $[Sb^{3+}]^3 = [S^{2-}]^2$
d. $3 \times [Sb^{3+}] = 2 \times [S^{2-}]$
e. $[Sb^{3+}] = 1/3 \times [S^{2-}]$
f. $[Sb^{3+}] = 1.5 \times [S^{2-}]$

17. If the solubility product for antimony sulfide $K_{sp} = 1.7 \times 10^{-93}$ at 25°C, what is the solubility of antimony sulfide at 25°C in moles of Sb_2S_3 per liter?

a. 1.1×10^{-19} M
b. 2.2×10^{-19} M
c. 3.3×10^{-19} M
d. 7.9×10^{-20} M
e. 3.7×10^{-19} M
f. 1.0×10^{-18} M

Questions 18, 19, 20, 21, and 22 refer to the following:

1.5 moles of an ideal (monatomic) gas is compressed isothermally and reversibly from state A (3.0 atm and 10.0 L) to half its volume. From state B, the temperature is increased at constant pressure to state C, where the volume is again what it was at state A.

It is recommended that you draw these changes on the diagram to the right.

R = 0.08206 L atm $K^{-1}mol^{-1}$ = 8.3145 J $K^{-1}mol^{-1}$

P

V

18. What is the pressure in atm at point B?

a. 1.0 atm
b. 1.5 atm
c. 3.0 atm
d. 4.0 atm
e. 4.5 atm
f. 6.0 atm
g. 7.5 atm
h. 9.0 atm

19. What is the temperature at points A and B?

a. −39°C
b. −29°C
c. −7°C
d. zero°C
e. 19°C
f. 26°C
g. 52°C
h. 77°C

20. How much work is done on the system in moving from A to B?

a. w = +1700 J
b. w = +1800 J
c. w = +1900 J
d. w = +2000 J
e. w = +2100 J
f. w = +2200 J
g. w = +2300 J
h. w = +2400 J

21. What work would be done if the system were to move back directly from C to A?

a. w = zero J
b. w = −1500 J
c. w = +1500 J
d. w = −1700 J
e. w = +1700 J
f. w = −2100 J
g. w = +2400 J
h. w = −2400 J

22. How much heat is liberated by the system in moving directly from C to A?

a. q = −1200 J
b. q = −2340 J
c. q = −3450 J
d. q = −4560 J
e. q = −5670 J
f. q = −6800 J
g. q = −7600 J
h. q = −6500 J

23. An entropy–driven reaction...

a. is spontaneous at low temperatures and high pressures
b. tends to proceed more rapidly than an enthalpy–driven reaction
c. produces great quantities of gas
d. is one in which the reactants are more ordered than the products
e. never occurs naturally
f. is extremely exothermic

24. What is the oxidation number of S in dithionic acid, $H_2S_2O_6$?

a. 0
b. +1
c. +2
d. +3
e. +4
f. +5
g. +6
h. +7

25. A 'breathalyzer' tests for alcohol by using an acidic dichromate solution to oxidize any alcohol in a person's breath. The dichromate solution is bright orange, which changes to green as the alcohol is oxidized. This color change is monitored by a photocell. Balance the equation for this redox reaction:

$$\_H^+(aq) + \_Cr_2O_7^{2-}(aq) + \_C_2H_5OH(l) \rightarrow \_Cr^{3+}(aq) + \_CO_2(g) + \_H_2O(l)$$

What is the sum of all six coefficients in the balanced equation?

a. 16 c. 19 e. 27 g. 36
b. 17 d. 22 f. 35 h. 38

26. Suppose you constructed a voltaic cell from the following two half cells:

E° ($SO_4^{2-}|H_2SO_3$) = +0.17 V

E° ($NO_3^-|NO$) = +0.96 V

What is the standard cell potential E°?

a. +0.17 V c. +0.79 V e. +0.96 V g. +1.13 V
b. –0.17 V d. –0.79 V f. –0.96 V h. –1.13 V

27. If the cell reaction (previous question) is represented by the equation:

$$3\,H_2SO_3(aq) + 2\,NO_3^-(aq) \rightarrow 3\,SO_4^{2-}(aq) + 2\,NO(g) + H_2O(l) + 4\,H^+(aq)$$

What is the value of ΔG° for the reaction?

a. –234 kJ c. –374 kJ e. –432 kJ
b. –283 kJ d. –388 kJ f. –457 kJ

28. How long will it take to plate out 63.55 grams of copper metal from an aqueous solution of Cu^{2+} if you were to use a current of 10 amps?

a. 26.2 sec c. 182 min e. 643 min
b. 161 min d. 322 min f. 26.8 hr

29. Which of the following square-planar or octahedral complexes exhibit geometrical *cis-trans* isomerism?

1 $[Pd(NH_3)_2Cl_2]$ 3 $[Pd(NH_3)_4]^{2+}$
2 $[Pd(NH_3)_3Cl]^+$ 4 $[Ru(NH_3)_4Br_2]^+$

a. 1 and 2 c. 1 and 4 e. 2 and 4 g. 1, 2, and 3
b. 1 and 3 d. 2 and 3 f. 3 and 4 h. 1, 3, and 4

30. In the octahedral complex ion $[Fe(H_2O)_6]^{2+}$, what is the LFSE (ligand field stabilization energy)? Ignore any possible contribution from pairing energies P).

a. –4 Dq c. –8 Dq e. –12 Dq g. –20 Dq
b. –6 Dq d. –10 Dq f. –16 Dq h. –24 Dq

31. Which list (row) of elements matches the following descriptions (in order):

1. The Group 5 element that exists normally as a gas
2. The Group 3 element that exhibits the greatest covalent character in its compounds
3. The Group 7 element that is most difficult to oxidize
4. The Group 2 element that is most similar in its chemical behavior to aluminum

	1	2	3	4
a.	N	B	I	Be
b.	P	Al	F	Mg
c.	As	Ga	Cl	Mg
d.	N	Al	Br	Be
e.	P	Ga	F	Mg
f.	As	B	Cl	Be
g.	N	Ga	I	Mg
h.	P	B	I	Be
i.	As	Al	F	Mg
j.	N	B	F	Be

The last four questions (32, 33, 34, and 35) *are written in pairs* (*a* and *b*). *You may choose which of each pair you answer. The first question of each pair is written specifically for students attending the 2:40 pm lectures by Prof. McHarris. You may, however, answer either one of each pair regardless of which lecture you attend. Answer only one of each pair. In other words, do not make more than one mark on the line on your answer sheet corresponding to each of these questions—if you do the question will be marked incorrect.*

Answer either 32a or 32b:

32*a*. A convenient way to prepare radioactive ^{204}Bi is by the reaction:

$^{206}\text{Pb}(p,3n)^{204}\text{Bi}$

What types of radiation would you expect to observe in the decay of ^{204}Bi?

a. α, γ
b. n, γ
c. e^+, α
d. e^-, γ
e. x-rays, γ
f. γ, fission products

32*b*. Based upon the n/p ratio, which of the following nuclides is most likely to be a positron emitter?

a. ^{238}U
b. ^{235}U
c. ^{3}H
d. ^{16}O
e. ^{15}O
f. ^{60}Co

Answer either 33a or 33b:

33*a*. Which of the following would probably *not* be a result of global warming?

a. Europe could become much colder.
b. Much of South Florida could be inundated by the ocean.
c. Famines in Africa could become much worse.
d. Polar bears could become extinct.
e. Hurricanes could become more frequent and more severe.
f. All of the above might well be consequences of global warming.

33*b*. What is the EAN (effective atomic number) of the nickel in the complex ion $[Ni(CN)_4]^{2-}$?

a. 26
b. 28
c. 30
d. 32
e. 33
f. 34
g. 36
h. 38

Answer either 34a or 34b:

34*a*. Which of the following is *not* an example of a fractal?

a. the coastline, say, of Great Britain
b. the Mandelbrot set
c. a figure with a dimension of 2.6
d. a sphere in four dimensions
e. a bifurcation diagram of a map in its chaotic region
f. a self-similar figure

34*b*. What is the value of ΔG at equilibrium?

a. the standard free energy change for the reaction
b. the same value that it has at any point during the reaction
c. the slope of the free energy curve at its maximum value
d. zero
e. it depends upon whether the change is exothermic or endothermic
f. the change in the enthalpy of the system when pure reactants are completely changed to pure products.

Answer either 35a or 35b:

35*a*. What is the "Butterfly Effect"?

a. a name for the extreme sensitivity of chaotic systems to initial conditions
b. a name for optical isomers that have a mirror plane of symmetry
c. a name indicating that many fractals resemble butterflies
d. a special breed of butterfly having a fractional dimension
e. a name suggested by the fact that many flowers resemble fractals
f. a name conjured up by Benoit Mandelbrot for his Mandelbrot set

35*b*. When the same element is both oxidized and reduced in a chemical reaction, the reaction is described as

a. disproportionation
b. disintegration
c. delocalization
d. disassociation
e. discrimination
f. disorientation

FINAL EXAMINATION

CHEMISTRY 142 FALL 2007

Friday December 14th 2007

1. At one stage in the decomposition reaction shown below, the ClO was produced at a rate of 36 mol L^{-1} min^{-1}. At what rate was the oxygen produced at this stage?

$$6\ ClO_3F(g) \rightarrow 2\ ClF(g) + 4\ ClO(g) + 7\ O_2(g) + 2\ F_2(g)$$

a. 24 mol L^{-1} m^{-1}	c. 40 mol L^{-1} m^{-1}	e. 48 mol L^{-1} m^{-1}	g. 56 mol L^{-1} m^{-1}
b. 36 mol L^{-1} m^{-1}	d. 42 mol L^{-1} m^{-1}	f. 49 mol L^{-1} m^{-1}	h. 63 mol L^{-1} m^{-1}

2. A compound undergoing a second-order decomposition decreases in concentration from 64 mol L^{-1} to 8.0 mol L^{-1} in 28 minutes. How long does it take for the 8.0 mol L^{-1} to decrease by a factor of four to 2.0 mol L^{-1}?

a. 8 min	c. 24 min	e. 48 min	g. 72 min	i. 168 min
b. 16 min	d. 32 min	f. 64 min	h. 96 min	j. 224 min

3. In the reaction of pyridine (C_5H_5N) with methyl iodide (CH_3I) in benzene solution, the following set of initial reaction rates was observed at 25°C for different initial concentration of the two reactants. What is the overall reaction order?

	Initial $[C_5H_5N]$	*Initial* $[CH_3I]$	*Initial rate*
Experiment 1	1.00×10^{-4} M	1.00×10^{-4} M	7.5×10^{-7} mol L^{-1} s^{-1}
Experiment 2	2.00×10^{-4} M	2.00×10^{-4} M	3.0×10^{-6} mol L^{-1} s^{-1}
Experiment 3	2.00×10^{-4} M	4.00×10^{-4} M	6.0×10^{-6} mol L^{-1} s^{-1}

a. −4	c. −2	e. 0	g. 2	i. 4
b. −3	d. −1	f. 1	h. 3	j. 5

4. Consider the following mechanism for a reaction in the gas state:

$$H_2O_2 \rightarrow H_2O + O$$
$$O + CF_2Cl_2 \rightarrow ClO + CF_2Cl$$
$$ClO + O_3 \rightarrow Cl + 2\ O_2$$
$$Cl + CF_2Cl \rightarrow CF_2Cl_2$$

Identify the reaction intermediate(s).

a. H_2O	f. CF_2Cl
b. CF_2Cl_2	g. O and CF_2Cl
c. O	h. CF_2Cl, Cl, and O
d. O_3	i. ClO and Cl
e. Cl	j. O, ClO, Cl, and CF_2Cl

5. A catalyst changes the rate of a chemical reaction because...

   a. it changes the equilibrium constant K
   b. it changes the coefficients in the balanced reaction
   c. it changes the temperature
   d. it changes the route of the reaction
   e. it makes the reaction more spontaneous (ΔG more negative)

6. Given that the equilibrium constants for the two reactions below are as shown:

$$XeF_6(g) + H_2O(g) \rightleftharpoons XeOF_4(g) + 2\,HF(g) \qquad K_1$$
$$XeO_4(g) + XeF_6(g) \rightleftharpoons XeOF_4(g) + XeO_3F_2(g) \qquad K_2$$

What is the equilibrium constant K_3 for the reaction:

$$XeO_4(g) + 2\,HF(g) \rightleftharpoons XeO_3F_2(g) + H_2O(g) \qquad K_3$$

a. K_1	c. $K_1 \times K_2$	e. K_2 / K_1	g. $K_2 - K_1$	i. $\sqrt{(K_2 / K_1)}$
b. K_2	d. K_1 / K_2	f. $\sqrt{K_1}$	h. $K_1 + K_2$	j. $\sqrt{(K_1 \times K_2)}$

7. Which one of the following electrolytes is weakly basic in aqueous solution?

a. $HClO_4$	c. KCN	e. KNO_3	g. LiI	i. NaCl
b. H_2SO_3	d. NH_4Br	f. $Ca(ClO_4)_2$	h. H_3PO_4	j. NH_4ClO_4

8. The net ionic equation representing the equilibrium established when potassium hydride is added to water is shown. Label the species present as either acid or base according to Brønsted-Lowry theory.

	H^- +	H_2O $\rightleftharpoons$	H_2 +	OH^-
a.	acid	base	acid	base
b.	acid	acid	base	base
c.	acid	base	base	acid
d.	base	acid	base	acid
e.	base	acid	acid	base
f.	base	base	acid	acid

9. In which direction will the following gas-phase equilibrium shift if the indicated changes are made? The reaction is exothermic in the forward direction. *Choose the row in which all answers are correct.*

$$2\,N_2O(g) + 3\,O_2(g) \rightleftharpoons 4\,NO_2(g)$$

	increase pressure	*add* O_2	*remove* NO_2	*increase temperature*
a.	shift left	shift left	shift left	shift left
b.	shift left	shift right	shift right	shift right
c.	shift left	shift right	shift right	shift left
d.	shift right	shift right	shift left	shift left
e.	shift right	shift right	shift right	shift left
f.	shift right	shift left	shift right	shift right

10. When acetic acid, CH_3CO_2H ($K_a = 1.8 \times 10^{-5}$) is neutralized by the addition of ammonia, NH_3 ($K_b = 1.8 \times 10^{-5}$) what is the value of the equilibrium constant for the neutralization reaction that occurs?

    a. 1.0×10^{-14} c. 1.8×10^{-14} e 1.8×10^{-19} g. 3.2×10^{-10}
    b. 3.2×10^{4} d. 3.1×10^{10} f. 1.0×10^{14} h. 1.8×10^{9}

11. Calculate the pH of a solution prepared by adding 50 mL 0.10 M HCl to 100 mL 0.10 M aqueous NH_3. (K_b for $NH_3 = 1.8 \times 10^{-5}$)

    a. 2.31 b. 4.75 c. 5.22 d. 8.78 e. 9.26 f. 11.1

12. What happens to the pH of the solution, after the titration described in the previous question, upon addition of 150 mL of water?

    a. The pH increases
    b. The pH decreases
    c. The pH stays the same

13. In an aqueous solution of potassium hydrogen phosphate K_2HPO_4, which of the following species is the strongest base present in the solution?

    a. H_3O^+
    b. H_2O
    c. K^+
    d. HPO_4^{2-}
    e. $H_2PO_4^-$

14. Oxalic acid ionizes in two stages in aqueous solution:

$$H_2C_2O_4 + H_2O \rightleftharpoons H_3O^+ + HC_2O_4^- \qquad K_{a1} = 5.9 \times 10^{-2} \quad pK_{a1} = 1.23$$
$$HC_2O_4^- + H_2O \rightleftharpoons H_3O^+ + C_2O_4^{2-} \qquad K_{a2} = 6.4 \times 10^{-5} \quad pK_{a2} = 4.19$$

    If 50 mL of a 0.1 M solution of sodium hydroxide (NaOH) is added to 50 mL of a 0.1 M solution of oxalic acid, is the solution that results at the equivalence point acidic, basic, or neutral?

    a. acidic
    b. basic
    c. neutral

15. Consider again the system described in the previous question. What is the pH of the solution after 75 mL of the sodium hydroxide solution has been added?

    a. 1.23 c. 2.71 e. 4.19 g. 5.42 i. 8.38
    b. 1.52 d. 1.89 f. 2.37 h. 3.89 j. 4.55

16. Silver chromate Ag_2CrO_4 (molar mass 331.73 g mol^{-1}) is a red solid that dissolves only slightly in water—to the extent of only 0.029 g per liter at 25°C. Calculate the solubility product K_{sp}.

    a. 8.7×10^{-5} c. 1.3×10^{-12} e. 3.3×10^{-13}
    b. 7.6×10^{-9} d. 2.7×10^{-12} f. 6.7×10^{-13}

17. The solubility products K_{sp} for some sparingly soluble carbonates are listed below. Which carbonate has the highest molar solubility?

a. silver carbonate Ag_2CO_3 $K_{sp} = 6.2 \times 10^{-12}$
b. barium carbonate $BaCO_3$ $K_{sp} = 8.1 \times 10^{-9}$
c. lead carbonate $PbCO_3$ $K_{sp} = 3.3 \times 10^{-14}$
d. strontium carbonate $SrCO_3$ $K_{sp} = 1.6 \times 10^{-9}$
e. calcium carbonate $CaCO_3$ $K_{sp} = 8.7 \times 10^{-9}$

Questions 18 through 22 refer to the following system.

The pressure on a system containing 2.0 moles of an ideal monatomic gas decreases from 15.0 atm to 5.0 atm as the system moves from state A to state B at a constant volume of 4.0 L. The system then expands at a constant pressure of 5.0 atm to reach state C at which point the temperature is again what it was at state A.

It is recommended that you draw these changes on the diagram.

$R = 0.08206$ L atm $K^{-1}mol^{-1} = 8.3145$ J $K^{-1}mol^{-1}$

P

V

18. What is the temperature at state B?

a. 89 K
b. 108 K
c. 122 K
d. 165 K
e. 186 K
f. 224 K
g. 298 K
h. 366 K

19. How much work is done by the system in moving at constant pressure from B to C?

a. w = –40 J
b. w = –60 J
c. w = –80 J
d. w = –2030 J
e. w = –4050 J
f. w = –6080 J
g. w = –8100 J
h. w = –9450 J

20. How much heat is absorbed by the system, or liberated by the system, in moving at constant pressure from B to C?

a. q = –60 J
b. q = +60 J
c. q = +100 J
d. q = –100 J
e. q = +6080 J
f. q = –6080 J
g. q = +10,130 J
h. q = –10,130 J

21. What is the change in the internal energy of the system ΔE as it moves from A to C?

a. ΔE = zero J
b. ΔE = +40 J
c. ΔE = –4050 J
d. ΔE = –6080 J
e. ΔE = +10,130 J
f. ΔE = +14,180 J
g. ΔE = –16,210 J
h. ΔE = –20,130 J

22. What is the entropy change in the system as it moves from state A to state C via state B?

a. 2.45 J K^{-1}
b. 4.47 J K^{-1}
c. 7.12 J K^{-1}
d. 9.83 J K^{-1}
e. 12.5 J K^{-1}
f. 15.7 J K^{-1}
g. 18.3 J K^{-1}
h. 23.1 J K^{-1}

23. The equilibrium constant K_a for the ionization of a weak acid at 25°C is 4.50×10^{-6}. What is the value of ΔG° for this ionization reaction?

a. –301 J
b. –935 J
c. –13.3 kJ
d. 2.31 kJ
e. 17.7 kJ
f. 22.4 kJ
g. 30.5 kJ
h. 36.7 kJ

24. The magnitude and sign of the calculated $\Delta G°$ in the previous question indicates that...

a. the reaction is very much reactant-favored
b. the ionization process is not spontaneous
c. the ionization process is exothermic
d. the equilibrium position lies almost completely on the product side of the equation
e. it takes a long time for the system to reach equilibrium

25. What are the oxidation numbers of O in the following compounds?

	Na_2O_2	O_2F_2	$K_2Cr_2O_7$
a.	−1	+1	−1
b.	−1	+1	−2
c.	−1	+2	−1
d.	−1	−2	−2
e.	−2	+1	−1
f.	−2	+1	−2
g.	−2	−2	−2
h.	−2	−2	−1

26. The oxidation of silver by chromate in basic conditions to produce silver(I) sulfide and chromium(III) hydroxide is represented by the equation:

$$Ag(s) + HS^-(aq) + CrO_4^{2-}(aq) \rightarrow Ag_2S(s) + Cr(OH)_3(s)$$

Balance the equation. How many water molecules are on which side of the balanced equation?

a. 1 on the left
b. 2 on the left
c. 3 on the left
d. 4 on the left
e. 5 on the left
f. 1 on the right
g. 2 on the right
h. 3 on the right
i. 4 on the right
j. 5 on the right

27. Given the following reduction potentials, calculate the potential (E°) for the following cell at standard conditions:

$$Zn(s)|Zn^{2+}(aq)||MnO_4^-(aq),Mn^{2+}(aq)|\textit{inert electrode}$$

$E° (Zn|Zn^{2+}) = -0.76$ V
$E° (MnO_4^-|Mn^{2+}) = +1.49$ V

a. −0.73 V
b. +0.73 V
c. −0.76 V
d. +0.76 V
e. −1.49 V
f. +1.49 V
g. −2.25 V
h. +2.25 V

28. If the cell reaction (previous question) is represented by the equation:

$$5\ Zn(s) + 2\ MnO_4^-(aq) + 16\ H^+(aq) \rightarrow 5\ Zn^{2+}(aq) + 2\ Mn^{2+}(aq) + 8\ H_2O(l)$$

What is the value of $\Delta G°$ for the reaction?

a. −235 kJ
b. −434 kJ
c. −704 kJ
d. −1085 kJ
e. −1556 kJ
f. −2171 kJ

29. A galvanic cell consists of a zinc anode immersed in a solution of zinc sulfate $ZnSO_4$ and a cadmium cathode immersed in a solution of cadmium sulfate $CdSO_4$. A salt bridge connects the two half-cells. A current of 1.45 amp flows for 2.60 hours. What is the change in the mass of the cadmium cathode?

a. +2.1 grams
b. –2.1 grams
c. +4.0 grams
d. –4.0 grams
e. +7.9 grams
f. –7.9 grams
g. +11.3 grams
h. –11.3 grams
i. +15.8 grams
j. –15.8 grams

30. Of the following block of elements, which one forms bonds with other elements with the greatest covalent character?

a. lithium
b. sodium
c. potassium
d. beryllium
e. magnesium
f. calcium
g. boron
h. aluminum
i. gallium

31. Using either your knowledge of its structure, or your understanding of common oxidation states, determine the subscript *n* for the magnesium in the formula for talc, $Mg_n(OH)_2Si_4O_{10}$.

a. 1 b. 2 c. 3 d. 4 e. 5 f. 6 g. 8

32. The low melting point of gallium and the reluctance of bromine to exhibit its highest oxidation state is attributed to the fact that

a. these elements are on the right side of the Periodic Table
b. both elements are in Period 4
c. both elements follow the filling of the first d orbital set with an increase of 10+ on the nucleus
d. both elements are p block elements
e. neither element is found in nature as a native element
f. both elements have an unusually high ionization energies

33. How many geometrical isomers of the octahedral complex $[MX_2Y_2Z_2]$ are there? The ligands X, Y, and Z are simple monodentate ligands.

a. 1
b. 2
c. 3
d. 4
e. 5
f. 6
g. 7
h. 8

34. What is the effective atomic number (EAN) of the transition element (Fe) in the octahedral coordination compound $[Fe(OH)(H_2O)_5](NO_3)_2$?

a. 23
b. 24
c. 26
d. 35
e. 36
f. 37
g. 38
h. 56

35. What is the ligand field stabilization energy (LFSE) for the cobalt ion in the octahedral complex ion $[Co(CN)_6]^{3-}$? Do not include any pairing energy P.

a. –4 Dq
b. –6 Dq
c. –8 Dq
d. –12 Dq
e. –16 Dq
f. –18 Dq
g. –20 Dq
h. –24 Dq

FINAL EXAMINATION

CHEMISTRY 142 SPRING 2008

Wednesday April 30th 2008

1. Consider the reaction

$$2\,N_2O_5(g) \rightarrow 4\,NO_2(g) + O_2(g)$$

If $NO_2(g)$ is formed at a rate of x Ms^{-1}, what are the corresponding rates for $N_2O_5(g)$ and $O_2(g)$?

	$\Delta[N_2O_5]/\Delta t$	$\Delta[O_2]/\Delta t$
a.	$2x$	$4x$
b.	$-2x$	$4x$
c.	$2x$	$-4x$
d.	$(1/2)x$	$(1/4)x$
e.	$-(1/2)x$	$(1/4)x$
f.	$(1/2)x$	$-(1/4)x$

2. A reactant involved in a reaction in which its order is 0 (zero-order) decreases in concentration from 48 mol L^{-1} to 12 mol L^{-1} in 60 seconds. How long does it take for its concentration to decrease from 12 mol L^{-1} to 3 mol L^{-1}?

a. 5 s
b. 10 s
c. 12 s
d. 15 s
e. 24 s
f. 30 s
g. 1 min
h. 2 min
i. 4 min
j. 8 min

3. Initial rate data are given below for the reaction

$$2\,NO(g) + 2\,H_2(g) \rightarrow N_2(g) + 2\,H_2O(l)$$

Experiment	*Initial [NO]*	*Initial [H_2]*	*Initial rate of formation of N_2*
1	0.420 M	0.122 M	0.136 Ms^{-1}
2	0.210 M	0.122 M	0.0339 Ms^{-1}
3	0.210 M	0.244 M	0.0678 Ms^{-1}
4	0.105 M	0.488 M	??

The rate equation for this reaction is

a. Rate = $k[NO][H_2]$
b. Rate = $k[NO]$
c. Rate = $k[NO][H_2]^2$
d. Rate = $k[NO]^2[H_2]$
e. Rate = $k[NO]^2$
f. Rate = $k[H_2]$
g. Rate = $k[NO]^{-1}[H_2]$
h. Rate = $k[NO]^{-1}[H_2]^{-1}$

4. What would the initial rate of the formation of $N_2(g)$ be in Experiment 4, described in the previous question, where $[NO]_{init}$ = 0.105 M and $[H_2]_{init}$ = 0.488 M?

a. 0.0168 Ms^{-1}
b. 0.0340 Ms^{-1}
c. 0.0678 Ms^{-1}
d. 0.136 Ms^{-1}
e. 0.272 Ms^{-1}
f. 0.544 Ms^{-1}

5. In a series of elementary steps making up the mechanism for a chemical reaction, a species involved in the reaction is formed in one step but then reused in a subsequent step. Such a species is referred to as a

a. reactant
b. catalyst
c. product
d. transition state
e. intermediate
f. radical

6. In the reaction profile shown on the right,
__ represents the energy of reaction,
__ represents the activated complex,
__ represents the reactants, and
__ represents the activation energy for the forward reaction.

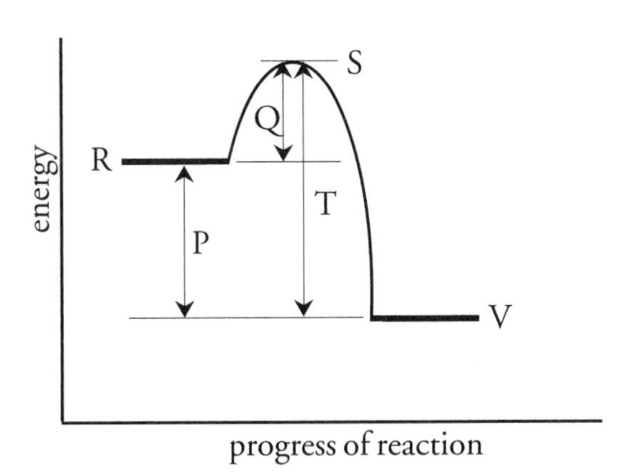

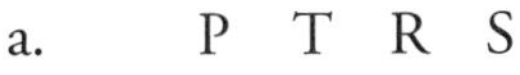

a. P T R S
b. T S R Q
c. S P Q R
d. P S R T
e. T S P Q
f. P S R Q
g. T Q R S

7. Classify the solutes given below based on the type of aqueous solution that they form when dissolved in water. Choose the row where all the responses are correct.

	Acidic	*Basic*	*Neutral*
a.	$MgCl_2$	Na_2HPO_4	LiBr
b.	$NaNO_3$	$NaNO_3$	KCl
c.	KCN	HCN	Na_3PO_4
d.	NaH_2PO_4	$NaNO_3$	NH_4Cl
e.	CH_3CO_2H	NaOH	KCN
f.	HF	CH_3NH_2	NH_4NO_3
g.	NH_4Cl	KCH_3CO_2	KBr

8. The net ionic equation for the equilibrium established when Na_2CO_3 is added to water is shown below. Classify the participants in the reaction as acid or base according to the definitions of Brønsted and Lowry:

	CO_3^{2-} +	H_2O $\rightleftharpoons$	HCO_3^- +	OH^-
a.	acid	base	acid	base
b.	base	acid	base	acid
c.	base	acid	acid	base
d.	acid	base	base	acid
e.	base	base	acid	acid
f.	acid	acid	base	base

9. At equilibrium, a 1.0 L flask contains 2.0 moles of ammonia, 1.0 mole of chlorine, 3.0 moles of hydrazine N_2H_4, and an unknown concentration of hydrogen chloride HCl. The value of the equilibrium constant for the system at this temperature is 12. What is the concentration of HCl in the flask?

$$2\,NH_3(g) \quad + \quad Cl_2(g) \quad \rightleftharpoons \quad N_2H_4(g) \quad + \quad 2\,HCl(g)$$

a. 1.0 M
b. 2.0 M
c. 3.0 M
d. 4.0 M
e. 6.0 M
f. 8.0 M
g. 12 M
h. 16 M

10. What is the pH of a solution prepared by diluting 50.0 mL of a 0.010*M* HCl solution to 500 mL?

a. 2.0
b. 3.0
c. 4.0
d. 5.0
e. 6.0
f. 8.0
g. 9.0
h. 10.0
i. 11.0
j. 12.0

11. Oxalic acid ionizes in two stages in aqueous solution:

$H_2C_2O_4 + H_2O \rightleftharpoons H_3O^+ + HC_2O_4^-$ $\quad K_{a1} = 5.9 \times 10^{-2}$ $\quad pK_{a1} = 1.23$

$HC_2O_4^- + H_2O \rightleftharpoons H_3O^+ + C_2O_4^{2-}$ $\quad K_{a2} = 6.4 \times 10^{-5}$ $\quad pK_{a2} = 4.19$

What is the pH of an aqueous solution of sodium hydrogen oxalate $NaHC_2O_4$?

a. 1.23
b. 1.71
c. 2.71
d. 3.22
e. 4.19
f. 5.42
g. 6.71
h. 8.44

12. Consider again the solution of sodium hydrogen oxalate described in the previous question. What is equilibrium constant for the reaction:

$$HC_2O_4^- + HC_2O_4^- \rightleftharpoons H_2C_2O_4 + C_2O_4^{2-}$$

a. 5.9×10^{-2}
b. 1.1×10^{-3}
c. 3.2×10^{-4}
d. 6.4×10^{-5}
e. 5.4×10^{-6}
f. 7.3×10^{1}
g. 9.2×10^{2}
h. 1.4×10^{3}

13. What is the pH of a 0.20 M solution of the weak base ammonia?

K_b for $NH_3 = 1.8 \times 10^{-5}$

a. 0.70
b. 2.72
c. 5.44
d. 6.97
e. 8.56
f. 9.89
g. 11.28
h. 13.30

14. Given the following acid ionization constants, K_a, determine which salt will produce the most basic solution.

Acid	K_a	*Acid*	K_a
Acetic acid	1.8×10^{-5}	Hydrocyanic acid	4.0×10^{-10}
Hydrofluoric acid	3.5×10^{-4}	Nitrous acid	4.3×10^{-4}
Hypochlorous acid	3.0×10^{-8}	Selenous acid	3.5×10^{-2}
Hydrogen sulfate	1.1×10^{-2}	Formic acid	1.8×10^{-4}

a. KCH_3CO_2
b. NaF
c. KClO
d. K_2SO_4
e. NaCN
f. KNO_2
g. $KHSeO_3$
h. $KHCO_2$

15. At one point during the titration of acetic acid (CH_3CO_2H) *vs.* NaOH, it was found that the concentration of the acid was 0.103 M and the concentration of its conjugate base ($CH_3CO_2^-$) was 0.068 M. What is the pH of the solution at this point?

K_a for $CH_3CO_2H = 1.8 \times 10^{-5}$

a. 1.57
b. 2.87
c. 3.44
d. 4.56
e. 4.74
f. 4.93
g. 5.22
h. 5.47
i. 6.37
j. 7.61

16. Which two solutes mixed together in aqueous solution would *not* act as a buffer solution?

a. $KHSO_4$ and K_2SO_4
b. HCN and NaCN
c. NH_3 and NH_4Cl
d. $KHSO_3$ and $MgSO_3$
e. KF and NaF
f. Na_2HPO_4 and NaH_2PO_4
g. Na_2CO_3 and $NaHCO_3$
h. $HClO_2$ and $NaClO_2$

17. What is the solubility (in mol L^{-1}) of silver hydroxide in water at 25°C?
The solubility product K_{sp} for AgOH = 1.5×10^{-8}.

a. 1.5×10^{-8}
b. 1.6×10^{-3}
c. 1.2×10^{-4}
d. 2.5×10^{-3}
e. 2.3×10^{-16}
f. 6.7×10^{7}

18. What is the solubility (in mol L^{-1}) of silver hydroxide in an aqueous solution buffered at a pH of 9.00?

a. 1.5×10^{-8}
b. 1.2×10^{-6}
c. 8.7×10^{-5}
d. 1.2×10^{-4}
e. 3.9×10^{-2}
f. 1.5×10^{1}
g. 1.5×10^{-3}
h. 1.5×10^{-13}

19. What is the sign of ΔG for a system approaching equilibrium from the product side (i.e. in a direction opposite to that represented by the equation for the reaction?

a. ΔG is negative
b. ΔG is zero
c. ΔG is positive

20. Determine the oxidation number of Xe in the following compounds:

	XeF_2	XeO_6^{4-}	$XeOF_4$
a.	0	+8	+4
b.	0	+12	+4
c.	+1	+2	+5
d.	+2	+12	+6
e.	+2	+4	+4
f.	+2	+8	+6
g.	+4	+8	+10
h.	+4	+4	+6

21. Balance the following redox reaction under basic conditions

$$Tl_2O_3(aq) + NH_2OH(aq) \rightarrow TlOH(s) + N_2(g)$$

What is the stoichiometric coefficient of water in the balanced equation? Indicate also whether water appears as a reactant or product in the balanced equation.

a. 1 reactant
b. 2 reactant
c. 3 reactant
d. 4 reactant
e. 6 reactant
f. 1 product
g. 2 product
h. 3 product
i. 5 product
j. 6 product

For the following questions (22–25), consider the following system

A 1.0 mol sample of an ideal monatomic gas at a pressure of 2.5 atm occupies a volume of 10.0 L. Call this state of the system A. Suppose that the gas is then cooled at constant volume to reach state B, which has a pressure of 1.0 atm. Subsequently, the the gas is allowed to expand at constant pressure to reach state C, where the temperature is the same as at state A.

P

V

22. How much heat is absorbed or released by the system as it moves *reversibly* from state A to state C?

a. 0.0 kJ
b. –1.01 kJ
c. 1.01 kJ
d. –1.52 kJ
e. 1.52 kJ
f. –2.32 kJ
g. 2.32 kJ
h. 6.21 kJ

23. How much work is done on the system as it moves from state A to state C along the path ABC?

a. 0.0 kJ
b. –1.01 kJ
c. 1.01 kJ
d. –1.52 kJ
e. 1.52 kJ
f. –2.32 kJ
g. 2.32 kJ
h. 5.72 kJ

24. How much heat is absorbed or liberated by the system as it moves from state A to state C along the path ABC?

a. 0.0 kJ
b. –1.01 kJ
c. 1.01 kJ
d. –1.52 kJ
e. 1.52 kJ
f. –2.32 kJ
g. 2.32 kJ
h. 5.72 kJ

25. Consider a fourth state of the system, D, which has the same pressure as state A and the same volume as state C. How does the work done as the system moves from state A to state C along the path ADC compare to the work done when following the path ABC?

a. The work done is the same for both paths
b. The work done along path ADC is greater than the work done along the path ABC
c. The work done along path ADC is less than the work done along the path ABC
d. It depends upon how much heat is liberated or absorbed
e. More information is needed

26. Ethanol (C_2H_5OH) has a melting point of –114.1°C and a boiling point of 78.5°C at 1.0 atm. What are the signs of ΔG, ΔH, and ΔS for the freezing and vaporization of ethanol at 20°C and 1.0 atm?

	freezing of ethanol at 20°C			vaporization of ethanol at 20°C		
	ΔG	ΔH	ΔS	ΔG	ΔH	ΔS
a.	–	+	+	+	–	–
b.	+	–	–	+	+	+
c.	+	+	+	–	–	–
d.	–	–	+	–	–	–
e.	+	+	–	+	–	+
f.	–	–	–	+	+	+
g.	–	–	+	+	–	–
h.	+	–	–	–	+	+

27. Calculate the equilibrium constant for the reaction at 600 K

$$H_2O(g) + \tfrac{1}{2} O_2(g) \rightleftharpoons H_2O_2(g)$$

using the following data for the same temperature

$$H_2(g) + O_2(g) \rightleftharpoons H_2O_2(g) \qquad K = 2.3 \times 10^{6} \qquad \Delta H° = -143 \text{ kJ}$$
$$2\,H_2(g) + O_2(g) \rightleftharpoons 2\,H_2O(g) \qquad K = 1.8 \times 10^{37} \qquad \Delta H° = -484 \text{ kJ}$$

a. 2.36×10^{-19} c. 7.89×10^{-16} e. 5.42×10^{-13} g. 2.15×10^{-12}
b. 1.02×10^{-25} d. 1.28×10^{-31} f. 9.76×10^{24} h. 4.24×10^{18}

28. Using data included in the previous question, calculate ΔS° for the reaction at 600 K:

$$H_2O(g) + \tfrac{1}{2} O_2(g) \rightleftharpoons H_2O_2(g)$$

a. -21 JK^{-1} c. -43 JK^{-1} e. -70 JK^{-1} g. -91 JK^{-1} i. -126 JK^{-1}
b. -35 JK^{-1} d. -55 JK^{-1} f. -79 JK^{-1} h. -102 JK^{-1} j. -133 JK^{-1}

29. Find the standard cell potential E° at 25°C for a cell in which Cd is oxidized to Cd^{2+} at the anode and Ni^{2+} is reduced to Ni at the cathode.

$E^{\circ}_{red}(Cd^{2+}|Cd) = -0.403$ V
$E^{\circ}_{red}(Ni^{2+}|Ni) = -0.257$ V

a. +0.146 V c. +0.257 V e. +0.403 V g. +0.660 V
b. –0.146 V d. –0.257 V f. –0.403 V h. –0.660 V

30. A layer of chromium metal is electroplated on an automobile bumper by passing a constant current of 200 A through a cell containing Cr^{3+}*(aq)*. How many minutes are required to deposit 125 g of chromium?

a. 19.3 c. 174 e. 3480 g. 69,600
b. 58.0 d. 1160 f. 23,200 h. 101,000

31. Which of the following statements are true?

I Graphite and the fullerenes are allotropes of carbon in which the carbon is sp^3 hybridized.

II S is more electronegative than O.

III One of the principal differences between the elements C, N, and O and their cogeners Si, P, and S is the ability of C, N, and O to form π bonds using p orbitals.

IV Group 3 elements, such as B and Al, form electron-deficient (<octet) compounds.

V Silicon forms strong bonds with oxygen.

a. III and IV	d. IV and V	g. II, IV and V	j. all are true
b. I and V	e. I, II and V	h. III, IV and V	
c. II and V	f. II, III and IV	i. I, III, IV and V	

32. A predominant reason for the great abundance of organic compounds is

a. the presence of carbon dioxide in the atmosphere
b. the ability of carbon form π bonds with nitrogen
c. the ability of carbon to form covalent bonds with hydrogen
d. the natural abundance of carbon on earth
e. the strong bond that carbon forms with oxygen
f. the ability of carbon to form single and double covalent bonds with itself

33. In the orange coordination compound $[Co(en)_3]^{3+}$, where en is the strong ligand ethylenediamine $NH_2CH_2CH_2NH_2$, the cobalt ion has ______ unpaired electrons, the ligand is ______, and the ligand field stabilization energy (LFSE) is ______. Ignore any pairing energy P.

a.	0	monodentate	–24 Dq	f.	1	tridentate	–4 Dq
b.	4	bidentate	–4 Dq	g.	4	bidentate	0 Dq
c.	0	bidentate	0 Dq	h.	0	bidentate	–12 Dq
d.	2	tridentate	–8 Dq	i.	6	monodentate	–4 Dq
e.	0	bidentate	–24 Dq	j.	6	bidentate	–24 Dq

34. Use valence bond theory to predict the hybridization of the transition metal's bonding orbitals and the geometry of the diamagnetic coordination complex $[Pt(NH_3)_4]^{2+}$. Note that Pt is in the same group as Ni.

a.	sp^3	tetrahedral	e.	dsp^3	square pyramidal
b.	dsp^2	tetrahedral	f.	dsp^3	trigonal bipyramidal
c.	sp^3	square planar	g.	d^2sp^3	octahedral
d.	dsp^2	square planar			

35. Which particle is produced in the following radioactive decay?

$$^{118}_{54}Xe \rightarrow ^{118}_{53}Au + ?$$

a. $^{1}_{1}p$ b. $^{4}_{2}\beta$ c. $^{1}_{0}n$ d. $^{0}_{-1}\alpha$ e. $^{0}_{+1}\beta$

FINAL EXAMINATION

CHEMISTRY 142 FALL 2008

Monday December 8th 2008

1. The following reaction was studied in a series of experiments with different initial concentrations of reactant and the following initial rates were observed. What is the order of the reaction with respect to I_2?

$$(C_2H_5)_2(NH)_2 + I_2 \rightarrow (C_2H_5)_2N_2 + 2\ HI$$

	$[(C_2H_5)_2(NH)_2]$	$[I_2]$	Initial Rate
Experiment 1	0.015 M	0.015 M	3.15 M s^{-1}
Experiment 2	0.015 M	0.045 M	9.45 M s^{-1}
Experiment 3	0.030 M	0.045 M	18.9 M s^{-1}

a. 0 c. 1 e. 3 g. −1 i. −3
b. 0.5 d. 2 f. 4 h. −2 j. −4

2. ^{223}Ra decomposes by a first-order mechanism. The rate constant for this process is 0.0606 day^{-1}. What is the half-life of ^{223}Ra?

a. 0.09 days c. 1.3 days e. 6.1 days g. 11.4 days
b. 0.4 days d. 3.7 days f. 9.2 days h. 14.2 days

3. A compound undergoing a second-order decomposition decreases in concentration from 100 M to 50 M in 10 minutes. How long does it take for the 50 M sample to decrease to 12.5 M?

a. 5 min c. 10 min e. 20 min g. 40 min i. 60 min
b. 7.5 min d. 15 min f. 30 min h. 50 min j. 90 min

4. A catalyst...

a. increases the amount of product at equilibrium
b. is not involved chemically in the reaction
c. changes the route the reaction takes between reactants and products
d. increases the activation energy required in the reaction
e. must be in the same phase as the reactants
f. shifts the equilibrium state toward the product side
g. must be continually replenished as it is used in the reaction

5. What is the value of K_c for the following equilibrium at 25°C given the equilibrium concentrations shown?

$$2\ SO_2(g) + O_2(g) \rightleftharpoons 2\ SO_3(g)$$

$[SO_2]$ = 0.344 M
$[O_2]$ = 0.172 M
$[SO_3]$ = 0.056 M

a. 0.15 c. 0.95 e. 3.24 g. 6.67
b. 0.31 d. 1.05 f. 4.55 h. 7.46

6. Calculate the equilibrium constant for: $4\ HBr(g) \rightleftharpoons 2\ H_2(g) + 2\ Br_2(g)$ $K = ???$

   Given the following information: $H_2(g) + Br_2(g) \rightleftharpoons 2\ HBr(g)$ $K = 7.9 \times 10^{11}$

   a. 1.6×10^{-24} c. 1.4×10^{-5} e. 6.2×10^{3} g. 8.9×10^{5}
   b. 1.3×10^{-12} d. 5.4×10^{2} f. 1.1×10^{5} h. 1.6×10^{12}

7. How many of the following solutes would you classify as a weak acid?

   H_3PO_4 HNO_3 HF CH_3CO_2H
   NH_3 KOH CdI_2 NaBr

   a. 0 c. 2 e. 4 f. 6 h. 8
   b. 1 d. 3 f. 5 g. 7

8. What is the pH of a 0.3 M solution of strontium hydroxide?

   a. 0.2 c. 1.7 e. 8.4 g. 13.5
   b. 0.5 d. 3.9 f. 11.9 h. 13.8

9. A Lewis acid...

   a. will donate an H^+
   b. will donate a pair of electrons
   c. will accept an H^+
   d. will accept a pair of electrons

10. Consider the following Bronsted-Lowry equilibrium:

    $OCl^- + H_2O \rightleftharpoons$ _____ $+ OH^-$

    Identify the missing product and decide if that product is acting as an acid or a base in this equilibrium.

    a. HOCl, acid c. OCl^{2-}, acid e. H_2OCl, acid g. H_3O^+, acid
    b. HOCl, base d. OCl^{2-}, base f. H_2OCl, base h. H_3O^+, base

11. What is the K_b for $HAsO_4^{2-}$ at 25°C?

    For H_3AsO_4: $K_{a1} = 2.5 \times 10^{-4}$ $K_{a2} = 5.6 \times 10^{-8}$ $K_{a3} = 3.0 \times 10^{-13}$

    a. 2.5×10^{-18} c. 3.0×10^{-13} e. 1.8×10^{-7} g. 0.033
    b. 1.6×10^{-15} d. 4.0×10^{-11} f. 2.5×10^{-4} h. 0.065

12. What is the pH of a 0.2 M solution of ammonium nitrate? (K_b for ammonia = 1.8×10^{-5})

    a. 1.80 c. 4.98 e. 7.40 g. 10.6
    b. 2.72 d. 5.56 f. 9.02 h. 11.3

13. A 0.1 M solution of formic acid HCOOH was titrated with 0.1 M sodium hydroxide. What is the pH of the solution when [HCOOH] = 0.03 M and [$HCOO^-$] = 0.07 M?

    The K_a for formic acid is 1.8×10^{-4}.

    a. 0.370 c. 1.52 e. 4.11 g. 5.91
    b. 1.15 d. 2.63 f. 4.60 h. 6.62

14. What is the value of the neutralization constant K_{neut} in a titration of hypoiodous acid HIO with ammonia?

K_a for HIO = 2.0×10^{-11}
K_b for NH_3 = 1.8×10^{-5}

a. 7.6×10^{-19}
b. 3.9×10^{-16}
c. 3.6×10^{-15}
d. 2.8×10^{-9}
e. 6.8×10^{-6}
f. 5.2×10^{-4}
g. 3.6×10^{-2}
h. 9.7×10^{2}

15. What is the solubility of tin(II) iodide if the K_{sp} for tin(II) iodide is 1.0×10^{-4}?

a. 2.5×10^{-5} M
b. 1.0×10^{-4} M
c. 0.010 M
d. 0.029 M
e. 0.046 M
f. 0.079 M
g. 1.06 M
h. 2.84 M

16. How will the addition of each of the following solutes affect the solubility of tin(II) iodide? The K_{sp} for tin(II) iodide is 1.0×10^{-4}. *(Choose the row in which all responses are correct.)*

	NaI	HI	HNO_3
a.	decrease	decrease	no change
b.	decrease	increase	no change
c.	decrease	no change	no change
d.	increase	increase	increase
e.	increase	decrease	decrease
f.	increase	no change	no change

17. 50 mL of a 1.0×10^{-5} M solution of magnesium nitrate is added to 50 mL of a 4.0×10^{-5} M solution of potassium phosphate K_3PO_4. The K_{sp} for magnesium phosphate is 1.0×10^{-24}. Which of the following statements is true regarding this solution?

a. $Q_{sp} = K_{sp}$; the system is at equilibrium
b. $Q_{sp} > K_{sp}$; precipitation occurs
c. $Q_{sp} > K_{sp}$; no precipitation occurs
d. $Q_{sp} < K_{sp}$; precipitation occurs
e. $Q_{sp} < K_{sp}$; no precipitation
f. $Q_{sp} < K_{sp}$; the system is at equilibrium

Questions 18 - 20 refer to the following system.

The pressure on a system containing 2.0 moles of an ideal monatomic gas decreases from 15.0 atm to 3.0 atm as the system moves from state A to state B at a constant volume of 4.0 L. The system then expands at a constant pressure of 3.0 atm to reach state C at which point the temperature is the same as it was at state A.

18. How much work is done on the system as it moves from A to B?

a. zero J
b. 48 J
c. 73 J
d. 1520 J
e. 2640 J
f. 3180 J
g. 4860 J
h. 6850 J

19. How much heat is absorbed of released when the system moves *reversibly* from A to C?

a. zero kJ
b. –4.9 kJ
c. –6.1 kJ
d. +9.8 kJ
e. +12 kJ
f. –7.3 kJ
g. +4.3 kJ
h. –9.8 kJ

20. What is the change in the internal energy of the system ΔE as it moves from B to C?

a. zero J
b. –7300 J
c. –3650 J
d. –6850 J
e. +7300 J
f. +6850 J
g. +3650 J
h. –12160 J

21. Imagine that you have two copper blocks, one with a mass of 359 grams at a temperature of 42.6°C, and another with a mass of 317 grams at a temperature of 112°C. The two blocks are brought into contact with each other in a insulated container. The specific heat of copper is 0.385 $JK^{-1}g^{-1}$. What is the entropy change for the heat transfer described?

a. 1.29 JK^{-1} c. 11.6 JK^{-1} e. 14.6 JK^{-1} g. 75.0 JK^{-1}
b. 3.91 JK^{-1} d. 12.7 JK^{-1} f. 29.3 JK^{-1} h. 76.1 JK^{-1}

22. What are the oxidation numbers for the element nitrogen in the following molecules? *(Choose the row in which all responses are correct.)*

	NH_3	N_2	NO_2
a.	+3	0	+2
b.	+3	−1	+2
c.	+3	0	+4
d.	−3	0	+2
e.	−3	−1	+4
f.	−3	0	+4

23. Balance the following redox reaction in acidic solution:

$$SO_3^{2-} + MnO_4^- \rightarrow SO_4^{2-} + Mn^{2+}$$

What is the coefficient for the water in the balanced equation? Also indicate if the water is on the reactant or product side of the equation.

a. 1 reactant c. 3 reactant e. 5 reactant g. 2 product i. 4 product
b. 2 reactant d. 4 reactant f. 1 product h. 3 product j. 5 product

24. Calculate the standard cell potential for the cell: $Cu \mid Cu^{2+} \parallel Ag^+ \mid Ag$

Given the half-cell standard reduction potentials: $E°(Ag^+ \mid Ag) = +0.799$ V
$E°(Cu^{2+} \mid Cu) = +0.337$ V

a. −0.125 V c. −0.462 V e. −1.136 V g. −1.473 V
b. +0.125 V d. +0.462 V f. +1.136 V h. +1.473 V

25. What is the value of $\Delta G°$ for the reaction described in question 24? *(Hint: Write the balanced equation for the cell!)*

a. −15 kJ c. −38 kJ e. −89 kJ g. −146 kJ
b. −27 kJ d. −45 kJ f. −95 kJ h. −224 kJ

26. Calculate the equilibrium constant K at 25°C for the following reaction:

$$2\ Cu + PtCl_6^{2-} \rightarrow 2\ Cu^+ + PtCl_4^{2-} + 2\ Cl^- \qquad E° = +0.16\ V$$

a. 6.23 c. 506 e. 2.6×10^5 g. 1.5×10^{15}
b. 12.4 d. 1260 f. 4.5×10^{12} h. 3.2×10^{64}

27. How long would it take to plate out 12.3 grams of copper metal from an aqueous solution of Cu^{2+} if you were to use a current of 1.89 amps?

a. 30 min c. 2 hours e. 4 hours g. 5 hours
b. 1 hour d. 3 hours f. 4.5 hours h. 5.5 hours

28. Which one of the following elements is described as being "electron perfect"?

a. hydrogen
b. nitrogen
c. oxygen
d. fluorine
e. helium
f. carbon
g. boron
h. lithium

29. Which one of the following elements is most similar in chemical reactivity to beryllium?

a. sodium
b. magnesium
c. aluminum
d. silicon
e. phosphorus
f. sulfur

30. Which of the following statements are true?

I Carbon is sp^3 hybridized in the diamond allotrope.
II It is possible for lithium to form a covalent bond with carbon.
III Unlike silicon, carbon has the ability to form π bonds using p orbitals.

a. I only
b. II only
c. III only
d. I and II
e. I and III
f. II and III
g. I, II, and III

31. How many geometrical isomers would you expect for the square planar complex $[Pt(NH_3)_2(CN)_2]$?

a. 1
b. 2
c. 3
d. 4
e. 5
f. 6
g. 7
h. 8

32. What is the ground state electron configuration of Ti^{3+}?

a. $[Ar]\ 4s^2\ 3d^5$
b. $[Ar]\ 3d^3$
c. $[Ar]\ 3d^5$
d. $[Ar]\ 4s^2\ 3d^3$
e. $[Ar]\ 4s^2\ 3d^1$
f. $[Ar]\ 4s^1$
g. $[Ar]\ 4s^2\ 3d^2$
h. $[Ar]\ 3d^1$

33. What is the effective atomic number (EAN) for the transition element in the octahedral coordination compound $[Mn(CN)_6]^{4-}$?

a. 23
b. 25
c. 35
d. 36
e. 37
f. 38
g. 39
h. 55

34. In the octahedral complex ion $[Fe(H_2O)_6]^{2+}$, what is the LFSE (ligand field stabilization energy)? Ignore any pairing energy, P, if applicable.

a. 0 Dq
b. –4 Dq
c. –6 Dq
d. –8 Dq
e. –12 Dq
f. –16 Dq
g. –20 Dq
h. –24 Dq

35. Balance the following nuclear reaction: $^{223}_{88}Ra \rightarrow \ ^{4}_{2}He + ?$
What is the missing isotope?

a. ^{227}Rn
b. ^{219}Rn
c. ^{227}Th
d. ^{219}Th
e. ^{238}U
f. ^{235}U
g. ^{206}Pb
h. ^{204}Pb

FINAL EXAMINATION

CHEMISTRY 142 SPRING 2009

Monday May 4th 2009

1. Consider the reaction:

$$3\ I^-(aq) + S_2O_8^{2-}(aq) \rightarrow I_3^-(aq) + 2\ SO_4^{2-}(aq)$$

If I^- reacts at a rate of x Ms^{-1}, the corresponding rates for formation of I_3^- and SO_4^{2-} are

	$\Delta[I_3^-]/\Delta t$	$\Delta[SO_4^{2-}]/\Delta t$
a.	$(1/3)x$	$(2/3)x$
b.	$(1/3)x$	$(3/2)x$
c.	$3x$	$(2/3)x$
d.	$3x$	$(3/2)x$
e.	x	$2x$
f.	x	$(1/2)x$
g.	$(1/3)x$	$2x$

For questions 2 and 3 consider the following reaction mechanism:

$$A + B \underset{k_{-1}}{\overset{k_1}{\rightleftharpoons}} C + D \qquad \textit{fast}$$

$$C + E \xrightarrow{k_2} F \qquad \textit{slow}$$

2. Identify the reactants, products, and intermediates:

	Reactant(s)	*Intermediate(s)*	*Product(s)*
a.	A, B	E	C, D, F
b.	A, B	C, E	D, F
c.	A, B, C, E	none	D, F
d.	A, B, C, E	D	F
e.	A, B, E	C	D, F
f.	A, B, E	none	C, D, F
g.	A, B, C	D, E	F
h.	A, B, C	E	D, F

3. What is the rate law expression for the reaction mechanism given above.

a. Rate = k[C][E]

b. Rate = k[A][B]

c. Rate = k[A][B][E]

d. Rate = k[A][B][E][D]$^{-1}$

d. Rate = k[A][B][E][D]$^{-1}$

e. Rate = k [E][D][A]$^{-1}$[B]$^{-1}$

f. Rate = k[A][B][C][E]

4. A compound undergoing a zero-order decomposition decreases in concentration from 96 mol L^{-1} to 24 mol L^{-1} in 30 minutes. How long does it take for the 96 mol L^{-1} to decrease to 6.0 mol L^{-1}?

a. 7.5 min
b. 20 min
c. 30 min
d. 37.5 min
e. 40 min
f. 80 min
g. 150 min
h. 200 min
i. 310 min
j. 630 min

5. What is the pH of a solution made by dissolving 1.5×10^{-2} mol of hypochlorous acid HClO in sufficient water to make 1.0 liter? The K_a for hypochlorous acid is 3.5×10^{-8}.

a. 1.82	c. 4.64	e. 6.23	g. 7.46
b. 3.73	d. 5.67	f. 7.00	h. 9.28

6. Label each of the species in the following aqueous equilibrium as either acid or base according to the Brønsted-Lowry theory.

	C_5H_5N +	H_2O ⇌	$C_5H_5NH^+$ +	OH^-
a.	base	acid	acid	base
b.	base	acid	base	acid
c.	acid	base	base	acid
d.	acid	acid	base	base
e.	acid	base	acid	base
f.	base	base	acid	acid

7. According to LeChatelier's Principle, in which direction will the following system shift if the specified changes are made? Choose the row in which all answers are correct.

$$PCl_3(g) + Cl_2(g) \rightleftharpoons PCl_5(g) \qquad DH° = \text{negative}$$

	increase T	*remove PCl_5*	*decrease volume*		*increase T*	*remove PCl_5*	*decrease volume*
a.	to left	to left	to left	e.	to right	to left	to left
b.	to left	to left	to right	f.	to right	to left	to right
c.	to left	to right	to left	g.	to right	to right	to left
d.	to left	to right	to right	h.	to right	to right	to right

8. Predict the pH of aqueous solutions containing the compounds given below.

	KNO_3	LiCN	NH_4Br	NaF
a.	neutral	basic	acidic	neutral
b.	basic	acidic	basic	neutral
c.	neutral	basic	acidic	basic
d.	neutral	basic	basic	basic
e.	basic	neutral	neutral	neutral
f.	acidic	basic	acidic	neutral
g.	neutral	basic	acidic	acidic

9. Using the acid ionization constants provided, select the best choice for preparing a pH = 6.5 buffer.

HF	$K_a = 6.6 \times 10^{-4}$
CH_3CO_2H	$K_a = 1.76 \times 10^{-5}$
H_2CO_3	$K_a = 4.3 \times 10^{-7}$
NH_4^+	$K_a = 5.6 \times 10^{-10}$
HCO_3^-	$K_a = 4.8 \times 10^{-11}$

a. $NaHCO_3$ and Na_2CO_3
b. CH_3CO_2H and NaOH
c. NH_4Cl and HCl
d. HF and KF
e. H_2CO_3 and $NaHCO_3$
f. NH_4Cl and NH_3
g. KCH_3CO_2 and CH_3CO_2H

10. Which one of the following electrolytes is classified as weak?

a. $KClO_4$ c. NaCl e. KCN g. LiI i. NH_4NO_3
b. H_3PO_4 d. NH_4Cl f. $Ba(OH)_2$ h. H_2SO_4 j. $HClO_3$

11. The K_a for hydrogen chromate $HCrO_4^-$ is 1.5×10^{-6}. Identify the conjugate base of $HCrO_4^-$ and determine its K_b.

a. CrO_4^{2-} 1.0×10^{-7} d. H_2CrO_4 1.0×10^{-7}
b. CrO_4^{2-} 1.5×10^{-8} e. H_2CrO_4 1.5×10^{-8}
c. CrO_4^{2-} 6.7×10^{-9} f. H_2CrO_4 6.7×10^{-9}

12. What is the value of K_c for the following equilibrium at 1500°C given the equilibrium concentrations shown?

$$CH_4(g) + H_2O(g) \rightleftharpoons CO(g) + 3\,H_2(g)$$

$[CH_4] = 0.40$ M
$[H_2O] = 0.10$ M
$[CO] = 0.30$ M
$[H_2] = 0.80$ M

a. 0.17 c. 0.38 e. 6.0 g. 17
b. 0.26 d. 3.8 f. 7.8 h. 26

13. Which of the following is the correct solubility product constant expression for the sparingly soluble barium iodate $Ba(IO_3)_2$?

a. $K_{sp} = [Ba^{2+}][IO_3^-]$
b. $K_{sp} = [Ba^{2+}]^2[IO_3^-]$
c. $K_{sp} = [Ba^{2+}][IO_3^-]^2/[Ba(IO_3)_2]$
d. $K_{sp} = [Ba^{2+}](2[IO_3^-])^2$
e. $K_{sp} = [Ba^{2+}][IO_3^-]^2$
f. $K_{sp} = [Ba^{2+}]^2[IO_3^-]/[Ba(IO_3)_2]$
g. $K_{sp} = [Ba][(IO_3)_2]$

For Questions 14 and 15, consider the diprotic acid carbonic acid H_2CO_3 in aqueous solution:

$H_2CO_3 + H_2O \rightleftharpoons HCO_3^- + H_3O^+$ $K_{a1} = 4.3 \times 10^{-7}$ $pK_{a1} = 6.37$
$HCO_3^- + H_2O \rightleftharpoons CO_3^{2-} + H_3O^+$ $K_{a2} = 4.8 \times 10^{-11}$ $pK_{a2} = 10.32$

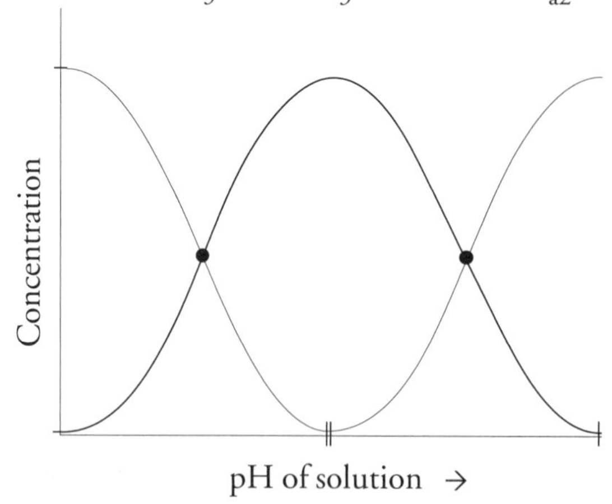

14. If 40 mL of 0.10 M NaOH is added to 40 mL of 0.10 M carbonic acid solution, the resulting solution will be

a. acidic b. basic c. neutral

15. If a total volume of 60 mL of 0.10 M NaOH is added to 40 mL of 0.10 M carbonic acid solution, the pH of the resulting solution will be

a. 3.19
b. 5.16
c. 6.37
d. 7.00
e. 8.35
f. 9.22
g. 10.32
h. 11.34
i. 12.78
j. 14.00

16. In a saturated solution of lead phosphate $Pb_3(PO_4)_2$, what is the relationship between the concentrations of the lead ions and the phosphate ions?

a. $[Pb^{2+}] = [PO_4^{3-}]$
b. $2 \times [Pb^{2+}] = 3 \times [PO_4^{3-}]$
c. $3 \times [Pb^{2+}] = 2 \times [PO_4^{3-}]$
d. $[Pb^{2+}]^2 = [PO_4^{3-}]^3$
e. $[Pb^{2+}]^3 = [PO_4^{3-}]^2$
f. $2 \times [Pb^{2+}]^3 = 3 \times [PO_4^{3-}]^2$

17. The solubility product for $Ca(OH)_2$ is $K_{sp} = 6.5 \times 10^{-6}$.
K_w for water $= 1.0 \times 10^{-14}$.
Calculate the equilibrium constant for the reaction at 25°C:

$$Ca(OH)_2(s) + 2\,HCl(aq) \rightleftharpoons Ca^{2+}(aq) + 2\,H_2O(l) + 2\,Cl^-(aq)$$

a. 6.5×10^{-6}
b. 6.5×10^{8}
c. 1.5×10^{-9}
d. 1.5×10^{5}
e. 6.5×10^{22}
f. 1.5×10^{-23}
g. 4.2×10^{-11}
h. 4.2×10^{17}
i. 4.2×10^{-25}

Questions 18 through 22 refer to the following system.

Consider a sample of an ideal monatomic gas at a pressure of 12.0 atm, a temperature of 27.0°C, occupying a volume of 20.0 liters. Call this state of the system A.

Suppose the gas is then cooled at constant volume to reach state B, where the pressure is 3.00 atm.

Subsequently, the gas is permitted to expand at constant pressure to reach state C, which has the same temperature as state A.

It is recommended that you draw these changes on the diagram to the right.

$R = 0.08206\ L\ atm\ K^{-1}mol^{-1} = 8.3145\ J\ K^{-1}mol^{-1}$

18. How many moles of gas are present in this system?

a. 1.07
b. 2.50
c. 4.30
d. 9.75
e. 15.0
f. 22.5
g. 45.0
h. 108
i. 45
j. 57

19. How much heat is absorbed by the system as it moves isothermally and reversibly from State A to State C?

a. q = zero J
b. q = +333 J
c. q = +3.03 kJ
d. q = +11.4 kJ
e. q = +22.7 kJ
f. q = +33.7 kJ
g. q = +42.1 kJ
h. q = +57.9 kJ

20. How much heat is released by the system as it moves from State A to State B?

a. q = zero J
b. q = –333 J
c. q = –6.33 kJ
d. q = –11.4 kJ
e. q = –27.4 kJ
f. q = –31.9 kJ
g. q = –45.6 kJ
h. q = –62.2 kJ

21. How much heat is absorbed by the system as it moves from State B to State C?

a. q = zero J
b. q = +333 J
c. q = +6.33 kJ
d. q = +11.4 kJ
e. q = +27.4 kJ
f. q = +31.9 kJ
g. q = +45.6 kJ
h. q = +62.2 kJ

22. Therefore determine how much work is done by or on the system as it moves from A to C along the path from A to B and then B to C.

a. w = zero J
b. w = –180 J
c. w = –14.3 kJ
d. w = –18.2 kJ
e. w = +27.4 kJ
f. w = +45.6 kJ
g. w = –57.2 kJ
h. w = –97.5 kJ

23. What is the oxidation number of chromium in sodium dichromate $Na_2Cr_2O_7$?

a. +7
b. +6
c. +5
d. +4
e. +3
f. +2
g. +1
h. 0
i. –3
j. –6

24. Balance the equation for the redox reaction in aqueous acidic solution shown below:

$$VO_2^{+} + Zn(s) \rightarrow VO^{2+} + Zn^{2+}$$

What is the stoichiometric coefficient for $H^+(aq)$ in the balanced equation. Does the $H^+(aq)$ appear on the left as a reactant or on the right as a product of the reaction?

a. 1 reactant
b. 2 reactant
c. 3 reactant
d. 4 reactant
e. 5 reactant
f. 1 product
g. 2 product
h. 3 product
i. 4 product
j 5 product

25. Consider the following reduction reactions and their half-cell standard reduction potentials:

$Ni^{2+}(aq) + 2e^- \rightarrow Ni(s)$ $\qquad E^\circ_{red}(Ni^{2+}|Ni) = -0.23$ V

$O_2(g) + 4H^+(aq) + 4e^- \rightarrow 2H_2O(l)$ $\qquad E^\circ_{red}(O_2|H_2O) = +1.23$ V

Suppose these reactions are used as the basis for the construction of a galvanic cell in which the reaction proceeds spontaneously. Identify the species oxidized and reduced in the cell.

	oxidized	*reduced*
a.	Ni^{2+}	H_2O
b.	Ni^{2+}	H^+
c.	Ni^{2+}	O_2
d.	Ni	O_2
e.	Ni	H^+
f.	Ni	H_2O
g.	H_2O	Ni^{2+}
h.	O_2	Ni
i.	H+	Ni

26. How long would it take to plate out 25.0 grams of copper metal from an aqueous solution of Cu^{2+} using a current of 5.27 amps and a cell potential of 6.0 volts?

a. 30 min	c. 2.0 hours	e. 4.0 hours	g. 5.0 hours	i. 6.0 hours
b. 1.0 hour	d. 3.0 hours	f. 4.5 hours	h. 5.5 hours	j. 6.5 hours

27. Some changes that occur in a system are path-independent. In other words, it doesn't matter what route is taken between the initial and final states of the system. Other changes depend upon the path taken. These are referred to as path-dependent. Which one of the following changes is path-dependent?

a. an entropy change
b. the change in internal energy of a system during an adiabatic expansion
c. an increase in the volume of the system
d. the heat absorbed in an isothermal expansion
e. a change in temperature
f. an increase in the enthalpy of the system
g. the decrease in free energy during a spontaneous process
h. a decrease in volume with no change in pressure

28. Some amount of hot water at 90.2°C is added to an insulated vessel containing 100 grams of ice at 0°C. As a result, 85 grams of the ice melted and 15 grams remained unmelted. What mass of hot water was added?

Latent heat of fusion of ice = 333 Jg^{-1}
Specific heat of water = 4.184 $JK^{-1}g^{-1}$

a. 45 g	c. 65 g	e. 85 g	g. 105 g	i. 125 g
b. 55 g	d. 75 g	f. 95 g	h. 115 g	j. 135 g

29. Consider again the system described in the previous question. What is the total change in the entropy of the system when the hot water is added to the ice and the system reaches equilibrium at 0°C?

a. +6.78 JK^{-1}	c. +14.1 JK^{-1}	e. +89.5 JK^{-1}	g. +193 JK^{-1}
b. +8.21 JK^{-1}	d. +17.9 JK^{-1}	f. +104 JK^{-1}	h. +212 JK^{-1}

30. Match the elements with the descriptions listed:

A an element similar in its chemical behavior to aluminum
B an element described as 'electron perfect'
C an element that exists as a diatomic molecule in its natural state
D an element that does not react with fluorine

	A	B	C	D
a.	sodium	silicon	sulfur	selenium
b.	beryllium	boron	nitrogen	neon
c.	gallium	carbon	phosphorus	xenon
d.	boron	manganese	oxygen	helium
e.	beryllium	carbon	nitrogen	helium
f.	magnesium	calcium	carbon	chlorine
g.	boron	germanium	hydrogen	iodine
h.	sodium	silicon	oxygen	xenon
i.	boron	helium	nitrogen	sodium

31. The mineral hardystonite has the empirical formula $Ca_nZnSi_2O_7$. What is the value of the subscript n?

a. 1 b. 2 c. 3 d. 4 e. 5 f. 6

32. How many geometrical isomers of the octahedral coordination compound $[Co(NH_3)_4ClBr]Cl$ are there?

a. 1 b. 2 c. 3 d. 4 e. 5 f. 6

33. The coordination compound $[Cu(NH_3)_4]^{2+}$ is paramagnetic. Use valence bond theory to determine the hybridization of the valence orbitals on the copper used in forming bonds with the ammonia ligands and hence determine the shape of the complex ion.

a. sp^2 trigonal planar
b. sp^3 tetrahedral
c. dsp^2 square planar
d. d^2sp^3 octahedral
e. d^4sp trigonal prismatic
f. dsp^3 trigonal bipyramidal
g. dsp^3 square pyramidal

34. In the octahedral complex ion $[Mn(CN)_6]^{3-}$, what is the LFSE (ligand field stabilization energy)? Ignore any pairing energy, P, if applicable.

a. 0 Dq
b. –4 Dq
c. –6 Dq
d. –8 Dq
e. –12 Dq
f. –16 Dq
g. –18 Dq
h. –20 Dq
i. –24 Dq
j. –30 Dq

35. What particle is produced in the following radioactive decay?

$$^{24}_{11}Na \rightarrow ^{24}_{12}Mg + ?$$

a. $^{4}_{2}\alpha$
b. $^{0}_{-1}\beta$
c. $^{1}_{0}n$
d. $^{1}_{1}p$
e. $^{0}_{1}\beta$

FINAL EXAMINATION

CHEMISTRY 142 FALL 2009

Monday December 17th 2009

1. Iodine atoms combine to form molecular iodine in the gas phase: $2\ I(g) \rightarrow I_2(g)$

   This reaction is second order and has a rate constant of $7.0 \times 10^9\ M^{-1}s^{-1}$ at 23°C. Calculate the half-life of the reaction when the initial concentration of $I(g)$ is 0.60 M.

   a. 9.9×10^{-11} s c. 1.4×10^{-10} s e. 7.0×10^{9} s
   b. 2.4×10^{-10} s d. 3.7×10^{-4} s f. 1.0×10^{10} s

2. Which of the following would be expected to increase the rate of a reaction?

   A. lower the reaction temperature
   B. increase the concentration of the reactants
   C. increase the activation energy
   D. add a catalyst

   a. A only c. C only e. A and B g. B and D i. B and C
   b. B only d. D only f. A and D h. A, B, and D j. A, B, C, and D

3. Consider the following mechanism for the reaction of nitric oxide and hydrogen:

   $2\ NO(g) \rightleftharpoons N_2O_2(g)$ *fast*
   $N_2O_2(g) + H_2(g) \rightarrow N_2O(g) + H_2O(g)$ *slow*
   $N_2O(g) + H_2(g) \rightarrow N_2(g) + H_2O(g)$ *fast*

   What is the rate law predicted by this mechanism?

   a. Rate = $k[N_2O_2]^{-2}[H_2]^2$ d. Rate = $k[NO]^2$
   b. Rate = $k[NO]^2[H_2]$ e. Rate = $k[H_2]$
   c. Rate = $k[NO][H_2]^2$ f. Rate = $k[NO]^2[H_2]^2[N_2O_2][N_2O]$

4. Consider the following reaction mechanism:

   $H_2O_2 + I^- \rightarrow H_2O + IO^-$ (rate determining step)
   $H_2O_2 + IO^- \rightarrow H_2O + O_2 + I^-$

   Identify the intermediate(s)?

   a. H_2O_2 c. IO^- e. O_2 g. H_2O_2 and I^-
   b. I^- d. H_2O f. I^- and IO^- h. H_2O, O_2, and I^-

5. What is the equation for the equilibrium that corresponds to the reaction quotient $Q_c = \dfrac{[NOBr]^2}{[NO]^2[Br_2]}$

   a. $2\ NO + Br_2 \rightleftharpoons 2\ NOBr$ d. $2\ NOBr \rightleftharpoons 2\ NO + Br_2$
   b. $NO + Br_2 \rightleftharpoons NOBr$ e. $NOBr \rightleftharpoons NO + Br_2$
   c. $NO + 2\ Br_2 \rightleftharpoons NOBr$ f. $NOBr \rightleftharpoons NO + 2\ Br_2$

6. Which one of the following statements is false regarding the equilibrium shown below?

$$2\ NaHCO_3(s) \rightleftharpoons Na_2CO_3(s) + H_2O(g) + CO_2(g)$$

a. Removing $CO_2(g)$ from the system will cause the equilibrium to shift to the right.
b. Adding $NaHCO_3(s)$ to the system will cause the equilibrium to shift to the right.
c. Adding a catalyst to the system will cause no shift in the equilibrium.
d. Adding $H_2O(g)$ to the system will cause the equilibrium to shift to the left.
e. Removing $Na_2CO_3(s)$ from the system will cause no shift in the equilibrium.

7. The following system has an equilibrium constant $K_c = 2.18 \times 10^6$. If a 10.0 L flask initially contains 3.2 moles of $HBr(g)$ and no $H_2(g)$ or $Br_2(g)$, what is the concentration of $H_2(g)$ once equilibrium is established?

$$H_2(g) + Br_2(g) \rightleftharpoons 2\ HBr(g) \qquad K_c = 2.18 \times 10^6$$

a. 2.2×10^{-4} M
b. 1.8×10^{-4} M
c. 1.4×10^{-3} M
d. 0.27 M
e. 0.32 M
f. 0.34 M
g. 0.99 M
h. 15 M

8. Which of the following statements is/are correct?

I. The lower the pOH, the more basic the solution.
II. The pH of pure water can change depending on the temperature.
III. The pH of a 1.0×10^{-8} M solution of HNO_3 is 8.0.
IV. The strongest acid that can exist in aqueous solution is the hydronium ion.

a. I only
b. II only
c. III only
d. IV only
e. II and III
f. I and IV
g. III and IV
h. I, II, and IV
i. II, III, and IV
j. I, II, III, and IV

9. How many of the following substances can act as a Lewis acid?

CN^- NH_3 Co^{3+} BF_3 CO F^-

a. 0
b. 1
c. 2
d. 3
e. 4
f. 5
g. 6

10. Decide if aqueous solutions of the following compounds will be acidic, basic, or neutral. Choose the row in which all answers are correct.

	NH_4Cl	$Mg(OH)_2$	$NaClO_4$	KCN
a.	basic	basic	acidic	acidic
b.	acidic	neutral	neutral	neutral
c.	basic	basic	basic	neutral
d.	neutral	basic	neutral	basic
e.	basic	basic	acidic	basic
f.	acidic	basic	neutral	basic

11. A 0.2 M solution of phosphoric acid H_3PO_4 was titrated with 0.2 M NaOH. For phosphoric acid, $K_{a1} = 7.5 \times 10^{-3}$, $K_{a2} = 6.2 \times 10^{-8}$, and $K_{a3} = 4.2 \times 10^{-13}$. What is the pH of the solution at the second equivalence point?

a. 1.41
b. 2.12
c. 4.67
d. 6.54
e. 7.21
f. 9.79
g. 12.38
h. 12.84

12. What is the pH of a 0.5 M aqueous solution of KF? The K_a for HF is 7.1×10^{-4}.

a. 1.72 b. 5.58 c. 6.91 d. 7.41 e. 8.42 f. 10.4 g. 10.9 h. 12.3

13. Calculate the equilibrium constant for: $2\ SO_3(g) \rightleftharpoons 2\ SO_2(g) + O_2(g)$ K_p = ???

given the following information: $SO_3(g) \rightleftharpoons SO_2(g) + 1/2\ O_2(g)$ $K_p = 1.32$

a. 0.015 b. 0.66 c. 1.15 d. 1.74 e. 2.64 f. 4.5 g. 114 h. 414

14. Phenol C_6H_5OH is a weak acid. The K_a for phenol at 25°C is 1.3×10^{-10}. What is the value of the neutralization constant K_{neut} in a titration of phenol with sodium hydroxide?

a. 7.7×10^{23} b. 1.3×10^{4} c. 5.6×10^{3} d. 1.3×10^{-2} e. 7.7×10^{-5} f. 1.9×10^{-7} g. 1.0×10^{-14} h. 1.3×10^{-24}

15. What is the solubility of aluminum hydroxide $Al(OH)_3$ if the solubility product K_{sp} is 1.8×10^{-33}.

a. 6.7×10^{-35} M b. 1.8×10^{-33} M c. 2.4×10^{-17} M d. 1.7×10^{-15} M e. 3.9×10^{-15} M f. 5.7×10^{-11} M g. 2.9×10^{-9} M h. 1.8×10^{-7} M

16. Suppose 25 mL of 5.0×10^{-3} M solution of calcium chloride $CaCl_2$, is added to 25 mL of 5.0×10^{-3} M solution of sodium fluoride NaF. Will precipitation of calcium chloride CaF_2 occur and what is the relationship between Q_{sp} and K_{sp}? The K_{sp} for CaF_2 is 5.3×10^{-9}.

a. yes, $Q_{sp} < K_{sp}$ b. yes, $Q_{sp} > K_{sp}$ c. yes, $Q_{sp} = K_{sp}$ d. no, $Q_{sp} < K_{sp}$ e. no, $Q_{sp} > K_{sp}$ f. no, $Q_{sp} = K_{sp}$

17. How will the addition of each of the following solutes affect the solubility of silver bromide? The K_{sp} for silver bromide is 5.0×10^{-13} and the K_f for $[Ag(NH_3)_2]^+$ is 1.6×10^{7}.

	NH_3	HNO_3	KBr
a.	increase	decrease	decrease
b.	decrease	no change	increase
c.	increase	no change	increase
d.	increase	no change	decrease
e.	decrease	increase	no change
f.	increase	increase	no change

18. Which of the following statements is/are true about a system at equilibrium?

I. At equilibrium, ΔG is equal to zero.
II. If $\Delta G° < 0$ (a negative value), then the reaction will have an equilibrium constant less than 1.
III. At equilibrium, entropy is at a maximum.
IV. The value of $\Delta G°$ at equilibrium is the same as the value of $\Delta G°$ at any other point in the reaction.

a. I only b. II only c. III only d. IV only e. I and III f. I and IV g. II and III h. II and IV i. I, III, and IV j. II, III, and IV

Questions 19 and 20 refer to the following system.

The pressure on a system containing 0.50 mole of an ideal monatomic gas increases from 2.0 atm to 8.0 atm as the system moves from state A to state B at a constant volume of 12.0 L. The system then decreases in volume at a constant pressure of 8.0 atm to reach state C. The temperature at state C is the same as the temperature at state A.

P

V

19. How much work is done on the system as it moves from A to B?

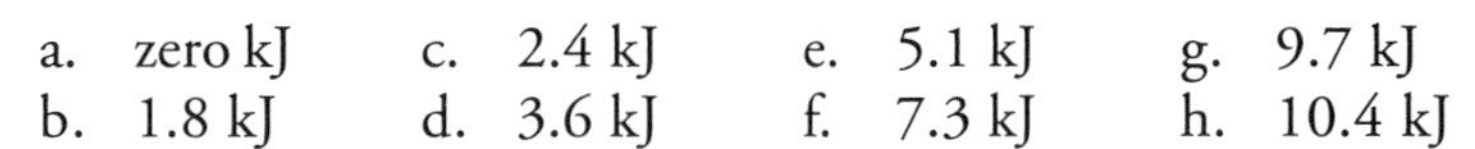

a. zero kJ	c. 2.4 kJ	e. 5.1 kJ	g. 9.7 kJ
b. 1.8 kJ	d. 3.6 kJ	f. 7.3 kJ	h. 10.4 kJ

20. How much heat is absorbed or released when the system moves reversibly from A to C?

a. −0.033 kJ	c. −2.7 kJ	e. −3.4 kJ	g. −5.8 kJ
b. +0.033 kJ	d. +2.7 kJ	f. +3.4 kJ	h. +5.8 kJ

21. How much heat is required to melt a 15.0 g block of ice initially at 0°C and warm the resulting water to 25°C?

Specific heat of water = 4.184 $J\ g^{-1}\ K^{-1}$
Latent heat of fusion for water = 333 $J\ g^{-1}$

a. 1569 J	c. 2090 J	e. 4995 J	g. 6564 J
b. 2064 J	d. 3321 J	f. 5374 J	h. 7992 J

22. What is the change in entropy for the process described in Question 21?

a. − 5.5 $J\ K^{-1}$	c. −18 $J\ K^{-1}$	e. −24 $J\ K^{-1}$	g. −42 $J\ K^{-1}$
b. +5.5 $J\ K^{-1}$	d. +18 $J\ K^{-1}$	f. +24 $J\ K^{-1}$	h. +42 $J\ K^{-1}$

23. Balance the following redox reaction under basic conditions:

$$MnO_4^-(aq) + CN^-(aq) \rightarrow MnO_2(s) + OCN^-(aq)$$

How many water molecules are on which side of the balanced equation?

a. 1, product	c. 2, product	e. 3, product	g. 4, product
b. 1, reactant	d. 2, reactant	f. 3, reactant	h. 4, reactant

24. Given the following reduction potentials, calculate the potential (E°) for the following cell at standard conditions:

$$Al(s) \mid Al^{3+}(aq) \parallel Br_2(l) \mid Br^-(aq)$$

$E°(Al^{3+} \mid Al) = -1.676$ V
$E°(Br_2 \mid Br^-) = +1.065$ V

a. −0.611 V b. +0.611 V c. −2.741 V d. +2.741 V

25. What is the value of $\Delta G°$ for the cell described in Question 24?

a. −1587 kJ	c. −793 kJ	e. −529 kJ	g. −354 kJ
b. +1587 kJ	d. +793 kJ	f. +529 kJ	h. +354 kJ

26. How many grams of zinc are deposited at the cathode by the passage of 2.15 A of current for 75 minutes in the electrolysis of an aqueous solution containing Zn^{2+} ions?

a. 0.055 g	c. 0.71 g	e. 3.3 g	g. 11.4 g
b. 0.23 g	d. 2.5 g	f. 9.7 g	h. 13.1 g

27. Which of the alkaline earth metals is most similar in chemical reactivity to aluminum?

a. Be b. Mg c. Ca d. Sr e. Ba

28. Which of the following is/are allotropes of carbon?

I. fullerene	III. diamond
II. quartz	IV. graphite

a. I only	c. III only	e. III and IV	g. I and IV	i. I, III, and IV
b. II only	d. IV only	f. II and III	h. III and IV	j. I, II, III, and IV

29. What is the oxidation state of nitrogen in each of the following molecules / ions?

	N_2O_3	N_2O	$N_2O_2^{2-}$	N_2O_5
a.	+3	−2	+1	+5
b.	+3	+1	−2	+5
c.	+5	+2	+3	−2
d.	+3	+1	+1	+3
e.	+2	+1	+3	+1
f.	+5	+2	−2	+3
g.	+5	−2	+5	+1
h.	+3	+1	+1	+5

30. What halogen is the most difficult to oxidize and which noble gas can form compounds with other elements? Choose the answer in which both questions are answered are correctly.

	halogen most difficult to oxidize	*noble gas that can form compounds with other elements*
a.	Cl_2	He
b.	F_2	Xe
c.	Br_2	Ar
d.	I_2	Ne
e.	F_2	He
f.	I_2	Xe

31. What is the ground state (lowest energy) electron configuration for Fe^{2+}?

a. [Ar] $4s^2\ 3d^6$	c. [Ar] $4s^2\ 3d^4$	e. [Ar] $4s^1\ 3d^5$
b. [Ar] $4s^2\ 3d^8$	d. [Ar] $3d^6$	f. [Ar] $3d^5$

32. How many of the following are considered weak field ligands?

H_2O Cl^- CO NH_3 OH^- CN^-

a. 1 c. 3 e. 5
b. 2 d. 4 f. 6

33. What is the effective atomic number (EAN) of the transition element iron in the octahedral coordination compound $[Fe(OH)(H_2O)_5](NO_3)_2$?

a. 23 c. 26 e. 36 g. 38
b. 24 d. 35 f. 37 h. 56

34. What hybridization of valence orbitals on the nickel atom in $Na_2[Ni(CN)_4]$ is necessary to accommodate the pairs of electrons donated from the cyanide ion?

a. sp b. dsp^2 c. sp^3 d. sp^2 e. sp^3d f. d^2sp^3

35. What is the ligand field stabilization energy (LFSE) for the cobalt ion in the octahedral complex $[CoF_6]^{3-}$? Do not include any pairing energy.

a. –4 Dq c. –8 Dq e. –16 Dq g. –20 Dq
b. –6 Dq d. –12 Dq f. –18 Dq h. –24 Dq

FINAL EXAMINATION

CHEMISTRY 142 SPRING 2010

Monday May 4th 2010

1. Butane burns in air to form carbon dioxide and water according to the equation:

$$C_4H_{10}(g) \quad + \quad 6\tfrac{1}{2}\, O_2(g) \quad \rightarrow \quad 4\, CO_2(g) \quad + \quad 5\, H_2O(g)$$

After 2.0 moles of C_4H_{10} and 13.0 moles of oxygen were placed in a 1.0 L flask, the concentrations of the reactants and products were monitored as a function of time. The graph shown on the right was obtained. Identify the components A, B, C, and D:

	A	B	C	D
a.	O_2	CO_2	C_4H_{10}	H_2O
b.	O_2	C_4H_{10}	CO_2	H_2O
c.	CO_2	O_2	C_4H_{10}	H_2O
d.	CO_2	H_2O	O_2	C_4H_{10}
e.	C_4H_{10}	O_2	CO_2	H_2O
f.	H_2O	C_4H_{10}	O_2	CO_2
g.	C_4H_{10}	O_2	H_2O	CO_2
h.	C_4H_{10}	H_2O	O_2	CO_2

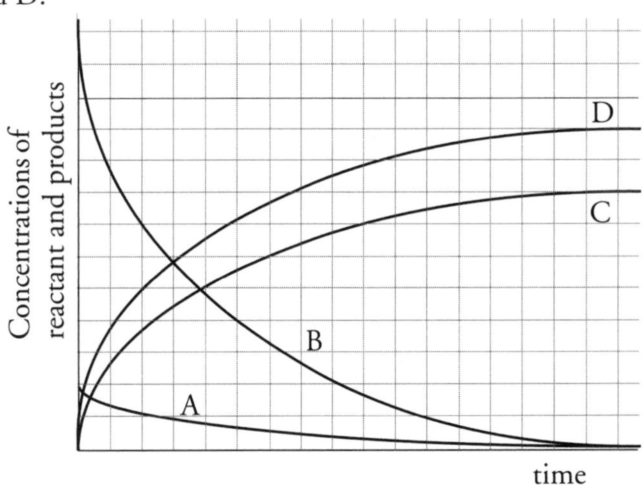

2. The concentration of a reactant in a reaction changed as follows as the reaction proceeded:

At time zero (initial concentration)	96 mol L^{-1}
After 12 minutes	48 mol L^{-1}
After *another* 24 minutes	12 mol L^{-1}
One hour after the start of the reaction	3 mol L^{-1}

What is the order of the reaction with respect to this reactant?

a. 0 b. 1 c. 2 d. 3 e. 4 f. 5

3. The reaction of a substance A with another substance B to form C was investigated in a series of experiments with different initial concentrations of the two reactants. The following initial rates were observed. What is the overall reaction order?

	Initial [A]	*Initial [B]*	*Initial rate of formation of C*
Experiment 1	0.40 M	0.20 M	1.2×10^{-1} mol L^{-1} s^{-1}
Experiment 2	0.20 M	0.20 M	6.0×10^{-2} mol L^{-1} s^{-1}
Experiment 3	0.10 M	0.40 M	1.2×10^{-1} mol L^{-1} s^{-1}
Experiment 4	0.20 M	0.10 M	1.5×10^{-2} mol L^{-1} s^{-1}

a. 4 c. 2 e. 0 g. −2 i. −4
b. 3 d. 1 f. −1 h. −3 j. −5

4. A radioactive isotope decomposes via a first-order reaction. What is the half-life of the isotope if 2/3 of the isotope decomposed in 12.0 days?

a. 4.41 days
b. 6.27 days
c. 7.57 days
d. 8.11 days
e. 9.39 days
f. 10.1 days
g. 20.1 days
h. 29.8 days

5. Calculate the equilibrium constant K_c for the following equilibrium if the concentrations of the various components at equilibrium are as shown:

$$Cl_2(g) + 2\ NO_2(g) \rightleftharpoons 2\ NO_2Cl(g)$$

Concentrations at equilibrium:
$[Cl_2(g)] = 3.0$ mole/L
$[NO_2(g)] = 2.0$ mole/L
$[NO_2Cl(g)] = 36.0$ mole/L

a. 3.0
b. 6.0
c. 12
d. 18
e. 24
f. 36
g. 72
h. 108
i. 216
j. 15,500

6. Consider the following disturbances on the equilibrium system shown. In which, if any, direction will the system shift to restore equilibrium in each case?

$$2\ FeCl_3(s) + 3\ H_2O(g) \rightleftharpoons Fe_2O_3(s) + 6\ HCl(g) \qquad \Delta H° = \text{negative (exothermic)}$$

	decrease temperature	*increase volume*	*add* $FeCl_3(s)$	*add* $H_2O(g)$
a.	right	right	right	right
b.	right	left	left	right
c.	right	left	right	right
d.	left	right	no change	left
e.	left	right	left	no change
f.	right	right	no change	right
g.	right	left	right	no change

7. According to the Brønsted-Lowry definitions of acids and bases, the conjugate acid of the hydrogen sulfate ion HSO_4^- is

a. H_3SO_3 b. HSO_3^- c. SO_4^{2-} d. HSO_4^- e. H_2SO_4 f. H_3O^+

8. The weak base hydrazine (N_2H_4) has an ionization constant K_b equal to 9.1×10^{-9}. Calculate the pH of a 0.010 M solution of hydrazine.

a. 1.22
b. 2.44
c. 6.32
d. 7.21
e. 8.98
f. 9.32
g. 10.29
h. 11.56

9. At 50°C, the value of K_w is 5.5×10^{-14}. What is the pOH of a neutral solution at 50°C?

a. 5.23
b. 5.50
c. 6.63
d. 6.70
e. 7.00
f. 7.32
g. 7.37
h. 7.89

10. The net ionic equation representing the equilibrium established when lithium hydride dissolves in water is shown. Label the species present as either acid or base according to Brønsted-Lowry theory.

	H^-	+ H_2O	H_2	+ OH^-
a.	acid	base	base	acid
b.	acid	acid	base	base
c.	acid	base	acid	base
d.	base	acid	base	acid
e.	base	acid	acid	base
f.	base	base	acid	acid

For Questions 11 through 14 (5 points each), consider the following distribution diagram for a titration of a 0.010 M solution of the diprotic acid H_2SO_3 vs. a 0.010 M solution of sodium hydroxide in aqueous solution:

$H_2SO_3 + H_2O \rightleftharpoons H_3O^+ + HSO_3^-$ $K_{a1} = 1.3 \times 10^{-2}$ $pK_{a1} = 1.89$

$HSO_3^- + H_2O \rightleftharpoons H_3O^+ + SO_3^{2-}$ $K_{a2} = 6.3 \times 10^{-8}$ $pK_{a2} = 7.20$

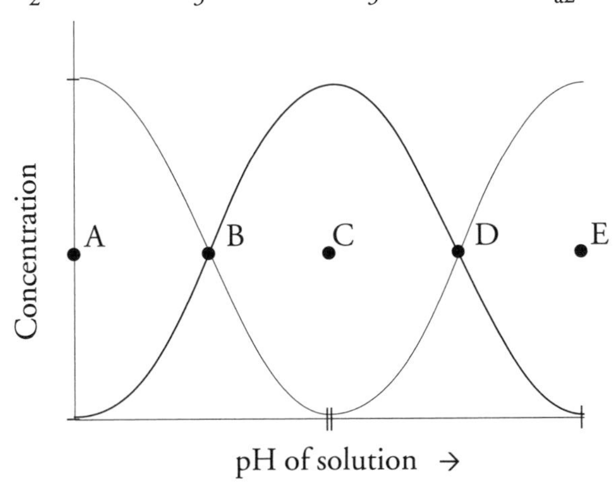

Use the following key:

a. 0.76
b. 1.89
c. 2.33
d. 4.55
e. 5.21
f. 6.59
g. 6.72
h. 7.20
i. 7.68
j. 9.09

11. What is the pH at point B?

12. What is the pH at point C?

13. What is the pH at point D?

14. What is the pH at a point halfway between C and D?

15. Which one of the following equimolar solutions will act as a buffer in aqueous solution?

a. KBr and KCl
b. Na_2CO_3 and K_2CO_3
c. LiF and HF
d. NaCl and NaClO
e. KCN and LiCN
f. H_2SO_4 and $NaHSO_4$
g. $NaClO_3$ and $HClO_3$

16. What is the value of the neutralization constant K_{neut} in a titration of hydrofluoric acid (HF) with potassium hydroxide?

$K_a(HF) = 3.5 \times 10^{-4}$

a. 3.5×10^{-4}
b. 4.3×10^{-7}
c. 2.0×10^{-15}
d. 2.9×10^{-11}
e. 2.9×10^{3}
f. 3.5×10^{10}
g. 2.9×10^{-5}
h. 3.7×10^{4}

17. The molar concentration of copper(II) phosphate $Cu_3(PO_4)_2$ in an aqueous solution is 1.6×10^{-17} M. What is the concentration of copper ions in the solution?

a. 3.2×10^{-17} M
b. 1.6×10^{-17} M
c. 1.2×10^{-17} M
d. 4.8×10^{-17} M
e. 6.4×10^{-17} M
f. 5.3×10^{-18} M

18. Which of the following is the correct solubility product constant expression for the sparingly soluble copper(II) phosphate $Cu_3(PO_4)_2$?

a. $K_{sp} = [Cu^{2+}][PO_4^{3-}]$
b. $K_{sp} = [Cu^{2+}]^2[PO_4^{3-}]$
c. $K_{sp} = [Cu^{2+}]^3[PO_4^{3-}]^2$
d. $K_{sp} = [Cu^{2+}]^2[PO_4^{3-}]^3$
e. $K_{sp} = [Cu^{2+}][PO_4^{3-}]^3$
f. $K_{sp} = [Cu^{2+}]^3[PO_4^{3-}]^2/[Cu_3(PO_4)_2]$
g. $K_{sp} = [Cu^{2+}]^3[PO_4^{3-}]$

19. 500 mL of a 0.200 M solution of ammonium phosphate $(NH_4)_3PO_4$ is added to 500 mL of a 0.300 M copper(II) nitrate $Cu(NO_3)_2$ solution. Copper(II) phosphate precipitates. What is the molar concentration of copper(II) phosphate in the solution after the precipitation.

K_{sp} for copper(II) phosphate = 1.40×10^{-37}.

a. 3.3×10^{-7} M
b. 2.7×10^{-4} M
c. 1.8×10^{-4} M
d. 1.1×10^{-6} M
e. 6.7×10^{-7} M
f. 1.67×10^{-8} M

20. Suppose a vessel contained 10 molecules of a gas. What is the probability that nine of the molecules would be in the lefthand half of the vessel with only one molecule in the righthand half??

a. 1/16
b. 1/26
c. 1/32
d. 1/51
e. 1/64
f. 1/102
g. 1/256
h. 1/333

21. Which is an expression of the second law of thermodynamics?

a. ΔE_{univ} = zero
b. S = zero at 0 K
c. $\Delta H = q_p$
d. ΔS_{univ} > zero for all spontaneous processes
e. At equilibrium, $\Delta G°$ equals zero
f. $\Delta E = q + w$

22. What is the oxidation number of the underlined element in each of the following compounds. Add them up using the correct signs. What is the sum?

$(NH_4)_2\underline{C}O_3$ $\underline{Br}F_5$ $\underline{Be}Cl_2$ $\underline{C}_6H_6$

a. +4	c. +8	e. +10	g. +12	i. +15
b. +5	d. +9	f. +11	h. +14	j. +17

For Questions 23 through 32 (5 points each), refer to the following system.

On the diagram draw the three points described below.

The initial state A of the system consists of 1.00 mole of an ideal gas at a pressure of 16.0 atm and a volume of 2.00 L. The gas expands isothermally to a new state of the system where the prerssure is 4.0 atm.

An alternative adiabatic expansion to the same volume as B leads to a state C of the system where the pressure is 1.59 atm.

pressure

volume

Use the following key for questions 23 and 24:

a. 79 K	c. 135 K	e. 178 K	g. 331 K	i. 401 K
b. 102 K	d. 155 K	f. 241 K	h. 390 K	j. 427 K

23. What is the temperature at point A?

24. What is the temperature at point C?

Use the following key for questions 25 through 29:

a. zero J	c. +1330 J	e. +3880 J	g. –4500 J	i. –1330 J
b. –44 J	d. –2930 J	f. –3880 J	h. +4500 J	j. –6970 J

25. How much work is done as the system moves down the isotherm from A to B, i.e. what is w?

26. How much work is done as the system moves down the adiabat from A to C, i.e. what is w?

27. How much work is done as the system moves from B to C, i.e. what is w?

28. What is the internal energy change ΔE as the system moves down the isotherm from A to B?

29. What is the internal energy change ΔE as the system moves down the adiabat from A to C?

Use the following key for questions 30 through 32:

a. zero JK^{-1} b. $+21.2\ JK^{-1}$ c. $-11.5\ JK^{-1}$ d. $-13.6\ JK^{-1}$ e. $+2.93\ JK^{-1}$ f. $-2.93\ JK^{-1}$ g. $+11.5\ JK^{-1}$ h. $+13.6\ JK^{-1}$ i. $+15.1\ JK^{-1}$ j. $-15.1\ JK^{-1}$

30 What is the entropy change ΔS as the system moves down the isotherm from A to B?

31. What is the entropy change ΔS as the system moves down the adiabat from A to C?

32. What is the entropy change ΔS as the system moves from B to C?

33. Balance the equation for the following reaction that occurs in aqueous solurion in the presence of acid:

$$\_Fe^{3+} + \_NH_3OH^+ \rightarrow \_N_2O + \_Fe^{2+}$$

Write any acid present as hydronium ions (not just H^+). What is the coefficient for H_2O in the balanced equation?

a. 1 b. 2 c. 3 d. 4 e. 5 f. 6 g. 7 h. 8 i. 9 j. 10

34. In an electrolysis cell, how long must a current of 5.00 amps be applied to produce 10.5 g of silver at the cathode from a solution of silver nitrate $AgNO_3$?

a. 36.2 s b. 2.11 min c. 5.22 min d. 6.73 min e. 15.6 min f. 24.1 min g. 31.3 min h. 42.7 min i. 1 hr 2.5 min j. 5 hr 7.9 min

35. Of the following block of elements, which one forms bonds with other elements with the greatest ionic character?

a. lithium b. sodium c. potassium d. beryllium e. magnesium f. calcium g. boron h. aluminum i. gallium

36. Sodium silicate, also known as liquid glass, is a metasilicate with the empirical formula Na_2SiO_n. What is the value of the subscript n?

a. 1 b. 2 c. 3 d. 4 e. 5 f. 6

37. Which of the following ligands is classified as weak?

a. NH_3 b. CO c. CN^- d. en (ethylenediamine) e. C_5H_5N f. H_2O

38. A predominant reason for the great abundance of organic compounds is

a. the presence of carbon dioxide in the atmosphere
b. the ability of carbon to form π bonds with nitrogen
c. the ability of carbon to form covalent bonds with hydrogen
d. the natural abundance of carbon on earth
e. the strong bond that carbon forms with oxygen
f. the ability of carbon to form single and double covalent bonds with itself

39. Which of the noble gases forms compounds with fluorine most readily?

a. He b. Ne c. Ar d. Kr e. Xe

40. What is the effective atomic number (EAN) for the transition metal in the octahedral coordination compound $K_4[Fe(CN)_6]$?

a. 30 c. 36 e. 42 g. 52
b. 32 d. 38 f. 48 h. 54

41. What is the ligand field stabilization energy (LFSE) for the iron ion in the octahedral coordination compound $K_4[Fe(CN)_6]$? Neglect any pairing energy.

a. –4 Dq c. –8 Dq e. –16 Dq g. –20 Dq
b. –6 Dq d. –12 Dq f. –18 Dq h. –24 Dq

42. How many geometrical isomers of the octahedral cobalt(III) coordination compound $[Co(NH_3)_2Br_2Cl_2]^-$ are there?

a. only 1 c. 3 e. 5 g. 7
b. 2 d. 4 f. 6 h. more than 7

FINAL EXAMINATION

CHEMISTRY 142 FALL 2010

Friday December 17th 2010

This exam consists of 40 questions. Questions are worth 10 points each, unless otherwise noted.

Kinetics

1. Consider the reaction

$$SO_2(g) + \tfrac{1}{3}O_3(g) \rightarrow SO_3(g)$$

When the instantaneous rate for consumption of SO_2, $\Delta[SO_2]/\Delta t$, is -0.100 mol $L^{-1}s^{-1}$, what are the corresponding rates for O_3 and SO_3?

	$\Delta[O_3]/\Delta t$	$\Delta[SO_3]/\Delta t$
a.	-3.33×10^{-2} mol $L^{-1}s^{-1}$	0.100 mol $L^{-1}s^{-1}$
b.	-0.133 mol $L^{-1}s^{-1}$	0.300 mol $L^{-1}s^{-1}$
c.	0.300 mol $L^{-1}s^{-1}$	-0.100 mol $L^{-1}s^{-1}$
d.	3.33×10^{-2} mol $L^{-1}s^{-1}$	-0.100 mol $L^{-1}s^{-1}$
e.	-0.300 mol $L^{-1}s^{-1}$	0.100 mol $L^{-1}s^{-1}$
f.	-0.100 mol $L^{-1}s^{-1}$	0.300 mol $L^{-1}s^{-1}$
g.	0.300 mol $L^{-1}s^{-1}$	0.100 mol $L^{-1}s^{-1}$

2. The initial rate for the oxidation of iron(II) by cerium(IV)

$$Ce^{4+}(aq) + Fe^{2+}(aq) \rightarrow Ce^{3+}(aq) + Fe^{3+}(aq)$$

is measured at several different initial concentrations of the two reactants.

$[Ce^{4+}]$	$[Fe^{2+}]$	Rate
1.1×10^{-5} M	1.8×10^{-5} M	2.0×10^{-7} mol $L^{-1}s^{-1}$
1.1×10^{-5} M	2.8×10^{-5} M	3.1×10^{-7} mol $L^{-1}s^{-1}$
3.4×10^{-5} M	2.8×10^{-5} M	9.5×10^{-7} mol $L^{-1}s^{-1}$

The rate law equation for this reaction is

a. rate = $k[Ce^{4+}]$
b. rate = $k[Fe^{2+}]$
c. rate = $k[Ce^{4+}][Fe^{2+}]$
d. rate = $k[Ce^{4+}]^2[Fe^{2+}]$
e. rate = $k[Ce^{4+}][Fe^{2+}]^2$
f. rate = $k[Ce^{4+}][Fe^{2+}]^{-1}$
g. rate = $k[Ce^{4+}]^{-1}[Fe^{2+}]$

3. Carbon dioxide reacts with itself to form carbon monoxide and oxygen according to the reaction

$$2\,CO_2(g) \rightarrow 2\,CO(g) + O_2(g)$$

Using the data provided below, determine the order of this reaction with respect to CO_2.

t, s	$[CO_2]$, M
0.0	0.100
7.8	0.080
17.9	0.060
32.2	0.040
42.2	0.030

a. –1 b. 0 c. 1 d. 2 e. 3 f. 4

4. At high temperature cyclobutane, C_4H_8, decomposes to ethylene

$$C_4H_8(g) \rightarrow 2\,C_2H_4(g)$$

The activation energy for this reaction is 260 $kJmol^{-1}$. At 800 K, this reaction has a rate constant k = 0.0315 s^{-1}. What is the rate constant for the reaction at 850 K?

a. $3.16 \times 10^{-3}\ s^{-1}$
b. $2.50 \times 10^{-2}\ s^{-1}$
c. $3.16 \times 10^{-2}\ s^{-1}$
d. $3.98 \times 10^{-2}\ s^{-1}$
e. $7.65 \times 10^{-2}\ s^{-1}$
f. $0.250\ s^{-1}$
g. $0.314\ s^{-1}$
h. $0.398\ s^{-1}$

5. Consider the following three-step reaction mechanism:

$$(CH_3)_3CBr \rightarrow (CH_3)_3C^+ + Br^-$$
$$(CH_3)_3C^+ + H_2O \rightarrow (CH_3)_3COH_2^+$$
$$(CH_3)_3COH_2^+ + Br^- \rightarrow (CH_3)_3COH + HBr$$

Identify the reactant(s), product(s), intermediate(s), and catalyst(s) in this mechanism.

	reactants	products	intermediates	catalysts
a.	$(CH_3)_3CBr$	$(CH_3)_3COH$, HBr	$(CH_3)_3C^+$, $(CH_3)_3COH_2^+$, Br^-	H_2O
b.	$(CH_3)_3CBr$, H_2O	$(CH_3)_3COH$, HBr	$(CH_3)_3C^+$, $(CH_3)_3COH_2^+$, Br^-	none
c.	$(CH_3)_3CBr$, H_2O, $(CH_3)_3C^+$	$(CH_3)_3COH_2^+$, Br^-, $(CH_3)_3COH$, HBr	none	none
d.	$(CH_3)_3CBr$, H_2O	$(CH_3)_3COH$, HBr	$(CH_3)_3C^+$, $(CH_3)_3COH_2^+$	Br^-
e.	$(CH_3)_3CBr$, H_2O	$(CH_3)_3COH$, HBr	$(CH_3)_3C^+$, Br^-	$(CH_3)_3COH_2^+$
f.	$(CH_3)_3CBr$	$(CH_3)_3COH$, HBr	$(CH_3)_3C^+$, $(CH_3)_3COH_2^+$, H_2O	Br^-
g.	$(CH_3)_3CBr$	Br^-, $(CH_3)_3COH$, HBr	$(CH_3)_3C^+$, $(CH_3)_3COH_2^+$	H_2O

Chemical Equilibria

6. Consider the reaction for the combustion of methane

$$CH_4(g) + 2\,O_2(g) \rightleftharpoons CO_2(g) + 2\,H_2O(g) \qquad \Delta H° = -803 \text{ kJ/mol}$$

Predict the direction in which equilibrium for this reaction will shift when the following changes are made to a reaction mixture at equilibrium. Select the row in which all of the answers are correct.

	increase temperature	*add catalyst*	*increase volume*	*remove* O_2
a.	shift to right	no effect	no effect	shift to left
b.	shift to left	shift to right	shift to left	shift to right
c.	shift to left	shift to right	no effect	shift to left
d.	shift to right	shift to right	no effect	shift to left
e.	shift to left	shift to right	shift to right	shift to left
f.	shift to left	no effect	no effect	shift to left
g.	shift to right	no effect	shift to right	shift to right
h.	shift to right	shift to right	shift to right	shift to right

7. Using the equilibrium constants for the following reactions:

$$SnO_2(s) + 2\,CO(g) \rightleftharpoons Sn(s) + 2\,CO_2(g) \qquad K_1 = 14$$

$$CO(g) + H_2O(g) \rightleftharpoons CO_2(g) + H_2(g) \qquad K_2 = 1.3$$

determine the equilibrium constant for the reaction

$$SnO_2(s) + 2\,H_2(g) \rightleftharpoons Sn(s) + 2\,H_2O(g)$$

a. 0.093 b. 0.12 c. 8.3 d. 11 e. 18 f. 24

Acid-Base Equilibria

8. Classify the solutions produced when the following compounds are dissolved in water as acidic, basic, or neutral. Choose the row in which all of the responses are correct.

	NH_3	NH_4Br	$KClO$	H_2CO_3
a.	basic	acidic	basic	acidic
b.	acidic	neutral	acidic	basic
c.	acidic	basic	neutral	neutral
d.	basic	neutral	neutral	acidic
e.	basic	acidic	neutral	acidic
f.	basic	neutral	basic	acidic
g.	acidic	neutral	basic	acidic
h.	acidic	basic	neutral	basic

9. Find the pH of a 0.80 M solution of hydrofluoric acid, HF ($K_a = 6.6 \times 10^{-4}$).

a. 0.10 c. 3.18 e. 7.00 g. 12.36
b. 1.64 d. 3.79 f. 10.21

Consider the following information when answering questions 10 and 11:
Salicylic acid is a weak, diprotic acid:

$$H_2C_7H_4O_3(aq) + H_2O(l) \rightleftharpoons HC_7H_4O_3^-(aq) + H_3O^+(aq) \qquad K_{a1} = 1.06 \times 10^{-3}$$
$$HC_7H_4O_3^-(aq) + H_2O(l) \rightleftharpoons C_7H_4O_3^{2-}(aq) + H_3O^+(aq) \qquad K_{a2} = 3.6 \times 10^{-14}$$

The figure given below represents the distribution of species present in solution when a 0.20 M salicylic acid solution is titrated using the strong base NaOH.

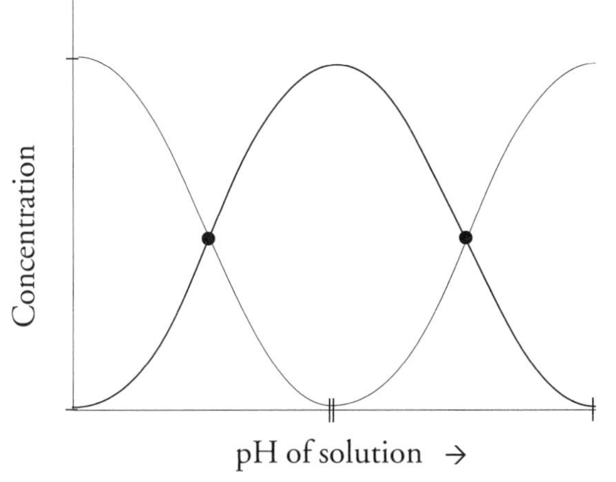

10. Identify the species present in solution when the pH of the solution is 10.0.

a. $H_2C_7H_4O_3$
b. $HC_7H_4O_3^-$
c. $C_7H_4O_3^{2-}$
d. $H_2C_7H_4O_3$ and $HC_7H_4O_3^-$
e. $H_2C_7H_4O_3$ and $C_7H_4O_3^{2-}$
f. $HC_7H_4O_3^-$ and $C_7H_4O_3^{2-}$
g. $H_2C_7H_4O_3$, $HC_7H_4O_3^-$, and $C_7H_4O_3^{2-}$

11. What is the pH of a solution containing 5.0×10^{-2} M $KHC_7H_4O_3$

a. 1.30 b. 2.97 c. 5.59 d. 8.21 e. 10.83 f. 13.44

12. K_a values are provided below for several weak acids. Identify the strongest acid and the acid which has the strongest conjugate base. Select the row in which the stongest acid and acid having the strongest conjugate base are correctly identified.

benzoic acid $K_a = 6.46 \times 10^{-5}$
chloroacetic acid $K_a = 1.4 \times 10^{-3}$
hydrocyanic acid $K_a = 6.17 \times 10^{-10}$
hydrosulfuric acid $K_a = 9.1 \times 10^{-8}$
nitrous acid $K_a = 4.6 \times 10^{-4}$

	strongest acid	*acid with strongest conjugate base*
a.	chloroacetic acid	hydrosulfuric acid
b.	benzoic acid	nitrous acid
c.	hydrocyanic acid	chloroacetic acid
d.	nitrous acid	hydrocyanic acid
e.	chloroacetic acid	hydrocyanic acid
f.	hydrosulfuric acid	chloroacetic acid
g.	hydrocyanic acid	nitrous acid
h.	chloroacetic acid	chloroacetic acid

13. Which of the following solutions will act as a buffer solution?

a. 0.75 M Na_3PO_4 and 0.80 M Na_2HPO_4
b. 1.0 M NaI and 1.0 M HI
c. 0.5 M NaOH and 0.5 M HCl
d. 0.45 M Na_2CO_3 and 0.45 M H_2CO_3
e. 0.30 M KCH_3CO_2 and 0.25 M HCO_2H
f. 1.0 M NH_3 and 1.0 M HNO_3

14. Find the pH of the solution that results when 200 mL of a 3.0×10^{-3} M NaOH solution is combined with 100 mL of a 5.0×10^{-3} HNO_3 solution.

a. 2.30 b. 3.30 c. 3.48 d. 4.00 e. 7.00 f. 10.00 g. 10.52 h. 10.70 i. 11.48

Solubility Equilibria

15. Find the equilibrium concentrations of Li^+ *(aq)* and CO_3^{2-} *(aq)* in a saturated aqueous solution of Li_2CO_3 ($K_{sp} = 8.2 \times 10^{-4}$).

	$[Li^+(aq)]$, M	$[CO_3^{2-}(aq)]$, M		$[Li^+(aq)]$, M	$[CO_3^{2-}(aq)]$, M
a.	1.5×10^{-1}	7.4×10^{-2}	e.	4.0×10^{-2}	2.0×10^{-2}
b.	5.9×10^{-2}	5.9×10^{-2}	f.	7.4×10^{-2}	7.4×10^{-2}
c.	7.4×10^{-2}	1.5×10^{-1}	g.	5.9×10^{-2}	1.2×10^{-1}
d.	1.2×10^{-1}	5.9×10^{-2}			

16. Which of the following will increase the solubility of $CuCO_3$ ($K_{sp} = 2.5 \times 10^{-10}$)? Note that ammonia complexes with copper(II) to form $[Cu(NH_3)_4]^{2+}$ ($K_f = 1.1 \times 10^{13}$).

    I. Addition of K_2CO_3
    II. Addition of HCl
    III. Addition of NH_3

    a. I only c. III only e. II and III g. I, II, and III
    b. II only d. I and II f. I and III

Thermodynamics

Questions 17 through 19 refer to the system described below. It is recommended that you draw a PV diagram to illustrate the changes described and/or make a table to keep track of P, V, T, and n for each of the states.

Consider a 2.00 mole sample of an ideal monatomic gas. In state A this gas sample has a pressure of 12.0 atm, volume of 4.10 L, and temperature of 300 K. This gas can undergo a reversible, isothermal expansion to reach state B which has a volume of 15.2 L. An alternative reversible, adiabatic expansion from state A to reach the same volume as state B leads to state C which has a pressure of 1.35 atm.

17. What is the change in the internal energy of the ideal monatomic gas in going from state B to state C?

    a. −7.28 kJ c. −4.37 kJ e. +4.37 kJ g. +7.28 kJ
    b. −6.54 kJ d. 0.00 kJ f. +6.54 kJ

18. How much work is done on the gas in the reversible, adiabatic expansion from state A to state C?

    a. −7.28 kJ c. −4.37 kJ e. +4.37 kJ g. +7.28 kJ
    b. −6.54 kJ d. 0.00 kJ f. +6.54 kJ

19. How much heat is transferred in going from state A to state C along the path ABC?

    a. −6.54 kJ c. −2.17 kJ e. 2.17 kJ g. 6.54 kJ
    b. −4.37 kJ d. 0.00 kJ f. 4.37 kJ h. 10.9 kJ

20. Ethanol has a boiling point of 78.5°C and a melting point of –114.1°C. What are the signs of ΔG, ΔH, and ΔS for the freezing of ethanol at –65°C and vaporization of ethanol at 95°C?

	Freezing at –65°C			Vaporization at 95°C		
	ΔG	ΔH	ΔS	ΔG	ΔH	ΔS
a.	–	–	–	+	+	+
b.	–	–	+	+	+	–
c.	+	+	–	–	–	+
d.	–	+	+	+	–	–
e.	+	+	–	–	+	+
f.	+	+	+	–	–	–
g.	+	–	–	–	+	+

21. Suppose that 20.0 g of steam at 100°C is bubbled through 225.0 g of water at 35°C in an insulated container. What is the temperature after the system has reached thermal equilibrium?

    Specific heat of water = 4.184 $J\ K^{-1}\ g^{-1}$
    Latent heat of fusion of water = 333 $J\ g^{-1}$
    Latent heat of vaporization of water = 2260 $J\ g^{-1}$

    a. 40.3°C
    b. 46.8°C
    c. 54.5°C
    d. 63.4°C
    e. 69.1°C
    f. 76.2°C
    g. 83.0°C
    h. 84.4°C

22. Find the entropy change for the process described in question 21.

    a. 0.00 $J\ K^{-1}$
    b. 5.51 $J\ K^{-1}$
    c. 15.3 $J\ K^{-1}$
    d. 18.8 $J\ K^{-1}$
    e. 119 $J\ K^{-1}$
    f. 136 $J\ K^{-1}$
    g. 258 $J\ K^{-1}$
    h. 261 $J\ K^{-1}$

23. Which of the following statements are true?

    I. Work, heat, and entropy are thermodynamic state functions.
    II. $\Delta S_{univ} < 0$ for spontaneous processes.
    III. $\Delta G°$ is minimized at equilibrium.
    IV. If $\Delta G° > 0$ for a reaction, the corresponding equilibrium constant $K < 1$.

    a. none are true
    b. I
    c. II
    d. III
    e. IV
    f. I and II
    g. III and IV
    h. II and III
    i. I and IV
    j. I, III, and IV

Electrochemistry

24. Determine the oxidation number for the atom underlined in each of the compounds given below. Add them together using the correct signs. What is the sum?

    $\underline{C}H_3OH$ $\quad$ $Mg_3(\underline{P}O_4)_2$ $\quad$ $\underline{Co}F_3$

    a. +3
    b. +4
    c. +5
    d. +6
    e. +7
    f. +8
    g. +9
    h. +10

25. Balance the following equation for an oxidation-reduction reaction occurring under basic conditions.

$$Ag(s) + HS^-(aq) + CrO_4^{2-}(aq) \rightarrow Ag_2S(s) + Cr(OH)_3(s)$$

What is the coefficient for the water in the reaction? Is water present in the balanced equation as a reactant or product?

a. 1, product
b. 2, product
c. 5, product
d. 7, product
e. 1, reactant
f. 2, reactant
g. 5, reactant
h. 7, reactant

26. A voltaic/galvanic cell is constructed from a standard $F_2(g)|F^-(aq)$ half cell (E°_{red} = 2.870 V) and a standard $Ti^{2+}(aq)|Ti(s)$ half cell (E°_{red} = –1.63 V). Identify the reactants in the spontaneous cell reaction.

a. $Ti(s)$, $F^-(aq)$
b. $F^-(aq)$, $F_2(g)$
c. $Ti(s)$, $F_2(g)$
d. $Ti^{2+}(aq)$, $F^-(aq)$
e. $Ti^{2+}(aq)$, $Ti(s)$
f. $Ti^{2+}(aq)$, $F_2(g)$

27. Use the standard reduction potentials given below to determine which of these species would be the best reducing agent.

$E^\circ_{red}(Au^+|Au) = +1.69$ V
$E^\circ_{red}(Ni^{2+}|Ni) = -0.23$ V
$E^\circ_{red}(Br_2|Br^-) = +1.09$ V
$E^\circ_{red}(Mn^{2+}|Mn) = -1.18$ V

a. Au^+
b. Au
c. Ni^{2+}
d. Ni
e. Br_2
f. Br^-
g. Mn^{2+}
h. Mn

28. In the electrolysis of manganese(IV) sulfate $Mn(SO_4)_2$, 12.0 g of manganese was deposited at the cathode over a period of 2.00 hours. If the same current for the same amount of time is passed through a separate electrolysis cell containing $AlCl_3$, how much aluminum will be deposited at the cathode?

a. 1.98 g
b. 3.96 g
c. 4.46 g
d. 5.84 g
e. 7.86 g
f. 11.9 g

Main Group Elements

29. *(5 points)* Which of the following properties is not characteristic of metallic elements?

a. malleability
b. high ionization energy
c. tendency to form cations
d. shiny, lustrous
e. low electronegativity
f. good conductors of electricity

30. *(5 points)* The carbon atoms in the molecule ethylene, $H_2C{=}CH_2$, are ______ hybridized. The double bond between the carbon atoms consists of ______ .

a. sp^2 one σ and one π bond
b. sp^3 two π bonds
c. sp^3 one σ and one π bond
d. sp one σ and one π bond
e. sp^2 two σ bonds
f. sp^2 two π bonds
g. sp two σ bonds
h. sp^3 two σ bonds
i. sp two π bonds

31. *(5 points)* Rank the halogens in order of increasing strength as an oxidizing agent.

a. $Br_2 < I_2 < F_2 < Cl_2$
b. $F_2 < Cl_2 < Br_2 < I_2$
c. $Cl_2 < F_2 < Br_2 < I_2$
d. $Br_2 < I_2 < Cl_2 < F_2$
e. $I_2 < F_2 < Br_2 < Cl_2$
f. $F_2 < Cl_2 < I_2 < Br_2$
g. $I_2 < Br_2 < Cl_2 < F_2$

From the list of elements provided below, select the correct response.

a. Ca c. Si e. Ne g. Xe
b. Na d. Mg f. Be h. O

32. *(4 points)* This metal forms phosphides having the formula M_3P.

33. *(4 points)* This element exhibits similar reactivity to Li because it has a comparable charge density/polarizing power.

34. *(4 points)* This element forms octet-deficient compounds.

35. *(4 points)* This element is largely unreactive except with O and F.

36. *(4 points)* Of the elements listed above, this element has the greatest electronegativity.

Coordination Compounds

37. *(5 points)* Identify the correct formula for the compound pentaamminehydroxoiron(III) chloride.

a. $Fe(NH_3)_5(OH)Cl_2$
b. $Fe(NH_3)_5(H_2O)Cl_3$
c. $[Fe(NH_3)_5(OH)]Cl_2$
d. $[Fe(NH_3)_5(H_2O)]Cl_2$
e. $[Fe(NH_3)_5(OH)]Cl_3$
f. $[Fe(NH_3)_5(H_2O)]Cl_3$

38. The coordination complex $[Ni(CN)_4]^{2-}$ is diamagnetic. Use valence bond theory to determine the hybridization of the orbitals on the nickel ion used to form bonds with the cyanide ligands and the shape of the coordination complex.

a. sp^2 trigonal planar
b. sp^3 tetrahedral
c. sp^3 square planar
d. dsp^2 tetrahedral
e. dsp^2 square planar
f. dsp^3 trigonal bipyramidal
g. d^2sp^3 octahedral
h. d^2sp^3 trigonal prism
i. d^4sp trigonal prism

39. How many geometric isomers are possible for the coordination complex $Mn(NH_3)_4Cl_2$?

a. 1 b. 2 c. 3 d. 4 e. 5 f. 6

40. Find the ligand-field stabilization energy (LFSE) for the cobalt ion in the octahedral coordination complex $[Co(H_2O)_6]^{3+}$. Ignore any contributions from the pairing energy P, if applicable.

a. 0 Dq c. –8 Dq e. –16 Dq g. –20 Dq
b. –4 Dq d. –12 Dq f. –18 Dq h. –24 Dq

FINAL EXAMINATION

CHEMISTRY 142 SPRING 2011

Wednesday May 4th 2011

1. The thermal decomposition of phosphine PH_3 into phosphorus and hydrogen is a first-order reaction:

$$4\,PH_3(g) \rightarrow P_4(g) + 6\,H_2(g)$$

At one point in the reaction, the hydrogen forms at a rate of 18.0 mol L^{-1} min^{-1}. What are the corresponding rates of change in the concentrations of phosphine PH_3 and phosphorus P_4?

	change in $PH_3(g)$	*change in $P_4(g)$*
a.	–4.0 mol L^{-1} min^{-1}	1.0 mol L^{-1} min^{-1}
b.	–6.0 mol L^{-1} min^{-1}	2.0 mol L^{-1} min^{-1}
c.	–8.0 mol L^{-1} min^{-1}	3.0 mol L^{-1} min^{-1}
d.	–12.0 mol L^{-1} min^{-1}	4.0 mol L^{-1} min^{-1}
e.	–18.0 mol L^{-1} min^{-1}	6.0 mol L^{-1} min^{-1}
f.	–8.0 mol L^{-1} min^{-1}	2.0 mol L^{-1} min^{-1}
g.	–12.0 mol L^{-1} min^{-1}	3.0 mol L^{-1} min^{-1}
h.	–18.0 mol L^{-1} min^{-1}	4.0 mol L^{-1} min^{-1}
i.	–4.0 mol L^{-1} min^{-1}	6.0 mol L^{-1} min^{-1}

2. The following data were collected for the reaction between hydrogen and nitric oxide at 700°C:

$$2\,H_2(g) + 2\,NO(g) \rightarrow N_2(g) + 2\,H_2O(g)$$

	Concentration of H_2	*Concentration of NO*	*Rate of reaction*
Expt 1	0.010 M	0.025	2.4×10^{-6} mol $L^{-1}s^{-1}$
Expt 2	0.0050 M	0.025	1.2×10^{-6} mol $L^{-1}s^{-1}$
Expt 3	0.010 M	0.0125	6.0×10^{-7} mol $L^{-1}s^{-1}$

What is the overall order of the reaction?

a. ½ b. 1 c. 1.5 d. 2 e. 2.5 f. 3 g. 4 h. 5 i. 6 j. 9

3. In one experiment, the concentration of a reactant decreased as follows as the reaction proceeded:

Initial concentration	224 mol L^{-1}
After 15 minutes	56 mol L^{-1}
1.25 hours after the start of the reaction	14 mol L^{-1}

How long did it take, from the beginning of the reaction, for the concentration of the reactant to reach 3.5 mol L^{-1}?

a. 2.0 hr 30 min
b. 3.0 hr
c. 3.0 hr 15 min
d. 3.0 hr 30 min
e. 4.0 hr
f. 4.0 hr 30 min
g. 5.0 hr 15 min
h. 5.0 hr 45 min
i. 6.0 hr

4. In the energy profile shown for an exothermic reaction, which labels correspond to the following?

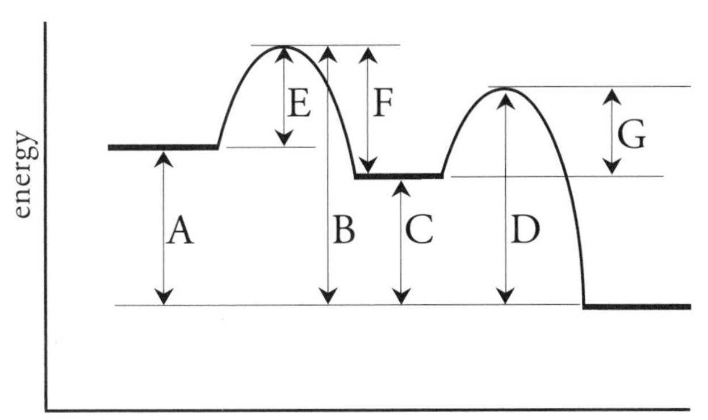

	Activation energy for the first step of the forward reaction	*The energy of reaction ΔE*	*Activation energy for the second step in the reverse direction*
a.	E	A	B
b.	F	C	E
c.	A	D	E
d.	G	B	E
e.	C	C	B
f.	E	A	F
g.	F	C	E
h.	A	D	E
i.	G	B	B
j.	C	A	G

5. Hydrogen can be prepared by heating natural gas (methane CH_4) in the presence of air at 1000°C. Calculate the equilibrium constant K_c if the concentration of the various components at equilibrium are as shown:

$$CH_4(g) + \frac{1}{2} O_2(g) \rightleftharpoons CO(g) + 2\,H_2(g)$$

Concentrations at equilibrium: $[CH_4(g)] = 1.0$ M, $[O_2(g)] = 4.0$ M, $[CO(g)] = 3.0$ M, $[H_2(g)] = 6.0$ M

a. 0.33 c. 8.0 e. 21 g. 33 i. 54
b. 3.3 d. 16 f. 27 h. 38 j. 76

6. The following equilibrium constants have been determined for oxalic acid at 25°C:

$$H_2C_2O_4(aq) + H_2O(l) \rightleftharpoons H_3O^+(aq) + HC_2O_4^-(aq) \qquad K_{a1} = 6.5 \times 10^{-2}$$

$$HC_2O_4^-(aq) + H_2O(l) \rightleftharpoons H_3O^+(aq) + C_2O_4^{2-}(aq) \qquad K_{a2} = 6.1 \times 10^{-5}$$

Calculate the equilibrium constant K_a for the reaction:

$$H_2C_2O_4(aq) + 2\,H_2O(l) \rightleftharpoons 2\,H_3O^+(aq) + C_2O_4^{2-}(aq) \qquad K_a = ??$$

a. 1.1×10^3 c. 2.5×10^5 e. 4.0×10^{-6} g. 9.4×10^{-4}
b. 3.3×10^3 d. 6.3×10^{10} f. 1.6×10^{-11} h. 1.2×10^{-8}

7. Heating sodium hydrogen carbonate in a closed vessel causes the following equilibrium to be established:

$$NaHCO_3(s) \rightleftharpoons Na_2CO_3(s) + H_2O(g) + CO_2(g)$$

Suppose that the equilibrium is disturbed by the following actions. In which direction, if any, will the system shift in order to restore equilibrium?

	add $NaHCO_3(s)$	*remove* $H_2O(g)$	*decrease volume*	*increase temperature*
a.	to right	to right	to left	to right
b.	to right	to left	to right	to right
c.	to right	no effect	to left	no effect
d.	no effect	to right	to left	to left
e.	no effect	to left	to right	no effect
f.	no effect	to right	to left	to right
g.	no effect	no effect	to left	to right
h.	to left	to right	to right	to left
i.	to left	to left	to right	to left
j.	to left	no effect	to left	to left

8. In which row are the solutes correctly described?

	Acidic solute	*Basic solute*	*Neutral solute*
a.	NaH_2PO_4	NH_3	$NaHCO_3$
b.	HF	NaOH	HCN
c.	NH_4Cl	KCN	CH_3CO_2H
d.	HNO_3	KF	KCN
e.	CH_3CO_2H	KOH	KH_2PO_4
f.	H_2SO_4	NH_3	KNO_3
g.	HCN	Li_2CO_3	NH_4NO_3
h.	H_2CO_3	HCN	NaCl
i.	$KHCO_3$	K_2CO_3	LiOH
j.	NH_3	NH_4Cl	KBr

9. Consider the following Brønsted-Lowry equilibrium:

$$HSO_3^- + HSO_3^- \rightleftharpoons H_2SO_3 + \underline{???????}$$

Identify the missing product, and decide whether that product is acting as a base or an acid in the equilibrium.

a. H_2SO_3 *acting as* an acid
b. OH^- *acting as* a base
c. SO_3^{2-} *acting as* a base
d. H_3O^+ *acting as* an acid
e. SO_4^{2-} *acting as* a base
f. HSO_3^- *acting as* an acid

10. A 1.00 M solution of sodium fluoride (NaF) has a pH = 8.57. What is the value of pK_a for hydrofluoric acid?

a. 1.65	c. 2.49	e. 4.27	g. 6.21	i. 8.54
b. 2.28	d. 3.14	f. 5.22	h. 7.68	j. 10.86

11. The net ionic equation for the equilibrium established when potassium hypochlorite KClO is added to water is shown below. Classify the participants in the reaction based on the Brønsted-Lowry definitions for acids and bases.

	$ClO^-(aq)$ +	$H_2O(l)$ ⇌	$HClO(aq)$ +	$OH^-(aq)$
a.	base	base	acid	acid
b.	base	acid	acid	base
c.	base	acid	base	acid
d.	acid	base	acid	base
e.	acid	acid	base	base
f.	acid	base	base	acid

12. As mentioned earlier, the two acid ionization constants for the diprotic acid oxalic acid at 25°C are:

$$H_2C_2O_4(aq) + H_2O(l) \rightleftharpoons H_3O^+(aq) + HC_2O_4^-(aq) \qquad K_{a1} = 6.5 \times 10^{-2}$$

$$HC_2O_4^-(aq) + H_2O(l) \rightleftharpoons H_3O^+(aq) + C_2O_4^{2-}(aq) \qquad K_{a2} = 6.1 \times 10^{-5}$$

What is the equilibrium constant for the following system that illustrates what happens when sodium hydrogen oxalate dissolves in water? The distribution diagram is illustrated for your convenience.

$$HC_2O_4^-(aq) + HC_2O_4^-(aq) \rightleftharpoons H_2C_2O_4(aq) + C_2O_4^{2-}(aq) \qquad K = ?$$

a. 1.1×10^3	c. 3.1×10^{-2}	e. 9.4×10^{-4}	g. 4.0×10^{-6}
b. 7.6×10^{-6}	d. 2.5×10^5	f. 3.3×10^{-5}	h. 2.6×10^5

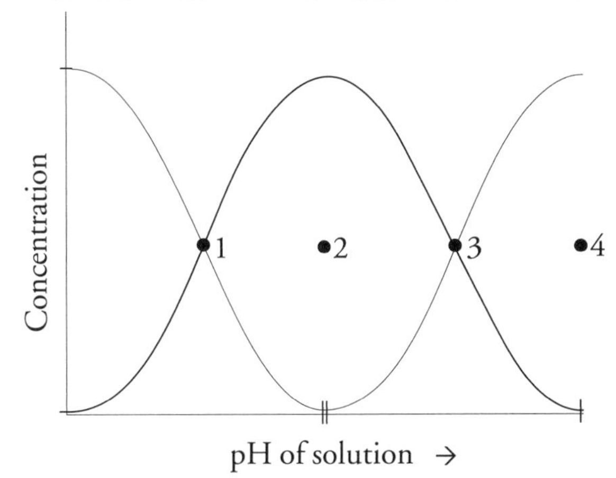

13. Referring to question 12, what is the pH at a point halfway between 1 and 2, where three-fourths of the oxalic acid has been neutralized by addition of sodium hydroxide to form sodium hydrogen oxalate ?

a. 1.19	c. 1.67	e. 2.10	g. 3.21	i. 4.21
b. 1.28	d. 1.95	f. 2.70	h. 3.67	j. 5.40

14. How many of the following equimolar aqueous solutions will act as a buffer solutions?

$NaCN$ and HCN
Na_2SO_3 and $NaHSO_3$
CH_3CO_2H and $LiCH_3CO_2$
K_2HPO_4 and Na_3PO_4
$NaNO_2$ and HNO_2
NaF and HF

a. 1 b. 2 c. 3 d. 4 e. 5 f. all of them

15. When silver carbonate Ag_2CO_3 dissolves in water, what is the relationship between the concentration of silver ions and the concentration of carbonate ions?

a. $[Ag^+] = 4 \times [CO_3^{2-}]$
b. $[Ag^+] = 2 \times [CO_3^{2-}]$
c. $[Ag^+] = [CO_3^{2-}]$
d. $[Ag^+] = (1/4) \times [CO_3^{2-}]$
e. $[Ag^+] = (1/2) \times [CO_3^{2-}]$

16. If the solubility product K_{sp} for silver carbonate Ag_2CO_3 is 8.1×10^{-12}, what is the molar solubility of silver carbonate in water at the same temperature?

a. 8.1×10^{-5} M
b. 2.1×10^{-3} M
c. 2.6×10^{-4} M
d. 5.7×10^{-6} M
e. 1.3×10^{-4} M
f. 2.0×10^{-4} M
g. 6.5×10^{-5} M
h. 4.3×10^{-5} M

17. If sodium sulfide is added to a system consisting of a saturated solution of silver carbonate and excess solid silver carbonate, and K_{sp} for silver sulfide is 6.0×10^{-51}, what is the ratio between the concentrations of sulfide ions and carbonate ions $[S^{2-}]/[CO_3^{2-}]$ in the solution at equilbrium?

a. 1.4×10^{39}
b. 5.2×10^{-18}
c. 3.7×10^{19}
d. 8.2×10^{-11}
e. 4.9×10^{-19}
f. 2.7×10^{-20}
g. 7.4×10^{-40}
h. 3.2×10^{-50}

18. When a system moves from State A to another state, State B, via two different paths, which of the following is the same for the two routes taken?

A. the change in the internal energy of system
B. the final temperature of the system
C. the work done by the system
D. the entropy change
E. the change in the mass, if any, of the system
F. the heat absorbed by the system.

a. only A
b. A and D
c. C
d. C and F
e. A and B
f. A, B, and D
g. A, B, D, and E
h. B and D
i. B, D, and E
j. A, D, and E

Questions 19 through 23 refer to the system described below. It is recommended that you draw a PV diagram to illustrate the changes described and/or make a table to keep track of P, V, T, and n for each of the states.

Consider an ideal monatomic gas at a pressure of 2.00 atm occupying a volume of 25.0 L at 31.5°C; call this State A. The gas is compressed adiabatically to a State B where the volume is 5.00 L and the pressure is 29.24 atm. The gas is then cooled at constant volume to a pressure of 10.0 atm; call this State C. The gas then expands isothermally and reversibly to return to State A.

19. How many moles of the ideal gas are present in the system?

a. 0.250 mol c. 0.750 mol e. 1.50 mol g. 3.00 mol i. 5.00 mol
b. 0.500 mol d. 1.00 mol f. 2.00 mol h. 4.00 mol j. 10.0 mol

20. How much work is done in going from State B to State C?

a. w = –4.56 kJ c. w = –10.1 kJ e. w = –14.6 kJ g. w = –26.8 kJ
b. w = –8.72 kJ d. w = –12.2 kJ f. w = –18.3 kJ h. w = zero

21. How much heat is lost by the system as is moves from State B to State C?

a. q = –4.56 kJ c. q = –10.1 kJ e. q = –14.6 kJ g. q = –26.8 kJ
b. q = –8.72 kJ d. q = –12.2 kJ f. q = –18.3 kJ h. q = zero

22. What is the entropy change in the system in going from State B to State C?

a. $-4.56\ JK^{-1}$ c. $-10.1\ JK^{-1}$ e. $-14.6\ JK^{-1}$ g. $-26.8\ JK^{-1}$
b. $-8.72\ JK^{-1}$ d. $-12.2\ JK^{-1}$ f. $-18.3\ JK^{-1}$ h. zero

23. What is the internal energy change ΔE of the system in going from State B to State C?

a. –4.56 kJ c. –10.1 kJ e. –14.6 kJ g. –26.8 kJ
b. –8.72 kJ d. –12.2 kJ f. –18.3 kJ h. zero

24. Suppose a vessel contained 4 molecules of an ideal gas. Imagine the vessel divided into two imaginary halves. What is the probability that one molecule would be in the right half and three molecules would be in the left half? Express the probability as a fraction of the total number of arrangements possible.

a. 1 in 2 d. 1 in 10 g. 1 in 64
b. 1 in 4 e. 1 in 16 h. 1 in 128
c. 1 in 8 f. 1 in 32 i. 1 in 256

25. Suppose all four molecules described in the previous question were confined to just one of the two halves of the container. Then the gas was allowed to expand isothermally so that the molecules occupy the entire container. What is the entropy change for this expansion?

a. 1.1×10^{-23} JK^{-1}
b. 3.3×10^{-24} JK^{-1}
c. 5.5×10^{-23} JK^{-1}
d. 3.8×10^{-23} JK^{-1}
e. 9.9×10^{-23} JK^{-1}
f. 3.3×10^{-22} JK^{-1}
g. 5.5×10^{-22} JK^{-1}
h. 3.3×10^{-21} JK^{-1}

26. What is the oxidation number of a chromium atom in ammonium dichromate $(NH_4)_2Cr_2O_7$?

a. −3
b. −1
c. 0
d. +1
e. +2
f. +3
g. +4
h. +5
i. +6
j. +7

27. Balance the equation for this redox reaction in acidic solution, using whole number coefficients:

$$Cu(s) + HNO_3(aq) \rightarrow Cu^{2+}(aq) + NO(g)$$

What coefficient is required for the $NO(g)$ in the balanced equation?

a. 1
b. 2
c. 3
d. 4
e. 5
f. 6
g. 7
h. 8
i. 9
j. 10

28. Use the standard half–cell reduction potentials to determine the equilibrium constant for the following reaction at 25°C:

$$5\ Fe^{2+}(aq) + MnO_4^-(aq) + 8\ H^+(aq) \rightarrow Mn^{2+}(aq) + 4\ H_2O(l) + 5\ Fe^{3+}(aq)$$

half-cell standard reduction potentials:

E°_{red} $MnO_4^-|Mn^{2+}$ = +1.51 V
E°_{red} $Fe^{3+}|Fe^{2+}$ = +0.77 V

a. 2.2×10^3
b. 8.7×10^6
c. 3.2×10^{12}
d. 1.0×10^{25}
e. 7.0×10^{48}
f. 3.5×10^{62}
g. 4.7×10^{86}
h. 5.1×10^{106}

29. How long does it take to produce 5.6 kg of aluminum from an electrolysis cell containing alumina (Al_2O_3) dissolved in cryolite (Na_3AlF_6) using a current of 10,000 amps?

a. 30 sec
b. 45 sec
c. 3.0 min
d. 10 min
e. 30 min
f. 45 min
g. 60 min
h. 100 min
i. 2.5 hours
j. 10 hours

30. Of the following block of elements, which one forms bonds with other elements with the greatest ionic character?

a. lithium
b. sodium
c. potassium
d. beryllium
e. magnesium
f. calcium
g. boron
h. aluminum
i. gallium

31. The following properties describe one of the elements in the Periodic Table. Which one?

    The element is a metal.
    The element forms an amphoteric oxide with the stoichiometry M_2O_3
    The element behaves in many respects just like beryllium.
    The element forms very strong bonds with oxygen.
    The element forms a chloride that is dimeric (M_2Cl_6) in the liquid state.

    a. Na
    b. Mg
    c. K
    d. Fe
    e. Sn
    f. Al
    g. Cu
    h. Zn
    i. Ca
    j. Li

32. Margarite, known otherwise as brittle mica, is a layered aluminosilicate with the empirical formula $[CaAl_n(OH)_2(Al_2Si_2O_{10})]$. What is the value of the subscript n?

    a. 1 b. 2 c. 3 d. 4 e. 5 f. 6

33. Which of the noble gases is most easily oxidized and forms compounds with oxygen and fluorine most readily?

    a. He b. Ne c. Ar d. Kr e. Xe

34. How many geometrical isomers of the octahedral cobalt(III) coordination compound triamminedibromochlorocobalt(III), $[Co(NH_3)_3Br_2Cl]$, are there?

    a. only 1
    b. 2
    c. 3
    d. 4
    e. 5
    f. 6
    g. 7
    h. more than 7

35. What is the ligand-field stabilization energy (LFSE) for the chromium ion in the octahedral coordination complex $[Cr(NH_3)_6]^{3+}$. Ignore any contributions from the pairing energy P, if applicable.

    a. 0 Dq
    b. –4 Dq
    c. –8 Dq
    d. –12 Dq
    e. –16 Dq
    f. –18 Dq
    g. –20 Dq
    h. –24 Dq

SOLUTIONS

You are strongly advised not to look at these solutions to the examination questions until you have tried to answer the questions yourself!

EXAMINATION 1

CHEMISTRY 142 SPRING 2006

Solutions

1. e. rate of NO_2 production = 4 × rate of O_2 production
   the rate of N_2O_5 use decreases

2. g. compare 1 and 2: order with respect to [CO] = 1
   compare 2 and 3: order with respect to $[Cl_2]$ = 1
   overall order = 1 + 1 = 2

3. g. the half-life doubles each half-life
   0.240 M to 0.120 M: 4 minutes
   0.120 M to 0.060 M: 8 minutes
   total time = 12 minutes

4. b. the intermediate (OI^-) is formed and then used
   the catalyst (I^-) is used and then reformed

5. b. k = 0.693/5270 = 0.000132
   $\ln([R]_t/[R]_0)$ = –kt
   ln(77/100) = –0.000132 × t
   t = 2000 yr

6. a. the increased energy available in the collisions is the the most important reason
   the molecules do collide more frequently but this is a minor effect

7. f. the yield of the product is the same with or without the catalyst
   the equilbrium composition is merely reached more quickly

8. c. $K = [NO_2]^4 / [N_2O]^2 \times [O_2]^3$
   $1.60 \times 10^6 = [NO_2]^4 / (4.0/20)^2 \times (2.0/20)^3$
   $[NO_2]^4 = 64$
   $[NO_2] = 2.8 \text{ mol L}^{-1}$

9. a. Δn is positive

10. e. reverse the equation and divide by two
    so, take the reciprocal of K and then take the square root

11. d. all the others are weak

12. d. the only one that is a base

13. c.

14. d. $H_2PO_4^-$ and HPO_4^{2-}

15. d. $pH = pK_{a2} = 7.21$

16. b. $pH = ½(pK_{a1} + pK_{a2}) = 4.67$

17. b. $K_b = K_w / K_a = 5.6 \times 10^{-10}$

18. c. calculate K_a and pK_a from the pH of the 0.20 M solution:
$K_a = [H_3O^+]^2/0.20$
$= (0.00191)^2/0.20 = 1.8 \times 10^{-5}$
$pK_a = 4.74$
$pH = pK_a + \log_{10}(\text{base/acid})$
$= 4.74 + \log_{10}(0.30/0.20)$
$= 4.74 + 0.176$
$= 4.92$

19. f. the conjugate base that is strongest
that is, the conjugate base of the weakest acid

20. d. $pH = pK_a + \log_{10}(\text{base/acid})$
if the [base] = [acid], then $pH = pK_a$
in this case the $pK_a = pK_{a1} = 4.1$

21. c. Lewis bases are electron-pair donors

22. g. $K_{neut} = K_b / K_w = 1.7 \times 10^{-6} / 1.0 \times 10^{-14}$
$= 1.7 \times 10^8$

EXAMINATION 1

CHEMISTRY 142 FALL 2006

Solutions

1. d. $-1/4\ \Delta[F_2]/\Delta t = -1/3\ \Delta[H_2O]/\Delta t = +1/6\ \Delta[HF]/\Delta t = +\Delta[OF_2]/\Delta t = +\Delta[O_2]/\Delta t$

   HF: Rate = 24.0 mol L^{-1} s^{-1} × (6 mol HF / 4 mol F_2) = 36.0 mol L^{-1} s^{-1}

   O_2: Rate = 24.0 mol L^{-1} s^{-1} × (1 mol O_2 / 4 mol F_2) = 6.0 mol L^{-1} s^{-1}

2. g. Rate = $k[C_2H_4]^x[H_2]^y$

   Compare experiments 1 and 3:
   $[H_2]$ is halved and the rate decreases by 4; $4 = 2^2$, therefore order with respect to $[H_2] = 2$

3. h. first-order decay: $\ln([R]_t/[R]_0) = -k \times t$
   $\ln(0.15/0.6) = -k \times (5.2)$
   $k = 0.27$

4. c. $\ln(k_1/k_2) = -E_A/R\ (1/T_1 - 1/T_2)$
   $\ln(1/2) = -(E_A/8.314) \times (1/298 - 1/377)$
   E_A = 8,200 J mol^{-1}

5. c. the intermediates are $(CH_3)_3C^+$ and Br^-

6. c. rate depends on the concentrations of the reactants in the slow step of mechanism
   the slow step is: $(CH_3)_2{=}CH_2 + HBr \rightarrow (CH_3)_3C^+ + Br^-$
   Rate = $k\,[(CH_3)_2{=}CH_2]\,[HBr]$

7. a.

8. e. HCN

9. c. $K = ([NOBr]^2)/([NO]^2[Br_2]) = 5^2/(1.8^2 \times 3) = 2.6$

10. g. Na^+ is a spectator ion; the net ionic equation is:
    $ClO^- + H_2O \rightleftharpoons HClO + OH^-$ $\quad K_b = K_w/K_a = 3.125 \times 10^{-7}$
    $(3.125 \times 10^{-7}) = x^2/1.75$ (ignoring x in the denominator)
    $x = 7.4 \times 10^{-4} = [OH^-]$
    $pOH = -\log[7.4 \times 10^{-4}] = 3.13$
    $pH = 14 - pOH = 14 - 3.13 = 10.87$

11. b. SCN^-, base — conjugate acid-base pairs differ by one H^+
    H_3O^+ is an acid, so SCN^- must be a base

12. d. a Lewis base must be able to donate an electron pair—CH_4 does not have an electron pair to donate

13. g.

14. h. $K_b = K_w/K_a = (10^{-14})/(1.2 \times 10^{-2}) = 8.3 \times 10^{-13}$

15. a. $K_p = (P_{CH3Cl})(P_{HCl}) / (P_{CH4})(P_{Cl2})$
$= (0.75 \times 0.75)/(0.5 \times 0.5) = 2.25$
$K_p = K_c(RT)^{\Delta n}$ but $\Delta n = 2 - 2 = 0$, so $K_p = K_c$

16. e.

$C(s) + CO_2(g) \rightleftharpoons 2CO(g)$	$K = 0.64$
$CO(g) \rightleftharpoons C(s) + ½\,O_2(g)$	$K = 1/(1 \times 10^3)$
add $CO_2(g) \rightleftharpoons CO(g) + ½\,O_2(g)$	$K = 0.64 / (1 \times 10^3) = 6.4 \times 10^{-4}$

17. e. $NaClO_4$ – Na^+ is a salt of NaOH (strong base), so no hydrolysis
ClO_4^- is a salt $HClO_4$ (strong acid), so no hydrolysis
Since no hydrolysis, no H_3O^+ or OH^- produced, so solution is neutral.

18. b. hydrolysis of HS^-:
$HS^- + H_2O \rightleftharpoons H_2S + OH^-$ $\quad K_b = K_w/K_{a1} = (10^{-14})/(9.5 \times 10^{-8}) = 1.05 \times 10^{-7}$
$K_b = [H_2S][OH^-]/[HS^-] = x^2/0.25 - x = x^2/0.25$ (ignore x in the denominator)
$[OH^-] = x = 1.62 \times 10^{-4}$
$pOH = -\log [OH^-] = 3.79$
$pH = 14 - 3.79 = 10.2$

19. e. $pH = pK_a + \log([CN^-]/[HCN])$
$= 9.3 + \log(0.25/0.8) = 8.8$

20. g.

$HClO + H_2O \rightleftharpoons ClO^- + H_3O^+$	$K_a = 2 \times 10^{-11}$
$NH_3 + H_2O \rightleftharpoons NH_4^+ + OH^-$	$K_b = 1.8 \times 10^{-5}$
$H_3O^+ + OH^- \rightleftharpoons 2H_2O$	$1/K_w = 1/10^{-14}$
$HClO + NH_3 \rightleftharpoons ClO^- + NH_4^+$	

$K_{neut} = (K_a \times K_b)/K_w = (2 \times 10^{-11})(1.8 \times 10^{-5})/(10^{-14}) = 3.6 \times 10^{-2}$

EXAMINATION 1

CHEMISTRY 142 SPRING 2007

Solutions

1. f. A and B are the products; C is the reactant (NH_3)
   A (H_2O) is formed faster than B (NO_2)

2. e. 60 mol CO_2 L^{-1} min^{-1} × (8 O_2 / 5 CO_2) = 96 mol^{-1} min^{-1}

3. g. decrease of 8 mol L^{-1} in 30 min
   decrease of 24 mol L^{-1} would take three times as long = 90 min

4. h. second-order half-lives double each half-life
   56 mol L^{-1} to 28 mol L^{-1} to 14 mol L^{-1} is two successive half-lives:
   first half-life = 4 min; second half-life = 8 min
   14 mol L^{-1} to 7 mol L^{-1} to 3.5 mol L^{-1} is two more successive half-lives:
   third half life = 16 min; fourth half-life = 32; total = 48 min

5. a.
$$\ln([R]_t/[R]_0) = -kt$$
$$\ln(0.75/1) = -(0.693/t_{1/2}) \times t$$
$$= -(0.693/5730 \text{ year}) \times t$$
$$t = 2380 \text{ yr}$$

6. f. In the energy profile for a reaction shown, D is the energy of the activated complex, A is the energy of the reaction, F is the energy of the products, and C is the activation energy for the reverse reaction.

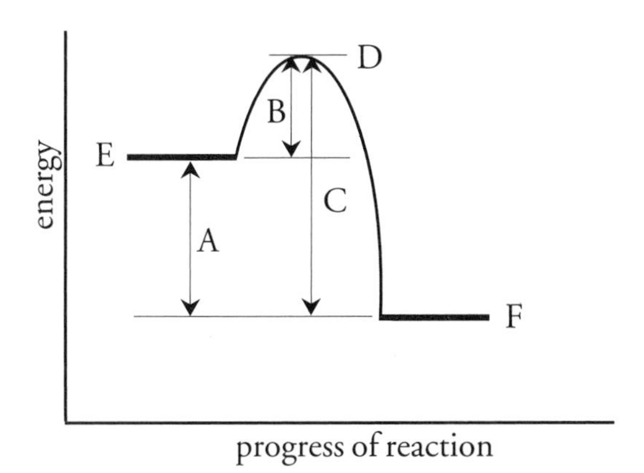

7. g. compare 2 and 3: order with respect to $[H_2]$ = 1
   compare 1 and 2: order with respect to [ICl] = 1
   overall order = 1 + 1 = 2

8. d. compare 4 with 1: (could compare 4 with any one of 1, 2, or 3):
   initial rate of formation of $I_2 = 3.7 \times 10^{-7} \times (4.7/1.5)^1 \times (2.7/1.5)^1 = 2.1 \times 10^{-6}$ mol L^{-1} s^{-1}

9. a. reactants are N_2 and H_2
   intermediates are FeN_2 and FeH_2
   catalyst is Fe
   product is ammonia NH_3

10. e.

11. a.

12. g. the rate is proportional to k and $k = A\ e^{-E_A/RT}$
comparing the two rates: $k_2/k_1 = e^{-E_{A2}/RT} - e^{-E_{A1}/RT} = e^{-(E_{A2}-E_{A1})/RT}$
$E_{A1} = 15{,}000$ J and $E_{A2} = 55{,}000$ J, so the ratio $k_2/k_1 = 10{,}000{,}000$

13. d.

14. h. KOH is a strong base, so $[OH^-] = 0.220 \times (10/250) = 8.8 \times 10^{-3}$ M
pOH = 2.06 and pH = 11.94

15. b. compare with HCl *(strong)* and then the series:
$HClO_4$ *(strong)*; $HClO_3$ *(strong)*; $HClO_2$ and HClO *(getting progressively weaker)*

16. c. redox equation

17. f. $K = 12.0 = (3.0) \times [HCl]^2 / (1.0)^2 \times (1.0)$ so $[HCl]^2 = 12.0 / 3.0 = 4.0$
[HCl] = 2.0 M; the number of moles in 2.0 L = 4.0 mol

18. e.

19. c.

20. c. NaH_2AsO_4 $pH = ½(pK_{a1} + pK_{a2}) = 5.43$ therefore acidic
Na_2HAsO_4 $pH = ½(pK_{a2} + pK_{a3}) = 9.89$ therefore basic
Na_3AsO_4 must be basic

21. e. $K = K_{a2}/K_{a1} = 5.6 \times 10^{-8} / 2.5 \times 10^{-4} = 2.2 \times 10^{-4}$

22. d. point C: $H_2AsO_4^-$ and $HAsO_4^{2-}$

23. g. HI and NaI —HI is strong—the acid or base used in a buffer must be weak

24. e.

25. g. $K_{neut} = K_a / K_w = 3.50 \times 10^{-4} / 1.0 \times 10^{-14} = 3.50 \times 10^{10}$

EXAMINATION 1

CHEMISTRY 142 FALL 2007

Solutions

1. c. 4 moles of methylhydrazine produce 25 moles of gas products (9 + 12 + 4)
   5000 liters of gas products would require 5000 liters × (4 mol CH_3NHNH_2 / 25 mol products)
   = 800 L CH_3NHNH_2 per second

2. e. sum of exponents = 2 – 1 = 1

3. a. 36.0 days = 3 half-lives, so 264 → 132 → 66 → 33 mg

4. j. second-order half-lives double each half-life
   144 mol L^{-1} to 72 mol L^{-1} to 36 mol L^{-1} is two successive half-lives:
   first half-life = 10 min; second half-life = 20 min
   36 mol L^{-1} to 18 mol L^{-1} to 9 mol L^{-1} is two more successive half-lives:
   third half-life = 40 min; fourth half-life = 80 min; total = 120 min = 2 hours

5. h. compare 1 and 2: order with respect to [A] = 1
   compare 2 and 4: order with respect to [B] = 2
   overall order = 1 + 2 = 3

6. c. more energy available in the collision

7. g. In the energy profile for a reaction shown,
   A is the energy of the reaction,
   D is the energy of the activated complex,
   F is the energy of the products, and
   B is the activation energy for the forward reaction.

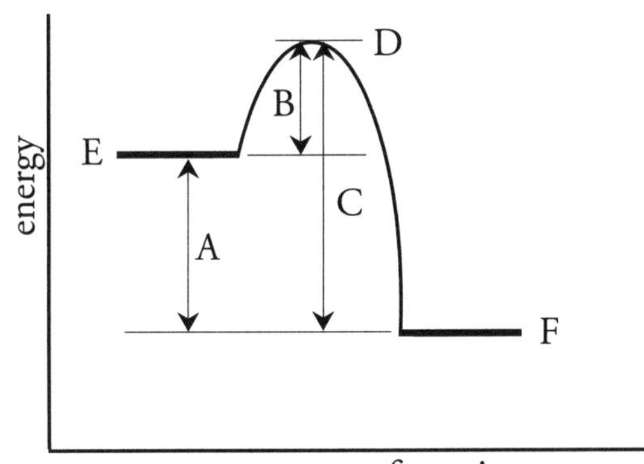

8. e. note both Br and H are intermediates

9. d. a catalyst changes the route (mechanism) of a reaction

10. h. $K = [SO_3]^2 / [SO_2]^2[O_2] = (1.5)^2 / (0.5)^2(2.0) = 4.5$

11. h.

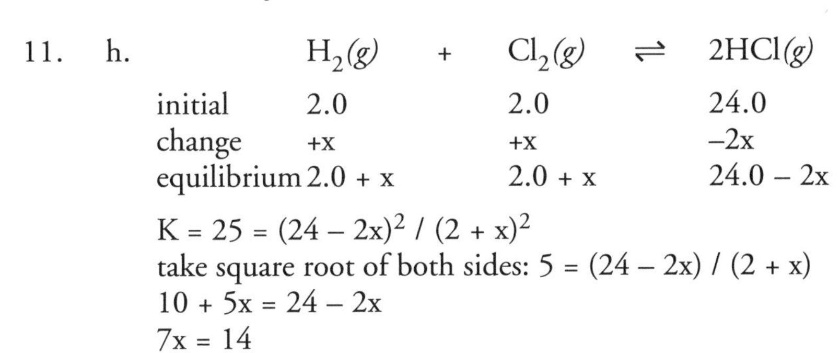

	$H_2(g)$	+ $Cl_2(g)$	⇌ $2HCl(g)$
initial	2.0	2.0	24.0
change	+x	+x	–2x
equilibrium	2.0 + x	2.0 + x	24.0 – 2x

$K = 25 = (24 - 2x)^2 / (2 + x)^2$
take square root of both sides: $5 = (24 - 2x) / (2 + x)$
$10 + 5x = 24 - 2x$
$7x = 14$
$x = 2$

12. a. $PCl_5 \rightleftharpoons PCl_3 + Cl_2$ $\Delta n = +1$

13. j. HNO_2 is a weak acid

14. b. reverse the first equation and divide by 2: (reciprocal *and* square root of K_1) $2V + W \rightleftharpoons R + S$ $K = 1/5$

reverse the second equation: (reciprocal of K_2) $2Q + P \rightleftharpoons W + 2V$ $K = 1/0.040$

add to obtain desired equation: $P + 2Q \rightleftharpoons R + S$ $K = 1 / (5 \times 0.040)$

15. e.

16. f.

17. e. K for the neutralization = $K_a \times K_b / K_w = 6.8 \times 10^{-4} \times 1.8 \times 10^{-5} / (1.0 \times 10^{-14}) = 1.2 \times 10^6$

18. d.

19. c. slightly acidic
recall that water itself produces a concentration of H_3O^+ of 10^{-7} M in pure water

20. b. K_{a1} is too large to neglect the x in the denominator, so solve by iteration
initially neglect the x, and $K_{a1} = 1.3 \times 10^{-2} = x^2 / (0.10)$ $x = 0.036$
make the denominator $0.10 - x = 0.064$ $K_{a1} = 1.3 \times 10^{-2} = x^2 / (0.064)$ $x = 0.029$
make the denominator $0.10 - x = 0.071$ $K_{a1} = 1.3 \times 10^{-2} = x^2 / (0.071)$ $x = 0.030$
make the denominator $0.10 - x = 0.070$ $K_{a1} = 1.3 \times 10^{-2} = x^2 / (0.070)$ $x = 0.030$
$x = [H_3O^+]$, $pH = -\log_{10}(0.030) = 1.52$

21. d. halfway to the equivalence point, [conjugate base] = [acid], so $pH = pK_{a1} = 1.89$

22. d. $pH = pK_{a1} + \log_{10}(3/1) = 1.89 + 0.477 = 2.37$

23. h. solution of HSO_3^-: $pH = ½(pK_{a1} + pK_{a2}) = ½(1.89 + 7.20) = 4.55$

24. g. halfway to the equivalence point, [conjugate base] = [acid], so $pH = pK_{a2} = 7.20$

25. h. $pH = pK_{a2} + \log_{10}(2/1) = 7.20 + 0.301 = 7.50$

EXAMINATION 1

CHEMISTRY 142 SPRING 2008

Solutions

1. c. PH_3 is *used* at 4 × the rate at which P_4 is formed, i.e. $-4\,x$
   H_2 is *produced* at 6 × the rate at which P_4 is formed, i.e. $6\,x$

2. d. I the activation energy does *not* depend upon the temperature
   II true
   III true—this is the predominant reason why reaction rates increase with increase in temperature
   IV catalysts *change* the mechanism

3. f.

4. g. second-order half-lives double each half-life
   16 mol L^{-1} to 8 mol L^{-1} to 4 mol L^{-1} is two successive half-lives:
   first half-life = 100 sec; second half-life = 200 sec
   4 mol L^{-1} to 2 mol L^{-1} to 1 mol L^{-1} is two more successive half-lives:
   third half-life = 400 sec; fourth half-life = 800 sec; total = 1200 sec = 20 min

5. b. compare 2 and 3: the $[NO_2^-]$ increases by 1.5 and the rate increases by 1.5
   order with respect to $[NO_2^-] = 1$
   compare 1 and 2: the $[NH_4^+]$ decreases by 2 and the rate decreases by 2
   order with respect to $[NH_4^+] = 1$
   Rate = $k\,[NH_4^+][NO_2^-]$

6. d. reactants: NO and H_2
   products: N_2 and H_2O
   intermediates are formed but then used in a subsequent step: N_2O_2 and N_2O

7. d. slowest step: Rate = $k_2\,[N_2O_2][H_2]$
   but from first step: $K = k_1/k_{-1} = [N_2O_2]/NO]^2$, so $[N_2O_2] = K\,[NO]^2$
   rate law: Rate = $k_2K\,[NO]^2[H_2] = k\,[NO]^2[H_2]$

8. e. for a first order decomposition, the half-life $t_{1/2}$ is constant,
   the change from 5.00 atm to 1.25 atm is two half-lives = 72 seconds
   $t_{1/2}$ = 36 seconds

9. e.

10. a. acidic: NH_4Cl, CH_3CO_2H, H_2SO_4
    basic: KCN, KOH, CH_3NH_2, KCH_3CO_2
    neutral: LiBr, KCl, $NaNO_3$

11. a. $Q = [CO][H_2]^3 / [H_2O][CH_4] = (0.15)(0.30)^3 / (0.035)(0.050) = 2.31$
Q is less than K (4.7), so the system moves to the right to increase [product]

12. g. $pOH = -\log_{10}[OH^-] = -\log_{10}(1.5 \times 10^{-4}) = 3.82$
$pH = 14 - pOH = 10.18$

13. a. I moves system to the right
II exothermic reaction, so moves system to the left
III left side occupies a larger volume (3 moles of gas), so moves system to the left

14. f. HN_3 acid and N_3^- conjugate base
H_2O base and H_3O^+ conjugate acid

15. a. $HCO_2H + H_2O \rightleftharpoons H_3O^+ + HCO_2^-$

$$K_a = \frac{[H_3O^+][HCO_2^-]}{[HCO_2H]} = \frac{x^2}{(0.2 - x)}$$

K_a is perhaps too large to neglect the x in the denominator, so check by iteration
initially neglect the x, and $K_a = 1.8 \times 10^{-4} = x^2 / (0.20)$ $x = 0.006$
make the denominator $0.20 - x = 0.194$ $K_a = 1.8 \times 10^{-4} = x^2 / (0.194)$ $x = 0.0059$
near enough...so
$x = [H_3O^+]$, $pH = -\log_{10}(0.0059) = 2.23$

16. c. HNO_2 is the stronger acid (K_a is larger), so its conjugate base is weaker

17. e. [conjugate base] = 0.022 M; [acid] = 0.066M
$pH = pK_a + \log_{10}([\text{conjugate base}]/[\text{acid}]) = 3.74 + \log_{10}(0.022/0.066) = 3.74 - 0.48 = 3.27$

18. d.

hydrolysis of Bz^-:	$Bz^- + H_2O \rightleftharpoons HBz + OH^-$	$K = K_w / K_a$
hydrolysis of NH_4^+:	$NH_4^+ + H_2O \rightleftharpoons H_3O^+ + NH_3$	$K = K_w / K_b$
ionization of water:	$H_3O^+ + OH^- \rightleftharpoons 2\,H_2O$	$K = 1 / K_w$
add:	$Bz^- + NH_4^+ \rightleftharpoons NH_3 + HBz$	$K = K_w / K_a \times K_b$

$K = (1 \times 10^{-14})/(6.5 \times 10^{-5})(1.8 \times 10^{-5}) = 8.55 \times 10^{-6}$

19. d. HBr is a strong acid

20. e. $HSeO_3^-$; the pH at point 2 is the average of the two pK_as = 4.38

EXAMINATION 1

CHEMISTRY 142 FALL 2008

Solutions

1. h. the rates of consumption and formation are proportional to coefficients in the balanced equation:
   coefficient for A: 0.172 / 0.086 = 2
   coefficient for B: 0.258 / 0.086 = 3
   coefficient for C: 0.086 / 0.086 = 1 therefore $2\,A + 3\,B \rightarrow C$

2. h. compare 1 and 2 to find the order with respect to mercury(II) chloride:
   $1.8 \times 10^{-5} / 7.1 \times 10^{-5} = 0.15^x / 0.3^x$ $x = 2$ —second order with respect to $HgCl_2$

   compare 2 and 3 to find the order with respect to the oxalate ion:
   $7.1 \times 10^{-5} / 3.5 \times 10^{-5} = 0.105^y/0.052^y$ $y = 1$ —first order with respect to $C_2O_4^{2-}$

   overall reaction order is 2 + 1 = 3

3. d. $k = 0.693 / 14.3\ \text{days} = 0.04846\ \text{day}^{-1}$
   $\ln([^{32}P] / 298\ \text{mg}) = -0.04846\ \text{day}^{-1} \times 23\ \text{days}$ $^{32}P = 97.8\ \text{mg}$

4. b. 12.0M to 6.0M takes 40 minutes
   6.0M to 3.0M takes 20 minutes
   3.0M to 1.5M takes 10 minutes
   1.5M to 0.75M takes 5 minutes
   3.0M to 0.75M will take 10 minutes + 5 minutes = 15 minutes

   alternative method:
   12.0M to 3.0M (decr of 9.0M) takes 1 hr
   so 3.0M to 0.75M (decr of 2.25M) takes
   1 hr × (2.25/9.0) = 15 minutes

5. g. II, III, and IV

6. e. $\ln(k_1/k_2) = -E_A / R \times (T_1^{-1} - T_2^{-1})$
   $\ln(0.0011\ \text{Lmol}^{-1}\text{s}^{-1} / 1.67\ \text{Lmol}^{-1}\text{s}^{-1}) = -E_A / 8.3145\ \text{JK}^{-1}\text{mol}^{-1} \times ((838\ \text{K})^{-1} - (1053\ \text{K})^{-1})$
   $E_A = 250\ \text{kJ}$

7. a. Start with the slow elementary step: Rate = k [Cl][$CHCl_3$]
   From the fast equilibrium: $K = [Cl]^2 / [Cl_2]$ rearrange: $[Cl] = \sqrt{K[Cl_2]}$
   Substitute into rate equation: Rate = $k \sqrt{K[Cl_2]}\ [CHCl_3]$
   Simplify: Rate = $k\ [Cl_2]^{1/2}\ [CHCl_3]$

8. b. Cl and CCl_3

9. f. $K_c = [ClF_3]^2 / [F_2]^3 \times [Cl_2]$ so $K_c = 3.0^2 / (2.0^3 \times 1.5) = 0.75$

10. d. reverse both equations, take the reciprocal of K:
    $2\,NO(g) + Br_2(g) \rightarrow 2\,NOBr(g)$ K = 1 / 0.014 = 71.429
    $2\,BrCl(g) \rightarrow Br_2(g) + Cl_2(g)$ K = 1 / 7.2 = 0.1389
    Add the equilibria, multiply K:
    $2\,NO(g) + 2\,BrCl(g) \rightarrow 2\,NOBr(g) + Cl_2(g)$ K = 71.43 × 0.1389 = 9.92

11. f.

	$H_2(g)$ +	$I_2(g)$ ⇌	$2\,HI(g)$	K = 54.3
initial:	1.0M	1.0M	6.0M	$Q = 6.0^2 / (1.0 \times 1.0) = 36$
change:	–x	–x	+2x	shift to the right to reach equil.
equilibrium:	1.0 – x	1.0 – x	6.0 + 2x	

$K = 54.3 = (6.0 + 2x)^2 / (1.0 - x)^2$ $x = 0.146$ and $[H_2] = 1.0 - 0.146 = 0.854$ M

12. j. acetic acid is a weak acid

13. a. $Sr(OH)_2 \rightarrow Sr^{2+} + 2\,OH^-$
$pOH = -\log(2 \times 0.00213) = 2.37$
$pH = 14 - 2.37 = 11.63$

14. f. H^+, Co^{3+}, SbF_5, BF_3, Ag^+

15. e. OBr^-, base

16. b. $H_2PO_4^-$ is the conjugate base of H_3PO_4
$K_b = K_w / K_{a1} = (1 \times 10^{-14}) / (7.1 \times 10^{-3}) = 1.4 \times 10^{-12}$

17. c.

	$CH_3CO_2^-$ + H_2O ⇌	CH_3CO_2H +	OH^-	$K_b = K_w/K_a$
initial:	0.15	0	0	$= 5.56 \times 10^{-10}$
change:	–x	+x	+x	
equilibrium:	0.15 – x	x	x	

$K_b = 5.56 \times 10^{-10} = x^2 / 0.15 - x$ $x = 9.13 \times 10^{-6}$ $pOH = -\log(9.13 \times 10^{-6}) = 5.04$
$pH = 14 - 5.04 = 8.96$

18. e. when $[^+H_3NCH_2CO_2^-] = [H_2NCH_2CO_2^-]$
$pH = pK_{a2}$ $pH = -\log(2.51 \times 10^{-10}) = 9.60$

when $[^+H_3NCH_2CO_2^-] = 0.1M$, all of the first H^+ has been removed (the first equivalence pt)
$pH = (pK_{a1} + pK_{a2}) / 2 = (2.34 + 9.60) / 2 = 5.97$

19. f. $pH = pK_{a2} + \log([\text{conjugate base}] / [\text{acid}]) = 9.60 + \log(0.025 / 0.075) = 9.12$

20. c. $NH_3 + H_2O \rightleftharpoons NH_4^+ + OH^-$ $K_b = 1.8 \times 10^{-5}$

0.15M 0.35M ??

$K_b = 1.8 \times 10^{-5} = (0.35 \times [OH^-]) / 0.15$ $[OH^-] = 7.714 \times 10^{-6}$
$pOH = -\log(7.714 \times 10^{-6}) = 5.11$
$pH = 14 - 5.11 = 8.89$

21. b. Addition of a strong acid will decrease the pH of the solution. Because you have more moles of both HF and KF in the solution than moles of acid added, the buffer capacity of the solution is not exceeded and the pH will only decrease slightly.

22. g. $K_{neut} = K_a \times K_b / K_w = (6.2 \times 10^{-10}) \times (1.8 \times 10^{-5}) / (1 \times 10^{-14}) = 1.12$

EXAMINATION 1

CHEMISTRY 142 SPRING 2009

Solutions

1. f. the decreasing concentration, C, is the reactant (NO_2)
   the two increasing concentrations, A and B, are the products
   but NO is formed at 2 × the rate at which O_2 is formed

2. e. the half-life $t_{½}$ is constant at 315 s; first order
   $k = 0.693 / t_{½} = 2.2 \times 10^{-3}\ s^{-1}$

3. e. overall order = 2 – 1 = 1
   the exponent is negative—increasing the $[O_2]$ will decrease the rate

4. e. B C 1 exothermic

5. i. second-order half-lives double each half-life
   96 mol L^{-1} to 48 mol L^{-1} to 24 mol L^{-1} is two successive half-lives:
   first half-life = 10 min; second half-life = 20 min
   24 mol L^{-1} to 12 mol L^{-1} to 6 mol L^{-1} to 6 mol L^{-1} is three more successive half-lives:
   third half-life = 40 min; fourth half-life = 80 min; fifth half-life = 160 min
   total = 10 + 20 + 40 + 80 + 160 min = 310 min

6. d. $\ln (k_1/k_2) = -E_A/R\ (1/T_1 - 1/T_2)$
   $\ln \{(20.1 \times 10^{-4})/(2.64 \times 10^{-4})\} = -(E_A/8.3145) \times (1/323.15 - 1/298.15)$
   $\ln (20.1/2.64) = +(E_A/8.3145) \times 25/(323.15 \times 298.15)$
   $E_A = 65{,}000\ J\ mol^{-1} = 65\ kJ\ mol^{-1}$

7. b. original concentration of $HClO_2$ = 0.60 M
   $[H_3O^+] = 7.91 \times 10^{-2}$ M
   equilibrium concentration of $HClO_2 = 0.60 - 7.91 \times 10^{-2}$ M = 0.521 M
   $K_a = [H_3O^+]^2/[HClO_2] = (7.91 \times 10^{-2})^2 / 0.521 = 1.20 \times 10^{-2}$

8. b. it helps to add the two equations to obtain the overall equation

reactants:	H_2O_2
products:	H_2O and O_2
catalysts used and then reformed:	Br_2
intermediates formed but then used:	H^+ and Br^-

9. c.

10. c. LiCl is neutral (7.0); $NaCH_3CO_2$ is basic (> 7.0); NH_4I is acidic (< 7.0); KSCN is basic (> 7.0

11. h. the three bases are IO^-, $CH_3CH_2CO_2^-$, and $C_6H_5NH_2$
equilibria lie on the weaker side, in these equations on the right side since $K > 1$

12. e. look for a K_a close to 10^{-5}; or a pK_a close to 5.0
the acetate buffer is closest; $pK_a = 4.75$
the buffer will have more base than acid present to raise the pH to 5.0

13. b. $Q_p = (22)^2 / (65)(36)^3 = 1.6 \times 10^{-4}$ which is less than K_p
the system will shift to the right (the product side) to increase the value of Q_p to K_p

14. e.

add N_2	right
reduce volume	right
add catalyst	no effect other than to reach equilibrium more quickly
remove H_2	left

15. b. remove a H^+ to form conjugate base: CO_3^{2-}
$K_b = K_w / K_a = 1.8 \times 10^{-4}$

16. e. $K_{neut} = 1 / K_b = K_a / K_w = 2.2 \times 10^{-5} / 1.0 \times 10^{-14} = 2.2 \times 10^9$

17. c. reverse the equation and divide by 2
so take the reciprocal of K and then the square root
$K = 6.3 \times 10^{-6}$

18. c. $\Delta n = -1$, the number of moles decrease by 1 as reactants go to products

19. d. all have lone pairs of electrons available: H_2O, NH_3, F^-

20. d. Henderson–Hasselbalch: $pH = pK_a + \log_{10}([base]/[acid])$
$pH = 7.46 + \log_{10}(0.40/0.50)$
$pH = 7.36$

21. d. $H_2C_2O_4$ and $HC_2O_4^-$; $pH = pK_{a1} = 1.19$

22. e. in the distribution diagram,
point (1) is reached after 25 mL NaOH (halfway to first neutralization)
the first neutralization point (2) is reached after 50 mL NaOH
another 25 mL would reach point (3)
at point 3, the two species present are $HC_2O_4^-$ and $C_2O_4^{2-}$

EXAMINATION 1

CHEMISTRY 142 FALL 2009

Solutions

1. c. ammonia is used at 4/6 the rate at which water is formed.
so the NH_3 will be used at $30 \times (4/6)$ mol L^{-1} min^{-1} = 20 mol L^{-1} min^{-1}

2. d. compare the two experiments to find the order with respect to N_2O_5
$5.45 \times 10^{-5} / 1.35 \times 10^{-5} = 3.15^x / 0.78^x$
x = 1, first order with respect to N_2O_5 and first order overall

3. c. $k = 0.693 / 56.3$ minutes = 0.01231 min^{-1}
$\ln([(CH_2)_2O] / 5.0\ g) = -0.01231\ min^{-1} \times 4.0\ hr \times (60\ min / 1\ hr)$
$[(CH_2)_2O] = 0.26\ g$

4. h. 0.010 M to 0.005 M takes 1630 seconds
0.005 M to 0.0025 M takes 3260 seconds
0.0025 M to 0.00125 M takes 6520 seconds
0.005 M to 0.00125 M takes 3260 seconds + 6520 seconds = 9780 seconds

5. h. A, B, and D

6. f. $\ln(k_1/k_2) = -E_A / R \times (T_1^{-1} - T_2^{-1})$
$\ln(4.5 \times 10^3\ s^{-1} / 1 \times 10^4\ s^{-1}) = -E_A / 8.3145\ JK^{-1}mol^{-1} \times ((274\ K)^{-1} - (283\ K)^{-1})$
$E_A = 57$ kJ

7. a. Start with the slow elementary step: Rate = k $[NH_3][HOCN]$
From the fast equilibrium: K = $[NH_3][HOCN] / [NH_4OCH]$
Rearrange: $[HOCN]$ = K $[NH_4OCN] / [NH_3]$
Substitute into rate equation: Rate = k $[NH_3]$ K $[NH_4OCN] / [NH_3]$
Simplify: Rate = k $[NH_4OCN]$

8. c. add the two equations together and cancel out the intermediate Cl(g)
$2\ NO_2Cl(g) \rightarrow 2\ NO_2(g) + Cl_2(g)$

9. b. $K_c = [NH_3]^2 / ([N_2] \times [H_2]^3)$
$= 0.002^2 / (0.10 \times 0.10^3) = 0.04$

10. e. reverse the equation and multiply by 1/2
so take the reciprocal of K and then the square root
K = 7.12

11. f.

increase temperature	shift left
decrease volume	no change
remove Fe(s)	no change
add a catalyst	no change (equilibrium will be established faster)

12. a.

	$H_2(g)$	+ $Br_2(g)$	$\rightleftharpoons$ $2\ HBr(g)$	$K = 2.18 \times 10^6$
initial:	0	0	0.26667 M	
change:	+x	+x	–2x	
equilibrium:	x	x	0.26667 – 2x	

$K = 2.18 \times 10^6 = (0.26667 - 2x)^2 / x^2$

$x = 1.8 \times 10^{-4} = [H_2]$

13. e. II and III

14. g. $[LiOH] = [OH^-] = 0.013$ M

$pOH = -\log(0.013) = 1.89$

$pH = 14 - pOH = 14 - 1.89 = 12.1$

15. e. F^-, NH_3, CO, and H_2O

16. c. $H_2BO_3^-$ is the conjugate base of H_3BO_3

$K_b = K_w / K_{a1} = (1 \times 10^{-14}) / (5.8 \times 10^{-10}) = 1.7 \times 10^{-5}$

17. b. NH_4NO_3 is acidic, KCH_3CO_2 is basic, HClO is acidic, LiCl is neutral

18. f.

	CN^-	+ H_2O $\rightleftharpoons$	HCN	+ OH^-	$K_b = K_w / K_a = 1.6 \times 10^{-5}$
initial:	0.3		0	0	
change:	–x		+x	+x	
equilibrium:	0.3 – x		x	x	

$K_b = 1.6 \times 10^{-5} = x^2 / 0.3 - x$

$x = 0.00219 = [OH^-]$

$pOH = -\log(0.00219) = 2.66$

$pH = 14 - 2.66 = 11.3$

19. d. when $[HCO_3^-] = 0.2$ M, all of the first H^+ has been removed (the first equivalence point)

$pH = (pK_{a1} + pK_{a2}) / 2$

$= (-\log(4.3 \times 10^{-7}) + -\log(5.6 \times 10^{-11})) / 2$

$= 8.31$

20. e. when $[HCO_3^-] = [CO_3^{2-}]$, $pH = pK_{a2} = -\log 5.6 \times 10^{-11} = 10.25$

21. h. $pH = pKa + \log([\text{conjugate base}] / [\text{acid}]) = -\log 1.8 \times 10^{-4} + \log(1.4 / 0.85) = 3.96$

22. a. strongest acid = largest K_a = thiocyanic acid

strongest conjugate base = weakest acid = smallest K_a = hypoiodous acid

EXAMINATION 1

CHEMISTRY 142 SPRING 2010

Solutions

1. e. products increase in concentration; & the coefficient for H_2O is greater than the coefficient for CO_2 similarly the reactants decrease in concentration; O_2 faster than propane by 5:1

2. g. the order is the exponent 2
the rate would increase by $3^2 = 9$ if the concentration of hypochlorite is tripled

3. c. 144 mol L^{-1} to 36 mol L^{-1} is two half-lives
36 mol L^{-1} to 18 mol L^{-1} is one more half-life = 16 minutes
18 mol L^{-1} to 9 mol L^{-1} is one more half-life = 60 – 12 – 16 = 32 minutes
the half lives are not the same; in fact, the half-lives are doubling (compare 16 min and 32 min) so second order

4. b. compare 1 and 2 to find the order with respect to [R] ([Q] is constant):
[R] doubles and so does the rate, so first order

compare 2 and 4 to find the order with respect to [Q] ([R] is constant):
[Q] doubles and the rate increases by 4,

$2^2 = 4$ so second order; and third order overall

5. f. use the Arrhenius equation: $\ln 2 = -(E_A/R)(1/645.15 - 1/623.15)$
$0.693 = -(E_A/8.3145)(1/645.15 - 1/623.15)$
$E_A = 105{,}000\ J = 105\ kJ$

6. d. use the integrated rate equation for a first order reaction:
$\ln(0.25/1) = -k\ t = -k \times 78$ (78 is the time taken in minutes)
$k = 0.01777$
$t_{½} = 0.693/0.01777 = 39$ min

7. b. more molecules have sufficient energy to reach the activated complex

8. b. reaction intermediates are formed in an early step and used in a subsequent step
they are neither reactants or products
N_2O and N_2O_2

9. c. substitute the equilibrium concentrations in the expression for the equilibrium constant
$K = [ClF_3]^2/([F_2]^3 \times [Cl_2]) = 6^2/(2^3 \times 3) = 1.5$

10. c. $K_p = K_c\ (RT)^{\Delta n} = K_c\ (8.3145 \times 298.15)^{-2}$
so K_p is smaller than K_c

11. a.

	2 HBr	$\rightleftharpoons$	H_2	+	Br_2	$K_c = 0.0081$
initial:	0.040		0		0	
change:	–2x		+x		+x	
equilibrium:	0.040 – 2x		x		x	

$K_c = x^2 / (0.040 - 2x)^2 = 0.0081$

take square root of both sides and solve for x = 0.0031

12. g.

13. c. remove an H^+; so HPO_4^{2-}

14. g. a solution is neutral when $[H_3O^+] = [OH^-]$

15. b. $K_w = [H_3O^+] \times [OH^-] = 3.8 \times 10^{-14}$; the pure water is neutral and $[H_3O^+] = [OH^-]$
$[H_3O^+] = \sqrt{3.8 \times 10^{-14}} = 1.95 \times 10^{-7}$
$pH = -\log_{10}[H_3O^+] = 6.71$

16. e. $K_b = [OH^-]^2 / 0.05 = 1.7 \times 10^{-9}$
$[OH^-] = \sqrt{8.5 \times 10^{-11}} = 9.22 \times 10^{-6}$
$pOH = -\log_{10}[OH^-] = 5.04$
$pH = 14 - pOH = 8.96$

17. e. H_2SO_4, HNO_3, $Mg(OH)_2$, HI, and NaOH

18. e.

19. c. reverse the first equation and add to the second equation to get the desired equation
the K for the autoionization of the HSO_3^- = $K_{a2}/K_{a1} = 4.85 \times 10^{-6}$

20. d. the average of the two $pK_a = 4.55$

21. e. a weak acid or base and its conjugate partner: HCN and KCN

22. b. $K_{neut} = K_a \times K_b / K_w = 4.9 \times 10^{-10} \times 1.8 \times 10^{-5} / 1.0 \times 10^{-14} = 8.82 \times 10^{-1}$

EXAMINATION 1

CHEMISTRY 142 FALL 2010

Solutions

1. a. $\Delta[ClO^-]/\Delta t = -5.4$ mol L^{-1} min^{-1} $\Delta[ClO_3^-]/\Delta t = 1.8$ mol L^{-1} min^{-1}

2. d. compare 1 and 3:
   $[O_2]$ constant, [NO] increases by × 2, the rate increases by × 4
   the order with respect to [NO] is 2
   compare 2 and 3:
   [NO] constant, $[O_2]$ decreases to ½, and the rate also decreases to ½
   the order with respect to $[O_2]$ is 1
   the rate equation for this reaction is Rate = $k[NO]^2[O_2]$

3. c. the rate constant k $= 0.693/t_{½}$
   $= 0.693\ /(3.92 \times 10^4)\ s^{-1}$
   $= 1.8 \times 10^{-5}\ s^{-1}$

4. c. 18.9 hours

5. h. 65.2 kJ mol^{-1}

6. e. reactants are NO_2 and CO
   products are NO, CO_2
   the intermediate is NO_3

7. b. the slow step is the first step, and the rate law is the rate equation for this first step
   Rate = $k[NO_2]^2$

8. g.

9. d. $K_c = 6.25 \times 10^{-3} = [H_2O]^2[O_2]\ /\ [H_2O_2]^2$
   $[O_2] = 1.56 \times 10^{-3}$ M

10. d. leave equation 1 the same
    multiply equation 2 by 2 and reverse
    add to obtain the desired equation
    therefore $K_3 = K_1/K_2^2$

11. a. Q = 0.248 and Q is less than K, so to reach equilibrium, reactants must be converted to products

12. a. $\Delta n = -1$ and $K_p = K_c(RT)^{\Delta n} = K_c(0.08206 \times 298)^{\Delta n} = K_c(24.5)^{-1} = K_c/24.5$
    K_p is smaller than K_c.

13. f.

14. b. $CH_3NH_2(aq)$ + $H_2O(l)$ ⇌ $CH_3NH_3^+(aq)$ + $OH^-(aq)$
base acid acid base

15. e. all except I

16. f. K < 1, and an equilibrium lies on the weaker side, in this case the reactant side, so...
HBrO*(aq)* is weaker than HSCN*(aq)*

17. f. 0.020 M NaOH excess, but the solution is diluted by a factor of 2 (from 500 mL to 1000mL)
$[OH^-] = 0.010$ M; pOH = 2.00; pH = 12.00

18. b. $K_a = 6.5 \times 10^{-5}$.= $[H^+][Bz^-] / [HBz] = [H^+]^2 / (0.25)$
$[H^+] = 4.03 \times 10^{-3}$ M
pH = 2.39

19. e. Na_2SO_3 basic; $HClO_4$ acidic; KNO_3 neutral; LiF basic

20. d. $pK_{a1} = 6.37$
$pK_{a2} = 10.25$
average = 8.31

21. c. $NaCH_3CO_2$ and CH_3CO_2H
KCN and HCN

22. g. $K_b = 1.8 \times 10^{-5}$ for NH_3
$K_a = (K_w/K_b) = (1.0 \times 10^{-14}/1.8 \times 10^{-5}) = 5.56 \times 10^{-10}$ for NH_4^+
$pK_a = 9.26$
$pH = pK_a + \log(\text{base/acid})$
$= 9.26 + \log(0.50/0.40)$
$= 9.26 + 0.10$
$= 9.36$

EXAMINATION 1

CHEMISTRY 142 SPRING 2011

Solutions

1. f. N_2O_5 is consumed at half the rate, i.e. 4.0 mol $L^{-1}min^{-1}$
and O_2 is produced at one quarter the rate, i.e. 2.0 mol $L^{-1}min^{-1}$

2. b. consider expts 1 and 2: the concentration of SO_3 decreases by a factor of 9
the rate increases by a factor of 3
$3 = (1/9)^{-½}$
the order with respect to SO_3 is –½

3. h. 128 mol L^{-1} to 32 mol L^{-1} is two half-lives, which = 18 min

first half-life	to 64 mmol L^{-1}	= 6 min	
second half-life	to 32 mmol L^{-1}	= 12 min	18 min total
third half-life	to 16 mmol L^{-1}	= 24 min	42 min total
fourth half-life	to 8 mmol L^{-1}	= 48 min	90 min total = 1.50 hr
fifth half-life	to 4 mmol L^{-1}	= 96 min	186 min total
sixth half-life	to 2 mmol L^{-1}	= 192 min	378 min total = 6 hr 18 min

4. d. first step is the rate-determining step
the overall rate depends upon the rate of this step

5. f. use the Arrhenius equation:

$$\ln(6.0 \times 10^{-5}/2.4 \times 10^{-6}) = -(E_A/8.3145)(1/630 - 1/575)$$
$$\ln(25) = -(E_A/8.3145)(-55/(630 \times 575))$$
$$3.219 = 1.826 \times 10^{-5}\ E_A$$
$$E_A = 176{,}280\ J = 176\ kJ$$

6. f.

7. c.

8. b. $K = (2.0^2 \times 4.0) / 3.0^3 = 16/27 = 0.59$

9. f. reverse equation 2 and multiply by 2, add to equation 1
$K_3 = K_1 / K_2^2$

10. c. $\Delta n = 7 - 6 = +1$
$K_p = K_c(RT)^{\Delta n} = K_c(0.08206 \times 298)^{\Delta n} = K_c \times 14.5$ — K_p is numerically larger

11. d.

12. f. HI is the only strong acid present

13. a. the pH decreases but the water remains neutral: $[H_3O^+] = [OH^-]$

14. c. the Lewis definitions characterize acid-base reactions in terms of the movement of electron pairs

15. c. the missing component is SO_4^{2-}, which acts as a base

16. e. pH = 4.85, therefore $[H_3O^+] = 1.41 \times 10^{-5}$
$K_a = [H_3O^+]^2 / 0.10 = 2.0 \times 10^{-9}$
$pK_a = -\log_{10}K_a = 8.70$

17. b.

18. e. salt of a strong acid (HNO_3) and a strong base (KOH)

19. f. $K_{neut} = 1/K_b = K_a/K_w = (3.5 \times 10^{-4})/(1.0 \times 10^{-14}) = 3.5 \times 10^{10}$

20. d. $H_2PO_3^- + H_2PO_3^- \rightleftharpoons H_3PO_3 + HPO_3^{2-}$ $K = K_{a2} / K_{a1} = 2.6 \times 10^{-7} / 1.0 \times 10^{-2}$

21. f. $pH = ½(pK_{a1} + pK_{a2}) = ½(2.00 + 6.59) = 4.30$

22. d. Na_2CO_3 and $KHCO_3$
NaF and HF
KCN and HCN
NaH_2PO_4 and Na_2HPO_4

EXAMINATION 2

CHEMISTRY 142 SPRING 2006

Solutions

1. c. $K_{sp} = [La^{3+}][IO_3^-]^3 = 27[La^{3+}]^4 = 6.1 \times 10^{-12}$
   $[M^{3+}] = 6.9 \times 10^{-4}$ M

2. b. decrease—the common ion suppresses the solubility

3. e. $[IO_3^-] = 2.0 \times 10^{-2}$ M
   $K_{sp} = [La^{3+}][IO_3^-]^3 = [La^{3+}][2.0 \times 10^{-2}]^3 = 6.1 \times 10^{-12}$
   $[La^{3+}] = 7.6 \times 10^{-7}$ M

4. e. In the one-liter mixture:
   $[Ba^{2+}] = 1.0 \times 10^{-5}$ M (diluted to ½)
   $[SO_4^{2-}] = 7.5 \times 10^{-6}$ M
   $Q = [Ba^{2+}][SO_4^{2-}] = 1.0 \times 10^{-5} \times 7.5 \times 10^{-6} = 7.5 \times 10^{-11}$
   Q is less than K_{sp}, and therefore no precipitation occurs.

5. c. $PbCl_2(s) \rightleftharpoons Pb^{2+} + 2\,Cl^-$ $\quad K_{sp} = 1.7 \times 10^{-5}$
   $PbF_2(s) \rightleftharpoons Pb^{2+} + 2\,F^-$ $\quad K_{sp} = 3.6 \times 10^{-8}$
   Reverse second equation and add:
   $PbCl_2(s) + 2\,F^- \rightleftharpoons PbF_2(s) + 2\,Cl^-$ $\quad K = 1.7 \times 10^{-5} / 3.6 \times 10^{-8} = 472$
   $K = [Cl^-]^2/[F^-]^2 = 472$, so $[Cl^-]/[F^-] = \sqrt{472} = 21.7$

6. g. at constant T, PV = constant
   if the pressure decreases from 8 to 2, the volume must increase × 4, to 20 L

7. d. PV = nRT
   8.0 atm × 5.0 L = 1.5 mol × 0.08206 × T
   T = 325 K = 52°C

8. g. $w = -P\Delta V = -2.0$ atm × 15 L = −30 L atm
   −30 L atm × 101.325 J L^{-1} atm^{-1} = −3040 J

9. f. $q_p = n\,C_p\,\Delta T$
   temperature at C is 325K and the temperature at B = 325 × (2/8) = 81.24 K
   $\Delta T = 243.7$ K
   $q_p = n\,C_p\,\Delta T = 1.5 \times (5R/2) \times 243.7 = 7600$ J

10. c. $\Delta E = q + w = 7600 - 3040$ J = +4560 J

11. d. Since ΔE from A to C = zero, ΔE from A to B must be −4560 J

12. f. $(\frac{1}{2})^5 = 1/32$

13. e. $\Delta S = nR\ \ln(V_f/V_i) = Nk\ \ln(V_f/V_i) = 5 \times 1.38 \times 10^{-23} \times 0.693 = 4.79 \times 10^{-23}\ JK^{-1}$

14. a. entropies are zero only at 0K (3rd law of thermodynamics)

15. f. $T = \Delta H°/\Delta S° = 5650\ J\ mol^{-1} / 28.9\ J\ K^{-1}\ mol^{-1} = 196\ K$

16. e. available energy (maximum amount of work possible)

17. b. autoionization more spontaneous, higher concentrations of H_3O^+ and OH^-, pH lower.

18. g. $\Delta G° = \Delta G_f^\circ(\text{products}) - \Delta G_f^\circ(\text{reactants})$
$= (-57 - 69 - 3 \times 95) - (-2 \times 51) = -309\ kJ$

19. a. $\Delta G° = -RT\ \ln K$
$-309{,}000\ J = -8.3145\ J\ K^{-1}\ mol^{-1} \times 298.15\ K \times \ln K$
$K = 1.4 \times 10^{54}$

20. f.
$2\ Al \rightarrow 2\ Al^{3+} + 6\ e^-$
$6\ e^- + 6\ H^+ \rightarrow 3\ H_2$

$2\ Al + 6\ H^+ \rightarrow 2\ Al^{3+} + 3\ H_2$
$\quad + 6\ OH^- \qquad + 6\ OH^-$

$2\ Al + 6\ H_2O \rightarrow 2\ Al^{3+} + 6\ OH^- + 3\ H_2$

or,

$2\ Al + 6\ H_2O + 2\ OH^- \rightarrow 2\ [Al(OH)_4]^- + 3\ H_2$

21. b. $+0.257\ V + 0.741\ V = 0.998\ V$

22. a. spontaneous $E° =$ positive, or $\Delta G° =$ negative

23. a. nickel metal is oxidized and is the reducing agent

24. $E = E° - (RT/nF)\ln([Zn^{2+}]/[Cu^{2+}])$
$[Cu^{2+}] = 0.50\ M$
$[Zn^{2+}] = 1.50\ M$
$E = 1.104\ V - (8.3145 \times 298.15 / 2 \times 96485)\ln(3) = 1.090\ V$

EXAMINATION 2

CHEMISTRY 142 FALL 2006

Solutions

1. e. $Sr_3(PO_4)_2 \rightarrow 3\ Sr^{2+} + 2\ PO_4^{3-}$
   $[Sr^{2+}] = 3/2 \times [PO_4^{3-}]$ because three Sr^{2+} ions are produced for every two PO_4^{3-} ions
   $K_{sp} = [Sr^{2+}]^3[PO_4^{3-}]^2 = 1.0 \times 10^{-31}$
   $K_{sp} = (3/2[PO_4^{3-}])^3 \times [PO_4^{3-}]^2 = 1.0 \times 10^{-31}$
   $27/8 \times [PO_4^{3-}]^5 = 1.0 \times 10^{-31}$
   $[PO_4^{3-}] = 4.95 \times 10^{-7}$ M and $[Sr^{2+}] = 3/2 \times [PO_4^{3-}] = 7.4 \times 10^{-7}$ M

   or

   $Sr_3(PO_4)_2 \rightarrow 3\ Sr^{2+} + 2\ PO_4^{3-}$
   3x 2x
   $K_{sp} = [Sr^{2+}]^3[PO_4^{3-}]^2 = (3x)^3 \times (2x)^2 = 1.0 \times 10^{-31}$
   $108\ x^5 = 1.0 \times 10^{-31}$
   $x = 2.47 \times 10^{-7}$
   and $[Sr^{2+}] = 3x = 7.4 \times 10^{-7}$ M

2. c. $K_{sp} = [Sr^{2+}]^3[PO_4^{3-}]^2 = 1.0 \times 10^{-31}$
   $[PO_4^{3-}] = 3.0 \times 10^{-2}$ M
   $(1 \times 10^{-31}) = [Sr^{2+}]^3(3.0 \times 10^{-2})^2$
   $[Sr^{2+}] = 4.8 \times 10^{-10}$

3. h. $K_{sp} = [Ag^+]^3[PO_4^{3-}]$
   $Q_{sp} = (2.0)^3(0.1) = 0.8$

4. e. the pressure at point B = the pressure at point A
   $P = nRT/V$ $P = (3)(0.08206)(298) / (2.5) = 29.3$ atm

5. g. the temperature increase from A to B must correspond directly to the volume increase:
   new temperature = 298 K × (11.5/2.5) = 1371 K

6. d. $w = -p\Delta V = -(4.5\text{ atm})(-9\text{ L}) = 40.5\text{ L atm} \times 101.325\text{ J L}^{-1}\text{ atm}^{-1} = 4104\text{ J} = 4.1\text{ kJ}$

7. c. A → B: $w = -p\Delta V = -(29.3\text{ atm})(9\text{ L}) = -263.7\text{ L atm} \times 101.325\text{ J L}^{-1}\text{ atm}^{-1} = -26.7\text{ kJ}$
   B → C: $w = -p\Delta V =$ zero
   C → D: $w = -p\Delta V = -(4.5\text{ atm})(-9\text{ L}) = 40.5\text{ L atm} \times 101.325\text{ J L}^{-1}\text{ atm}^{-1} = 4.1\text{ kJ}$
   Total for A → B → C → D: $w = -26.7\text{ kJ} + 0 + 4.1\text{ kJ} = -22.6\text{ kJ}$

8. a. $q_p = nC_p\Delta T = (3)(5/2)(8.314)(45.7 - 210) = -10245\text{ J} = -10.2\text{ kJ}$

9. d. must be no change; internal energy is a state function; 0 kJ

10. d.

11. d. $W_i = 1$, $W_f = 4$ so the change $\Delta W = 4 - 1 = 3$

12. g. $W_f = 9$, $W_i = 1$

$\Delta S = k \ln W_f - k \ln W_i = k \ln(W_f/W_i)$

$= (1.38 \times 10^{-23}) \ln(9/1)$

$= 3.03 \times 10^{-23}$ J K^{-1}

13. e. $\Delta S_v = \Delta H_v / T_b$

$\Delta H_v = 25.8$ kJ mol^{-1} = 25,800 J mol^{-1}

$\Delta S_v = 25{,}800$ J mol^{-1}/266.8 K = 96.7 J K^{-1} mol^{-1}

14. b.

15. b. $\Delta H° = [(3 \times -110.5 \text{ kJ}) + (1 \times -826 \text{ kJ})] - [(3 \times -393.5 \text{ kJ})]$

$= +23$ kJ

$= 23{,}000$ J

$\ln (K_2/K_1) = \Delta H°/R\ [(1/T_1) - (1/T_2)]$

$\ln (2/1) = (23{,}000/8.314)\ [(1/298) - (1/T_2)]$

$T_2 = 322$ K

16. e. $\Delta S° = \Sigma\ S°_{(products)} - S°_{(reactants)}$

$\Delta S° = [(3 \times 198 \text{ JK}^{-1}) + (1 \times 90 \text{ JK}^{-1})] - [(2 \times 27 \text{ JK}^{-1}) + (3 \times 214 \text{ JK}^{-1})] = -12 \text{ JK}^{-1}$

$\Delta G° = \Delta H° - T\Delta S°$

$\Delta G° = (23{,}000 \text{ J}) - (100 \text{ K})(-12 \text{ JK}^{-1}) = 24{,}200 \text{ J} = 24.2 \text{ kJ}$

17. b. nonspontaneous

18. e. $3\ CH_3OH(aq) + 8\ H^+(aq) + Cr_2O_7^{2-}(aq) \rightarrow 3\ CH_2O(aq) + 2\ Cr^{3+}(aq) + 7\ H_2O(l)$

19. g. change the sign of the reduction potential for the oxidation at the anode and add the two:

$E° = 0.73$ V + 2.37 V = 3.10 V

20. f. $[Cr^{3+}] = 1.5$ M and $[La^{3+}] = 0.5$ M

$E = E° - (RT/nF) \ln([Cr^{3+}]/[La^{3+}])$

$E = 3.100 \text{ V} - (8.314 \times 298)/(3 \times 96{,}485) \ln(1.5/0.5)$

$= 3.100 \text{ V} - 0.009 \text{ V} = 3.091 \text{ V}$

21. d. 0 V

22. d.

EXAMINATION 2

CHEMISTRY 142 SPRING 2007

Solutions

1. c. $K_{sp} = [Mg^{2+}]^3[PO_4^{3-}]^2$

2. f. the formula of the salt is $Mg_3(PO_4)_2$; $[Mg^{2+}] = 3/2 \times [PO_4^{3-}]$
three Mg^{2+} ions are produced for every two PO_4^{3-} ions

3. e. $Mg_3(PO_4)_2 \rightarrow 3\ Mg^{2+} + 2\ PO_4^{3-}$
$K_{sp} = [Mg^{2+}]^3[PO_4^{3-}]^2 = 1.0 \times 10^{-24}$
$K_{sp} = (3/2[PO_4^{3-}])^3 \times [PO_4^{3-}]^2 = 1.0 \times 10^{-24}$
$27/8 \times [PO_4^{3-}]^5 = 1.0 \times 10^{-24}$
$[PO_4^{3-}] = 1.24 \times 10^{-5}$ M and solubility of $Mg_3(PO_4)_2 = 1.24 \times 10^{-5}$ M $/2 = 6.21 \times 10^{-6}$ M

4. c. $K_{sp} = [Mg^{2+}]^3[PO_4^{3-}]^2 = 1.0 \times 10^{-24}$
$[PO_4^{3-}] = 0.010$ M
$(1.0 \times 10^{-24}) = [Mg^{2+}]^3(0.010)^2$
$[Mg^{2+}] = 2.15 \times 10^{-7}$

5. b. $K_{sp} = [Mg^{2+}]^3[PO_4^{3-}]^2 = 1.0 \times 10^{-24}$
$Q_{sp} = (5.0 \times 10^{-6})^3(2.0 \times 10^{-5})^2 = 5.0 \times 10^{-26}$ which is less than K_{sp} so no precipitation

6. e. $CuCl(s) \rightarrow Cu^{2+}(aq) + Cl^-(aq)$ $K_{sp1} = 1.7 \times 10^{-7}$
$CuCN(s) \rightarrow Cu^{2+}(aq) + CN^-(aq)$ $K_{sp2} = 3.5 \times 10^{-20}$
reverse equation 2 and add to obtain...
$CuCl(s) + CN^-(aq) \rightarrow CuCN(s) + Cl^-(aq)$ $K = K_{sp1} / K_{sp2} = 4.9 \times 10^{12}$

7. d.

8. b. $q = +1670$ J
$w = -P\Delta V = -0.880 \text{ atm} \times 4.6 \text{ L} \times 101.325 \text{ J L}^{-1} \text{ atm}^{-1} = -410$ J
$\Delta E = q + w = 1670 - 410 = 1260$ J

9. g. PV is constant at constant temperature; pressure at C = 5 atm × 12/3 = 20 atm

10. f. $T = PV / nR = (20 \times 3)/(2 \times 0.08206) = 365.6 \text{ K} = 92.4°C$

11. g. $q_v = n C_v \Delta T = 2 \times 3/2 \times 8.3145 \times (365.6 - 91.4) = 6840$ J

12. f. $w = -nRT \ln(V_f/V_i) = -2 \times 8.3145 \times 365.6 \times \ln 4 = -8430$ J

13. e. $\Delta S = q_{rev}/T = 8430 / 365.6 = 23 \text{ JK}^{-1}$ *or* $\Delta S = nR \ln(V_f/V_i) = 2 \times 8.3145 \times \ln 4 = 23 \text{ JK}^{-1}$

14. e. $\Delta S° = \Sigma\ S°_{(products)} - S°_{(reactants)} = 2 \times 240\ J\ K^{-1} - 304\ J\ K^{-1} = 176\ J\ K^{-1}$
$\Delta H° = 57.1\ kJ = 57{,}100\ J$
$\Delta G° = \Delta H° - T\Delta S° = 57100 - (176 \times 298.15)\ J = 4630\ J = 4.63\ kJ$

15. f. $\Delta G° = -RT \ln K_w = -8.3145 \times 298.15 \times \ln(1.0 \times 10^{-14}) = 79{,}912\ J = 80\ kJ$

16. c.

17. a. $Cl_2 + 2\,H_2O \rightarrow 2\,HClO + 2\,H^+ + 2\,e^-$
$Cl_2 + 2\,e^- \rightarrow 2\,Cl^-$
$2\,Cl_2 + 2\,H_2O \rightarrow 2\,HClO + 2\,H^+ + 2\,Cl^-$
divide by 2...
$Cl_2 + H_2O \rightarrow HClO + H^+ + Cl^-$

18. c. change the sign of the reduction potential for the oxidation at the anode and add the two:
$E° = -0.54\ V + 1.51\ V = 0.97\ V$

19. j. 10 electrons lost by the ten I^- ions and 10 electrons gained by the two Mn in MnO_4^-

20. a. E° is positive (and ΔG° is negative)

21. c. $\Delta G° = -nFE° = -10 \times 96485 \times 0.97 = -935{,}900\ VC = -935{,}900\ J = -936\ kJ$

22. e.

23. c. 5 mol Ag × (1 mol e^- / 1 mol Ag) × (1 F / 1 mol e^-) × (96485 C / 1 F) × (1 A s / 1 C) × (1 / 100 A) × (1 min / 60 s) × (1 hr / 60 min) = 1.34 hr

24a. f. An Inconvenient Truth

24b. f. ΔG° is the difference between the beginning and the end on the free energy diagram regardless of whether the system is at the beginning, the end, at equilibrium, or anywhere else on the diagram
ΔG is zero at equilibrium

25a. b.

25b. b.

26. h. $18\,H_2O(l) + 3\,As_2S_3(s) + 14\,HClO_3(aq) \rightarrow 6\,H_3AsO_4(aq) + 9\,H_2SO_4(aq) + 14\,HCl(aq)$

EXAMINATION 2

CHEMISTRY 142 FALL 2007

Solutions

1. c. $Ag_3PO_4 \rightarrow 3\ Ag^+ + PO_4^{3-}$
   $K_{sp} = [Ag^+]^3[PO_4^{3-}] = 8.9 \times 10^{-17}$
   $[Ag^+] = 3 \times [PO_4^{3-}]$ — i.e. three Ag^+ ions are produced for every PO_4^{3-} ion
   $K_{sp} = (3[PO_4^{3-}])^3 \times [PO_4^{3-}] = 8.9 \times 10^{-17}$
   $27 \times [PO_4^{3-}]^4 = 8.9 \times 10^{-17}$
   $[PO_4^{3-}] = 4.3 \times 10^{-5}$ M and the solubility of Ag_3PO_4 is the same

2. a. decrease—the addition of a common ion suppresses the solution process

3. b. $K_{sp} = [Ag^+]^3[PO_4^{3-}] = 8.9 \times 10^{-17}$
   $[PO_4^{3-}] = 5 \times 10^{-2}$ M
   $K_{sp} = [Ag^+]^3(5 \times 10^{-2}) = 8.9 \times 10^{-17}$
   $[Ag^+]^3 = 1.78 \times 10^{-15}$
   $[Ag^+] = 1.2 \times 10^{-5}$ M
   there are three Ag^+ ions in each Ag_3PO_4 formula unit, so
   the solubility of the salt $Ag_3PO_4 = 1.2 \times 10^{-5}$ M / 3 = 4.0×10^{-6} M

4. b. $K_{sp} = [Ag^+]^3[PO_4^{3-}] = 8.9 \times 10^{-17}$
   $[Ag^+] = 1.5 \times 10^{-4}$ M × 2/3 (the solution is diluted to 3/2 times the original volume)
   $[PO_4^{3-}] = 2.0 \times 10^{-5}$ M × 1/3 (the solution is diluted to 3 times the original volume)
   $Q_{sp} = (1.0 \times 10^{-4})^3(6.7 \times 10^{-6}) = 6.7 \times 10^{-18}$ which is less than K_{sp} so no precipitation

5. h. $AgCl(s) \rightarrow Ag^+(aq) + Cl^-(aq)$ $\quad K_{sp1} = 1.8 \times 10^{-10}$
   $AgI(s) \rightarrow Ag^+(aq) + I^-(aq)$ $\quad K_{sp2} = 8.3 \times 10^{-17}$
   reverse equation 2 and add to obtain...
   $AgCl(s) + I^-(aq) \rightarrow AgI(s) + Cl^-(aq)$ $\quad K = K_{sp1} / K_{sp2} = 2.2 \times 10^6$

6. b. the carbonate ion is the conjugate base of a very weak acid
   the carbonate ion reacts with acid to produce CO_2 and water, which causes more $MgCO_3$ to dissolve

7. f. PV is a constant at constant temperature: 12 atm × 3 L = 4 atm × 9 L

8. f. use PV = nRT, T = PV / nR = (12 × 3)/(1.5 × 0.08206) = 292.47 K = 19.32°C

9. h. the pressure decreases by a factor of 3, so must the temperature: 292.47 / 3 = 97.5 K

10. f. $w = -P\Delta V$ = 4.0 atm × 6.0 L = –24 L atm × 101.325 J / L atm = –2430 J

11. g. $q_p = n\ C_p\ \Delta T = 1.5 \times 5/2 \times 8.3145 \times (292.5 - 97.5) = +6080$ J

12. e. $\Delta E = q + w = 6080 - 2430 = +3650$ J

13. d. since ΔE = zero for A to C (isothermal), then ΔE for A to B must be –3650 J

14. f. $w = -nRT \ln(V_f/V_i) = -1.5 \times 8.3145 \times 292.5 \times \ln(3/9) = 4000$ J

15. d. q (heat) is not a state function—it is path dependent

16. f. $W_{initial} = 1$ and $W_{final} = 8$, the change = 7

17. d. $\Delta S = Nk \ln(V_f/V_i) = 3 \times 1.38066 \times 10^{-23} \times \ln 3 = 4.6 \times 10^{-23}\ JK^{-1}$

18. e. constant pressure heating from 298 K to 545 K:
$\Delta S = nC_p \ln(T_f/T_i) = 2.5 \times (5/2) \times 8.3145 \times \ln(545/298) = 31.4\ JK^{-1}$
or separate into two steps:
isothermal expansion from 3.5 L to 6.4 L:
$\Delta S = nR \ln(V_f/V_i) = 2.5 \times 8.3145 \times \ln(6.4/3.5) = 12.55\ JK^{-1}$
increase in temperature from 298 K to 545 K at constant volume:
$\Delta S = nC_v \ln(T_f/T_i) = 2.5 \times (3/2) \times 8.3145 \times \ln(545/298) = 18.82\ JK^{-1}$
total = $12.55\ JK^{-1} + 18.82\ JK^{-1} = 31.4\ JK^{-1}$ (same result)

19. c. this statement merely defines ΔG

20. f. +6 (for each Cr)

21. f. six electrons transferred

22. c. 4 on the left: $4\ H_2O + 4\ CrO_4^{2-} + 4\ PH_3 \rightarrow 4\ Cr(OH)_4^- + P_4 + 4\ OH^-$

23. c. the reaction is not spontaneous in the direction in which it is written

24. b. $E^\circ_{anode} + E^\circ_{cathode} = -1.104$ V
$E^\circ_{anode} - 0.7618\ V = -1.104$ V
$E^\circ_{anode} = -1.104\ V + 0.7618$ V
$= -0.342$ V —this is for oxidation at the anode
change sign for reduction: $E^\circ_{red}(Ag^+|Ag) = +0.342$ V

25. f. $\Delta G^\circ = -nFE^\circ = -2 \times 96{,}485 \times (-1.104\ V) = 213{,}000\ J = 213$ kJ

26. d. 50 g Cu × (1 mol Cu / 63.55 g Cu) × (2 mol e^- / 1 mol Cu) × (1 F / 1 mol e^-) × (96485 C / 1 F) × (1 A s / 1 C) × (1 / 15 A) × (1 min / 60 s) × (1 hr / 60 min) = 2.81 hr

EXAMINATION 2

CHEMISTRY 142 SPRING 2008

Solutions

1. d. $K_{sp} = [Fe^{3+}][OH^-]^3$

2. d. $K_{sp} = [Ag^+][OH^-] = 1.5 \times 10^{-8}$ $[Ag^+] = \sqrt{(1.5 \times 10^{-8})} = 1.23 \times 10^{-4}$ M
   $K_{sp} = [Fe^{2+}][OH^-]^2 = 1.6 \times 10^{-14}$ $[Fe^{2+}] = \sqrt[3]{(1.6 \times 10^{-14} / 4)} = 1.59 \times 10^{-5}$ M
   $K_{sp} = [Fe^{3+}][OH^-]^3 = 1.1 \times 10^{-36}$ $[Fe^{3+}] = \sqrt[4]{(1.1 \times 10^{-36} / 27)} = 4.49 \times 10^{-10}$ M
   $K_{sp} = [Mg^{2+}][OH^-]^2 = 1.2 \times 10^{-11}$ $[Mg^{2+}] = \sqrt[3]{(1.2 \times 10^{-11} / 4)} = 1.44 \times 10^{-4}$ M
   $K_{sp} = [Zn^{2+}][OH^-]^2 = 4.5 \times 10^{-17}$ $[Zn^{2+}] = \sqrt[3]{(4.5 \times 10^{-17} / 4)} = 2.24 \times 10^{-6}$ M

3. e. $K_{sp} = [Fe^{2+}][OH^-]^2 = 1.6 \times 10^{-14}$
   $[0.025\ M][OH^-]^2 = 1.6 \times 10^{-14}$
   $[OH^-]^2 = 1.6 \times 10^{-14} / 0.025 = 6.4 \times 10^{-13}$
   and therefore $[OH^-] = \sqrt{(6.4 \times 10^{-13})} = 8.0 \times 10^{-7}$ M

4. b. any acid—nitric acid is an example

5. b. the free energy of the system decreases toward equilibrium

6. e. $\Delta G°$

7. c. it doesn't change

8. b. the system is isolated and the entropy must increase if the reaction is spontaneous

9. b. Q is not yet equal to K; it increases as the system moves from left to right

10. c. I and III only
    II exothermic reactions ($\Delta H°$ negative) are frequently product–favored because a negative $\Delta H°$ makes $\Delta G°$ more likely negative—but the value of $T\Delta S°$ is also important! For example, the exothermic freezing of water is not product–favored at 25°C
    IV as in the example just given, if the T is high enough and $\Delta S°$ is (+), then $\Delta G°$ will be (–)

11. c.

12. f. constant pressure heating from 298 K to 447 K:
    $\Delta S = nC_p \ln(T_f/T_i) = 4.0 \times (5/2) \times 8.3145 \times \ln(447/298) = 33.7\ JK^{-1}$
    or separate into two steps:
    isothermal expansion from 5.0L to 7.5L:
    $\Delta S = nR \ln(V_f/V_i) = 4.0 \times 8.3145 \times \ln(7.5/5.0) = 13.5\ JK^{-1}$

increase in temperature from 298 K to 447 K at constant volume:
$\Delta S = nC_v \ln(T_f/T_i) = 4.0 \times (3/2) \times 8.3145 \times \ln(447/298) = 20.2\ JK^{-1}$
total = $13.5\ JK^{-1} + 20.2\ JK^{-1} = 33.7\ JK^{-1}$ (same result—entropy is a state function)

13. a. $\Delta S = Nk \ln(V_f/V_i) = 6.0 \times 1.38065 \times 10^{-23} \times \ln(2) = 5.74 \times 10^{-23}\ JK^{-1}$

14. i. one arrangement in 2^6, i.e. 1/64

15. b. use PV = nRT, T = PV/nR = (18 × 3)/(2.5 × 0.08206) = 263.22 K = –10°C

16. g. $w = -nRT \ln(V_f/V_i) = -2.5 \times 8.3145 \times 263.22 \times \ln(6/18) = 6011\ J = 6.0\ kJ$

17. a. no expansion work done — constant volume

18. g. constant pressure work, $w = -P\Delta V = -9.0$ atm × –12.0 L × 101.325 J / L atm = 10.9 kJ

19. h. ΔE = zero (isothermal) but $\Delta E = q + w$, so $q = -w$
w = 6.0 kJ (from question 16), so q = –6.0 kJ

20. e. ΔE = zero regardless of the path between A and C (the internal energy E is a state function)
if w = 10.9 kJ (from questions 17 & 18), then q = –w = –10.9 kJ
or, if you really want to calculate it...
$q = q_v$ (for A → B) + q_p (for B → C) = $n C_v \Delta T + n C_p \Delta T$
q = (2.5 × 3/2 × 8.3145 × 526.44) + (2.5 × 5/2 × 8.3145 × –526.44)
q = –2.5 × 8.3145 × 526.44 = –10.9 kJ

21. f. $\Delta E = q + w$ = (2.5 × 3/2 × 8.3145 × 526.44) + zero = 16.4 kJ

22. g. ΔE = zero for A → C, and ΔE for A → B = 16.4 kJ (question 21), so ΔE for B → C = –16.4 kJ
or, if you really want to calculate it...
$\Delta E = q + w$ = (2.5 × 5/2 × 8.3145 × –526.44) + 10,943 J (question 18) = –16,414 J = –16.4 kJ

23. g. ΔH and ΔS always both negative for condensation; ΔG depends on temperature, negative at 50°C and positive at 140°C

24. h. –3 –2 +3 +4 = +2

25. i. $4\ H_2O_2(aq) + Cl_2O_7(aq) + 2\ OH^-(aq) \rightarrow 2\ ClO_2^-(aq) + 4\ O_2(g) + 5\ H_2O(l)$

26. g. $E^\circ_{anode} + E^\circ_{cathode}$ = +0.763 V (change sign for oxidation) +0.770 V = +1.533 V

27. d. $Zn(s) + 2\ Fe^{3+}(aq) \rightarrow Zn^{2+}(aq) + 2\ Fe^{2+}(aq)$

$E = E^\circ - (RT/nF) \ln([Zn^{2+}][Fe^{2+}]^2/[Fe^{3+}]^2)$
$= 1.533\ V - (8.3145 \times 298)/(2 \times 96485) \times \ln[(0.726)(0.736)^2/(0.274)^2]$
$= 1.533\ V - 0.021\ V = 1.512\ V$

EXAMINATION 2

CHEMISTRY 142 FALL 2008

Solutions

1. f. $Cu(OH)_2 \rightarrow Cu^{2+} + 2\,OH^-$ $K_{sp} = 2.2 \times 10^{-20}$
   $K_{sp} = [Cu^{2+}][OH^-]^2 = [Cu^{2+}] \times (2[Cu^{2+}])^2 = 4[Cu^{2+}]^3$
   $[Cu^{2+}] = 1.8 \times 10^{-7}$

2. a. $Cu(OH)_2 \rightarrow Cu^{2+} + 2\,OH^-$ $K_{sp} = 2.2 \times 10^{-20}$
   pH = 12; pOH = 2; $[OH^-] = 10^{-2}$
   $K_{sp} = 2.2 \times 10^{-20} = [Cu^{2+}][10^{-2}]^2$
   $[Cu^{2+}] = 2.2 \times 10^{-16}$

3. b. $CuBr(s) \rightarrow Cu^+ + Br^-$ $K_{sp} = 4.2 \times 10^{-8}$
   add HBr: decrease solubility (addition of a common ion)
   add HCl: no change
   add NH_3: increase solubility (Cu^+ forms a complex with NH_3)
   add KBr: decrease solubility (addition of a common ion)

4. e. $Ag_2CrO_4 \rightarrow 2\,Ag^+ + CrO_4^{2-}$ $K_{sp} = 9.0 \times 10^{-12} = [Ag^+]^2[CrO_4^{2-}]$
   dilution of Ag^+: (100 mL)(5.0×10^{-4} M) = (175 mL)[Ag^+] [Ag^+] = 2.86×10^{-4} M
   dilution of CrO_4^{2-}: (75 mL)(1.0×10^{-4} M) = (175 mL)[CrO_4^{2-}] [CrO_4^{2-}] = 4.29×10^{-5} M
   $Q_{sp} = (2.86 \times 10^{-4})^2(4.29 \times 10^{-5}) = 3.51 \times 10^{-12}$ $Q_{sp} < K_{sp}$ therefore no precipitation

5. c. $\Delta E = q + w = +280\text{ J} - 130\text{ J} = +150\text{ J}$

6. b.

7. f. $T = (PV) / (nR) = (15\text{ atm} \times 4.0\text{ L}) / (2.0\text{ mol} \times 0.08206\text{ L atm K}^{-1}\text{ mol}^{-1}) = 365.6\text{ K} = 92.5\text{ °C}$

8. d. $w = -nRT \ln(V_f / V_i) = -2.0\text{ mol} \times 8.3145\text{ J K}^{-1}\text{ mol}^{-1} \times 365.6\text{ K} \times \ln(20/4)$
   $w = -9785\text{ J} = -9.8\text{kJ}$

9. g. $w = -P\Delta V = -3.0\text{ atm} \times (20\text{ L} - 4\text{ L}) = -48\text{ L atm} \times (101.325\text{ J} / 1\text{ L atm}) = -4860\text{ J}$

10. e. $q_p = n(5/2)R\Delta T = 2.0\text{ mol} \times (5/2) \times 8.3145\text{ J K}^{-1}\text{ mol}^{-1} \times (365.6\text{ K} - 73.12\text{ K}) = +12160\text{ J}$

11. b. $\Delta E = q + w$
    $q_v = n(3/2)R\Delta T = 2.0\text{ mol} \times (3/2) \times 8.3145\text{ J K}^{-1}\text{ mol}^{-1} \times (73.12\text{ K} - 365.6\text{ K}) = -7300\text{ J}$
    w = 0 (volume is constant)
    $\Delta E = -7300\text{ J} + 0 = -7300\text{ J}$

12. d. $\Delta S = nR \ln(V_f / V_i)$
$= (6 \text{ molecules}) \times (1 \text{ mol} / 6.022 \times 10^{23} \text{ molecules}) \times 8.3145 \text{ J K}^{-1} \text{ mol}^{-1} \times \ln(4/1)$
$= 1.15 \times 10^{-22} \text{ J K}^{-1}$

13. g. $\Delta S° = [2 \text{ mol} \times 187.0 \text{ J K}^{-1} \text{ mol}^{-1}] - [(1 \text{ mol} \times 131.0 \text{ J K}^{-1} \text{ mol}^{-1}) + (1 \text{ mol} \times 223.0 \text{ J K}^{-1} \text{ mol}^{-1})]$
$= +20 \text{ J K}^{-1}$

14. b. ΔG is negative (spontaneous reaction)
ΔH is negative (exothermic reaction)
ΔS is positive (gases have higher entropy than an ionic crystalline lattice)

15. e. $\Delta G = \Delta G° + RT \ln Q$
$Q_p = (P_{NO2})^2 / (P_{N2O4}) = 0.122^2 / 0.453 = 0.0329$
$\Delta G = 5400 \text{ J mol}^{-1} + (8.3145 \text{ J K}^{-1} \text{ mol}^{-1} \times 298 \text{ K} \times \ln(0.0329)) = -3060 \text{ J} = -3.1 \text{ kJ}$
$Q_p < K_p$ therefore shift right

16. c. oxygen is –2, hydrogen is +1, sulfur is +5

17. a. half reactions: $(H_2O + CN^- \rightarrow CNO^- + 2\,H^+ + 2\,e^-) \times 3$
$(3\,e^- + 4\,H^+ + MnO_4^- \rightarrow MnO_2 + 2\,H_2O) \times 2$
add: $3\,H_2O + 3\,CN^- + 8\,H^+ + 2\,MnO_4^- \rightarrow 3\,CNO^- + 6\,H^+ + 2\,MnO_2 + 4\,H_2O$
simplify: $3\,CN^- + 2\,H^+ + 2\,MnO_4^- \rightarrow 3\,CNO^- + 2\,MnO_2 + H_2O$
in basic solution: $3\,CN^- + 2\,H_2O + 2\,MnO_4^- \rightarrow 3\,CNO^- + 2\,MnO_2 + H_2O + 2\,OH^-$
simplify: $3\,CN^- + \underline{H_2O} + 2\,MnO_4^- \rightarrow 3\,CNO^- + 2\,MnO_2 + 2\,OH^-$

18. a. $E°_{anode} + E°_{cathode} = -1.36 \text{ V}$ (change sign for oxidation) $+ 1.61 \text{ V} = +0.25 \text{ V}$

19. e. $\Delta G° = -nFE° = -2.0 \text{ mol} \times 96458 \text{ C} \times 0.25 \text{ V} = -48242.5 \text{ J} = -48 \text{ kJ}$

20. g. $E = E° - (RT)/(nF) \times \ln Q$
$= 0.012 \text{ V} - (8.1345 \text{ J K}^{-1} \text{ mol}^{-1} \times 298 \text{ K})/(2 \text{ mol} \times 96485 \text{ C}) \times \ln(0.50/0.001) = -0.0680 \text{ V}$

21. f. $0.912 \text{ A} \times 18 \text{ hr} \times (3600 \text{ s} / 1 \text{ hr}) \times (1 \text{ C} / 1 \text{ As}) \times (1 \text{ mol } e^- / 96485 \text{ C}) \times (1 \text{ mol Mg} / 2 \text{ mol } e^-)$
$\times (24.31 \text{ g Mg} / 1 \text{ mol Mg}) = 7.45 \text{ g Mg}$

22. c. $\Delta G = \Delta H - T\Delta S$
at equilibrium, $\Delta G = 0$ so $\Delta H = T\Delta S$
$T = \Delta H/\Delta S = 10900 \text{ J mol}^{-1} / 39.1 \text{ J K}^{-1} \text{ mol}^{-1} = 278.8 \text{ K} = 5.6°C$

EXAMINATION 2

CHEMISTRY 142 SPRING 2009

Solutions

1. e. $K_{sp} = [Y^{3+}]^2[CO_3^{2-}]^3$

2. c. $3 \times [Y^{3+}] = 2 \times [CO_3^{2-}]$
three carbonate ions are produced for every two yttrium ions, i.e. $[CO_3^{2-}] = 3/2 \times [Y^{3+}]$

3. e. $K_{sp} = [Tl^+]^2[S^{2-}] = 6.0 \times 10^{-22}$
$[Tl^+]^2[0.10] = 6.0 \times 10^{-22}$
$[Tl^+]^2 = 6.0 \times 10^{-22} / 0.10 = 6.0 \times 10^{-21}$
and therefore $[Tl^+] = \sqrt{(6.0 \times 10^{-21})} = 7.7 \times 10^{-11}$ M

4. h. $BaF_2(s) \rightleftharpoons Ba^{2+}(aq) + 2\,F^-(aq)$ $\quad K_{sp} = 2.4 \times 10^{-5}$
$HF(aq) + H_2O(l) \rightleftharpoons H_3O^+(aq) + F^-(aq)$ $\quad K_a = 7.2 \times 10^{-4}$
$2\,H_2O(l) \rightleftharpoons H_3O^+(aq) + OH^-(aq)$ $\quad K_w = 1.0 \times 10^{-14}$
reverse the second equation and multiply by 2; multiply the third equation by 2, and add all three:
$BaF_2(s) \rightleftharpoons Ba^{2+}(aq) + \cancel{2\,F^-}(aq)$ $\quad K_{sp} = 2.4 \times 10^{-21}$
$2\,\cancel{H_3O^+}(aq) + \cancel{2\,F^-}(aq) \rightleftharpoons 2\,HF(aq) + 2\,\cancel{H_2O}(l)$ $\quad K = 1/(7.2 \times 10^{-4})^2$
$\cancel{4}^{2}\,H_2O(l) \rightleftharpoons 2\,\cancel{H_3O^+}(aq) + 2\,OH^-(aq)$ $\quad K_w = (1.0 \times 10^{-14})^2$

$BaF_2(s) + 2\,H_2O(l) \rightleftharpoons Ba^{2+}(aq) + 2\,HF(aq) + 2\,OH^-(aq)$
$K = 2.4 \times 10^{-5} \times 1/(7.2 \times 10^{-4})^2 \times (1.0 \times 10^{-14})^2 = 4.6 \times 10^{-27}$

5. b. pH = 13.33
pOH = 0.67, therefore [OH–] = 0.214 M
there are two OH^- ions for evergy Ba^{2+} ion, so $[Ba^{2+}] = 0.107$ M
$K_{sp} = [Ba^{2+}][OH^-]^2 = [0.107][0.214]^2 = 4.9 \times 10^{-3}$

6. d. ideal gas law: P V = n R T: $\quad$ 8 atm × 15.0 L = n × 0.08206 × 323.15 $\quad$ n = 4.5 moles

7. a. the volume at State C = init vol × (8.0/3.0) = 40 L
isothermal expansion: $\quad w = -nRT \ln(V_f/V_i) = -4.5 \times 8.3145 \times 323.15 \times \ln(40/15) = -11{,}860$ J
work = – 12 kJ

8. a. constant volume—no expansion work done

9. e. $w = -P\Delta V = -3.0 \text{ atm} \times (40-15) \text{ L} = -75 \text{ L atm}$
$w = -75 \text{ L atm} \times 101.325 \text{ J L}^{-1}\text{atm}^{-1} = -7600 \text{ J} = -7.6$ kJ

10. f. $\Delta E = q + w$
w = zero and $q = n\overline{Cv}\Delta T$
temperature at State A = 323.15 K
temperature at State B = 323.15 × (3/8) = 121.18 K; so ΔT = 201.97 K
$\Delta E = q = n\overline{Cv}\Delta T = 4.5 \times 3/2 \times 8.3145 \times 201.97 = 11{,}335$ J
$\Delta E = -11$ kJ

11. d. ΔE = zero for any route between A and C, so $q = -w = +7.6$ kJ
or $q = n\overline{Cv}\Delta T$ for A to B + $n\overline{Cp}\Delta T$ for B to C = −11,335 J + 18, 892 J = +7557 J = +7.6 kJ

12. j. all are false statements

13. c. work depends upon the path

14. e. $(5/25)^{20} = 1.0 \times 10^{-14}$

15. e. $\Delta S = k \ln(V_f/V_i)^N = N k \ln(V_f/V_i) = 20 \times 1.38 \times 10^{-23} \times \ln(25/5) = 4.4 \times 10^{-22}$

16. d. (2 × 1) + (2 × ?) + (4 × −2) = 0, so oxidation number of carbon = +3

17. h. heat lost by hot water: $q = 100 \text{ g} \times 4.184 \text{ JK}^{-1}\text{g}^{-1} \times 80 \text{ K} = 33{,}472$ J
mass of ice melted: mass = $33{,}472 \text{ J}/333 \text{ Jg}^{-1} = 100$ g
mass of ice added: mass = 100 g melted + 15 g remaining = 115 g total

18. a. for ice melting: $\Delta S = \Delta H/T = 33{,}472 \text{ J}/273.15 \text{ K} = 122.5 \text{ JK}^{-1}$
for hot water cooling: $\Delta S = m \times \text{sp ht} \times \ln(T_f/T_i) = 100 \text{ g} \times 4.184 \text{ JK}^{-1}\text{g}^{-1} \times \ln(273.15/353.15)$
$= -107.5 \text{ JK}^{-1}$
total: $\Delta S = +15 \text{ JK}^{-1}$

19. b.

oxidation:	$2\ S_2O_3^{2-}$	$\rightarrow S_4O_6^{2-} + 2\ e^-$
reduction:	$2e^- + I_3^-$	$\rightarrow 3\ I^-$
add:	$2\ S_2O_3^{2-} + I_3^-$	$\rightarrow S_4O_6^{2-} + 3\ I^-$

20. c. change the sign for the oxidation and add: $E° = -0.800 + 1.358 = 0.558$ V

21. b. $E° = (RT/nF) \ln K$
$0.15 = (8.3145 \times 298.15/2 \times 96485) \ln K$
$\ln K = 11.68$
$K = 1.2 \times 10^5$

22. d. 20.0 A × 1.00 hr × (3600 s/1 hr) × (1 C/1 As) × (1 mol e^-/96485 C) × (1 mol Al/3 mol e^-) × (26.98 g / 1 mol Al)
= 6.71 g Al

EXAMINATION 2

CHEMISTRY 142 FALL 2009

Solutions

1. h. $K_{neut} = K_b / K_w = (4.4 \times 10^{-4}) / (1 \times 10^{-14}) = 4.4 \times 10^{10}$

2. a. $K_{sp} = [Ca^{2+}][F^-]^2$ and $[F^-] = 2[Ca^{2+}]$
$5.3 \times 10^{-9} = [Ca^{2+}] \times (2[Ca^{2+}])^2 = 4[Ca^{2+}]^3$
$[Ca^{2+}] = 1.1 \times 10^{-3}$ M

3. d. $K_{sp} = [Ca^{2+}][F^-]^2$ and $[F^-] = 0.010$ M
$5.3 \times 10^{-9} = [Ca^{2+}] \times (0.010)^2$
$[Ca^{2+}] = 5.3 \times 10^{-5}$ M

4. d. $[Ca^{2+}]_{dilute} = [Ca^{2+}]_{conc} \times V_{conc} / V_{dilute} = (1.0 \times 10^{-5}\text{ M}) \times 25\text{ mL} / 50\text{ mL} = 5.0 \times 10^{-6}$ M
$[F^-]_{dilute} = [F^-]_{conc} \times V_{conc} / V_{dilute} = (1.5 \times 10^{-5}\text{ M}) \times 25\text{ mL} / 50\text{ mL} = 7.5 \times 10^{-6}$ M
$Q_{sp} = [Ca^{2+}][F^-]^2 = (5.0 \times 10^{-6}) \times (7.5 \times 10^{-6})^2 = 2.8 \times 10^{-16}$
$Q_{sp} < K_{sp}$, therefore no precipitation

5. a. NH_3 will increase solubility because the formation of the complex $[Ag(NH_3)_2]^+$ takes Ag^+ out of solution, shifting the equilibrium to the right.

6. c. $\Delta E = q + w$
$-180\text{ J} = -40\text{ J} + w$
$w = -140$ J

7. c. $P_aV_a = P_cV_c$
$V_c = (2.0\text{ atm} \times 12.0\text{ L}) / 8.0\text{ atm} = 3.0$ L

8. g. constant pressure, $w = -P\Delta V = -(8.0\text{ atm}) \times (3.0\text{ L} - 12.0\text{ L}) = 72$ L atm
72 L atm × (101.325 J / 1 L atm) × (1kJ / 1000J) = 7.3 kJ

9. c. find temperature at State A = temperature at State C:
$T_A = PV/nR = (2.0\text{ atm} \times 12.0\text{ L}) / (0.50\text{ mol} \times 0.08206\text{ L atm K}^{-1}\text{ mol}^{-1}) = 585$ K
$w = -nRT \ln(V_f/V_i) = -(0.5\text{mol}) \times (8.3145\text{ J K}^{-1}) \times (585\text{ K}) \times \ln(3.0\text{ L} / 12.0\text{L}) = 3371\text{ J} = 3.4$ kJ

10. b. $\Delta E = q + w$; $w = -P\Delta V$; volume is constant so $\Delta V = 0$ and $w = 0$; $\Delta E = q_v$
$T_B = PV/nR = (8.0\text{ atm} \times 12.0\text{ L}) / (0.50\text{ mol} \times 0.08206\text{ L atm K}^{-1}\text{ mol}^{-1}) = 2340$ K
$q_v = n(3/2)R\,\Delta T = (0.5\text{ mol}) \times (3/2) \times (8.3145\text{ J K}^{-1}) \times (2340\text{K} - 585\text{K}) = 10944\text{ J} = 10.9$ kJ

11. g. heat lost by steam condensing + heat lost by water cooling = heat gained by ice melting
$[10.0\text{ g} \times 2260\text{ J g}^{-1}] + [10.0\text{ g} \times 4.184\text{ J K}^{-1}\text{ g}^{-1} \times 100] = m_{ice} \times 333\text{ J g}^{-1}$
$m_{ice} = 80.4$ g ice melts

12. e. $\Delta S_{process} = \Delta S_{condense\ steam} + \Delta S_{cool\ water} + \Delta S_{melt\ ice}$
$\Delta S_{condense\ steam} = -(m \times \text{latent heat}) / T = (10\ g \times 2260\ J\ g^{-1}) / 373.15\ K = -60.57\ J\ K^{-1}$
$\Delta S_{cool\ water} = m \times c \times \ln(T_f/T_i) = 10\ g \times 4.184\ J\ K^{-1}\ g^{-1} \times \ln(273.15\ K/373.15\ K) = -13.05\ J\ K^{-1}$
$\Delta S_{melt\ ice} = (m \times \text{latent heat}) / T = (80\ g \times 333\ J\ g^{-1}) / 273.15\ K = +97.53\ J\ K^{-1}$
$\Delta S_{process} = -60.57\ J\ K^{-1} + -13.05\ J\ K^{-1} + 97.53\ J\ K^{-1} = +23.91\ J\ K^{-1}$

13. f. entropy and enthalpy are both state functions

14. c. $\Delta S° = \Sigma(\Delta S°_{prodcuts}) - \Sigma(\Delta S°_{reactants}) = (2 \times 192.5) - [(1 \times 191.6) + (3 \times 130.7)] = -198.7\ J\ K^{-1}$

15. d. If the temperature is increased, the value of $\Delta G°$ will become more *negative.*

16. f. I and IV are true
II. At equilibrium, ΔG is equal to zero.
III. At equilibrium, entropy is at a maximum.

17. a.

$HClO_3$	O is –2, H is +1, Cl must be +5
ClO_4^-	O is –2, Cl must be +7
Cl_2	an element by itself, oxidation number must be 0
HCl	H is +1, Cl must be –1

18. f. oxidation half-reaction: $6\ H_2O + Br_2 \rightarrow 2\ BrO_3^- + 12\ H^+ + 10e^-$ leave as is
reduction half-reaction: $2\ e^- + Br_2 \rightarrow 2\ Br^-$ multiply by 5
add: $6\ H_2O + Br_2 + 10\ e^- + 5\ Br_2 \rightarrow 2\ BrO_3^- + 12\ H^+ + 10e^- + 10\ Br^-$
adjust for basic conditions by adding 12 OH^- to both sides:
$6\ H_2O + 6\ Br_2 + 12\ OH^- \rightarrow 2\ BrO_3^- + 12\ H_2O + 10\ Br^-$
cancel the water and reduce the coefficients to the smallest whole number ratio:
$3\ Br_2 + 6\ OH^- \rightarrow BrO_3^- + 3\ H_2O + 5\ Br^-$

19. g. anode half-reaction is the oxidation half-reaction: $Cd(s) \rightarrow Cd^{2+}(aq) + 2e^-$
change the sign for the oxidation at the anode and add the two half-cell potentials:
$E° = +\ 0.40\ V + (-0.13\ V) = +0.27\ V$

20. b. $\Delta G = -nFE° = -2 \times 96485 \times 0.27 = -52102\ J = -52\ kJ$

21. d. $E° = +0.76\ V - 0.74\ V = +0.02\ V$
$E = E° - (RT / nF) \times \ln([Zn^{2+}]^3 / [Cr^{3+}]^2)$
$= 0.02 - [(8.3145 \times 298.15) / (6 \times 96485) \times \ln(0.0085^3 / 0.0100^2)]$
$= 0.04\ V$

22. b. $60.2\ g\ Al \times (1\ mol\ Al / 26.98\ g\ Al) \times (3\ mol\ e^- / 1\ mol\ Al) \times (96485\ C / 1\ mol\ e^-) \times (1As / 1C)$
$\times (1 / 50\ A) \times (1\ hr / 3600\ s) = 3.6\ hr$

EXAMINATION 2

CHEMISTRY 142 SPRING 2010

Solutions

1. e. $[Pb^{2+}] = 3.6 \times 10^{-4}$

   $[NO_3^-] = 2 \times [Pb^{2+}] = 7.2 \times 10^{-4}$

2. a. $K_{sp} = [Pb^{2+}][SO_4^{2-}]$

3. e. the solutions are diluted when mixed
   in the final solution the concentration of $SO_4^{2-} = [SO_4^{2-}] = ½ \times 5.4 \times 10^{-2}$ M $= 2.7 \times 10^{-2}$ M
   $K_{sp} = [Pb^{2+}][SO_4^{2-}] = 1.8 \times 10^{-8}$
   $[Pb^{2+}] = 1.8 \times 10^{-8} / 2.7 \times 10^{-2}$ M $= 6.7 \times 10^{-7}$ M

4. e.

$$MX_2 \rightleftharpoons M^{2+} + 2\,X^- \qquad K_{sp} = 6.0 \times 10^{-18}$$

$$M^{2+} + 4\,NH_3 \rightleftharpoons [M(NH_3)_4]^{2+} \qquad K_f = 5.0 \times 10^{10}$$

$$MX_2 + 4\,NH_3 \rightleftharpoons [M(NH_3)_4]^{2+} + 2\,X^- \qquad K = K_{sp} \times K_f = 3.0 \times 10^{-7}$$

5. c. the solutions are diluted when mixed
   concentration of $Ca^{2+} = [Ca^{2+}] = ½ \times 1.0 \times 10^{-4}$ M $= 5.0 \times 10^{-5}$ M
   concentration of $CO_3^{2-} = [CO_3^{2-}] = ½ \times 2.0 \times 10^{-4}$ M $= 1.0 \times 10^{-4}$ M
   $Q_{sp} = [Ca^{2+}][CO_3^{2-}] = (5.0 \times 10^{-5})(1.0 \times 10^{-4}) = 5.0 \times 10^{-9}$
   $Q_{sp} < K_{sp}$ therefore no precipitation

6. f. pressure at B = 4.00 atm
   temperature at B = P V / nR = 4.00 × 5.00 L / 1.00 × 0.08206 = 243.7 K (same as at A)

7. h. $w = -nRT \ln(V_f/V_i) = -1.00 \times 8.3145 \times 243.7 \times \ln(5/2) = -1857$ J

8. a. w = zero (constant volume change)

9. b. the same as the entropy change down the adiabatic expansion A → C = zero
   S, entropy, is a state function and ΔS is path-independent

10. c. the increase in entropy A → B must be balanced by the increase from B → C, since the sum = zero

sum:	$nR\ln(V_f/V_i) + n\overline{C}_v \ln(T_f/T_i) = 0$
$\overline{C}_v = (3/2)R$:	$nR\ln(V_f/V_i) + n \times (3/2) \times R \times \ln(T_f/T_i) = 0$
cancel n and R:	$\ln(V_f/V_i) + (3/2) \times \ln(T_f/T_i) = 0$
	$\ln(5/2) + (3/2) \times \ln(T_f/243.7) = 0$
	$\ln(T_f/243.7) = -0.6109$
	$T_f = 132.3$ K

11. e. ΔE (A → C) = q + w, but q = zero, so ΔE (A → C) = w
ΔE (A → B → C) must be the same (path-independent)
but ΔE (A → B) = zero (isothermal expansion)
and w for B → C = zero (constant volume)
ΔE (B → C) = q
so ΔE (A → C) = q for B → C
$= n\bar{C}_v \Delta T$
$= n\bar{C}_v (T_f - T_i)$
$= 1.00 \times (3/2) \times 8.3145 \times (132.3 - 243.7)$
$= -1389$ J

12. g. answer to a): probability = $(1/2)^8 = 1/256$
answer to b): each of the 8 molecules could be the one in the right half
probability = $8 \times 1/256 = 1/32$

13. b.

isothermal	ΔE = zero
constant volume	$\Delta E = q_v$
constant pressure	$\Delta E = q_p - P\Delta V$
adiabatic	$\Delta E = w$

14. e. it is ΔG that is zero at equilibrium, not $\Delta G°$

15. c. two, heat and work

16. g. $6 + (-1) + 0 + 2 = 7$

17. h. Li the metal with the most negative reduction potential
i.e., the metal that is oxidized most easily
note that, if anything, the positive ions would tend to be oxidizing agents

18. e. chlorine Cl_2 is oxidized to ClO_3^- (it is also the element being reduced to Cl^-)
for the oxidation, the change in the oxidation number is 0 to +5

19. b. $H_2O + 2\ MnO_4^- + 3\ SO_3^{2-} \rightarrow 2\ MnO_2 + 3\ SO_4^{2-} + 2\ OH^-$

20. b. change the sign for the oxidation at the anode and add the two half-cell potentials
$E° = 0.356V + 1.685V = +2.041V$

21. e. $\Delta G° = -nFE° = -2 \times 96{,}485 \times 2.041 = -393{,}852\ J = -394\ kJ$

22. d. 10 A × 24 hr
× (3600 s/1 hr) × (1 C/1 As) × (1 mol e^-/96485 C) × (1 mol Cu/2 mol e^-) × (63.55 g /1 mol Cu)
= 285 g Cu

EXAMINATION 2

CHEMISTRY 142 FALL 2010

Solutions

1. c. $Pb(OH)_2(s) \rightleftharpoons Pb^{2+}(aq) + 2\,OH^-(aq)$
   $K_{sp} = [Pb^{2+}][OH^-]^2$

2. a. $[Ag^+] = 3 \times [PO_4^{3-}]$
   $K_{sp} = [Ag^+]^3[PO_4^{3-}] = 3^3 \times [PO_4^{3-}]^4 = 1.8 \times 10^{-18}$
   $[PO_4^{3-}] = {}^4\sqrt{}(1.8 \times 10^{-18} / 27) = 1.6 \times 10^{-5}$ M

3. c. $K_{sp} = [Mg^{2+}][F^-]^2 = (0.050)[F^-]^2 = 6.4 \times 10^{-9}$
   $[F^-] = \sqrt{}(6.4 \times 10^{-9} / 0.050) = 3.6 \times 10^{-4}$ M

4. a. The concentrations are halved when the solutions are mixed
   $Q_{sp} = [Fe^{2+}][F^-]^2 = (5.0 \times 10^{-3})(1.2 \times 10^{-2})^2 = 7.2 \times 10^{-7}$
   No precipitation of FeF_2 because $Q < K_{sp}$

5. e. HNO_3 no effect (Br^- is the conjugate base of a strong acid)
   NaBr decreases (common ion effect)
   EDTA increases (the Pb^{2+} forms the EDTA complex ion)

6. c. q = –25 kJ (heat released by the system as it is cooled)
   w = +10 kJ (work done on the system)
   $\Delta E = q + w = -15$ kJ

7. f. According to the first law of thermodynamics, $\Delta E_{univ} = 0$
   The second law of thermodynamics indicates that for any spontaneous process, $\Delta S_{univ} > 0$

8. f. PV = nRT
   n = PV / RT = 6.80 × 14.3 / 0.08206 × (273.15 + 45) = 3.72 mol (5 points)

9. d. no change in volume; no work done

10. c. the expansion is isothermal, therefore ΔE = zero and the heat absorbed must equal the work done
   V_f (at point C) = 14.3 L × (6.80/2.44) = 39.85 L
   $w = -nRT \ln(V_f/V_i) = -3.72 \times 8.3145 \times 318 \times \ln(39.85/14.3) = -10{,}085$ J = –10.1 kJ
   q = +10.1 kJ (10.1 kJ heat is transferred from the surroundings to the gas sample)

11. b. $\Delta E = q + w = nC_p\Delta T - P\Delta V$
   $= 3.72 \text{ mol} \times (5/2)8.3145\ JK^{-1}mol^{-1} \times (-204K) - (2.44 \text{ atm} \times -25.6 \text{ L} \times 101.325\ JL^{-1}atm^{-1})$
   $= -15{,}774 + 6329 = -9.44$ kJ
   alternatively, more easily, for B → A, w = 0 and $q = nC_v\Delta T = 9.46$ kJ, so C → B must be –9.46 kJ

12. d. ΔE = zero regardless of the path (E is a state function and ΔE is path-independent)

13. b. $\Delta S = k \ln(W_f/W_i) = k \ln(V_f^N/V_i^N) = k \ln(V_f/V_i)^N = Nk \ln(V_f/V_i)$
$= 9 \times 1.38 \times 10^{-23} \times \ln(4/1) = 1.72 \times 10^{-22}$ J K^{-1}

14. f. the entropy of block A increases (increase in temperature)
the entropy of block B decreases (decrease in temperature)
the total entropy of the system increases (spontaneous process—it happens)

15. a. $\Delta G° = \Delta H° - T\Delta S°$
$\Delta H° = -74.9$ kJ $= -74{,}900$ J
$T\Delta S° = 298.15$ K $\times (186.3 - 2 \times 130.7 - 5.6)$ JK^{-1} $= 24{,}060$ J
$\Delta G° = -74{,}900$ J $+ 24{,}060$ J $= -50{,}800$ J $= -50.8$ kJ

16. g. $K = e^{-(\Delta G°/RT)} = 1.59 \times 10^7$

17. e. +5

18. a. $S_2O_6^{2-}(aq) + 2\,H_2O \rightarrow 2\,SO_4^{2-}(aq) + 4\,H^+ + 2\,e^-$ × 3
$2\,HClO_2(aq) + 6\,H^+ + 6\,e^- \rightarrow Cl_2(g) + 4\,H_2O$

$3\,S_2O_6^{2-}(aq) + 2\,H_2O + 2\,HClO_2(aq) \rightarrow 6\,SO_4^{2-}(aq) + 6\,H^+ + Cl_2(g)$

19. a. In a voltaic/galvanic cell, *oxidation* occurs at the anode and electrons flow from the *anode to cathode* in the external circuit.

20. c. Pt*(s)* | $Fe^{2+}(aq)$, $Fe^{3+}(aq)$ || $Cl^-(aq)$ | $Cl_2(g)$ | C*(s, graphite)*
E° = –0.77 + 1.36 = 0.59 V

21. c. $Cl_2(g)$

22. d. $3\,Pb^{2+}(aq) + 2\,Cr(s) \rightarrow 3\,Pb(s) + 2\,Cr^{3+}(aq)$
E = E° – (RT/nF) × ln([products]/[reactants])
E° = +0.74 V (oxidation) – 0.13 V (reduction) = 0.61 V
E = 0.61 V – (8.3145 × 298 / 6 × 96485) × ln([6.5 × 10^{-4}]^2/[8.9 × 10^{-1}]^3) V
= 0.61 V + 0.062 V
= 0.67 V

23. a. all answers for E in the preceding question are positive, so the reaction must be spontaneous

24. d. 20.0 A × 6.00 hr × (3600 s /1 hr) × (1 C /1 As) × (1 mol e^- / 96485 C) × (1 mol Cu /2 mol e^-)
× (63.55 g Cu /1 mol Cu) = 142 g Cu

EXAMINATION 2

CHEMISTRY 142 SPRING 2011

Solutions

1. c. three calcium ions are produced for every two phosphate ions

2. c. $K_{sp} = [Ca^{2+}]^3[PO_4^{3-}]^2$

3. c. $[Ca^{2+}] = (3/2) \times [PO_4^{3-}]$
   $K_{sp} = [Ca^{2+}]^3[PO_4^{3-}]^2$
   $K_{sp} = ((3/2) \times [PO_4^{3-}])^3[PO_4^{3-}]^2 = (27/8) \times [PO_4^{3-}]^5 = 1.2 \times 10^{-26}$
   $[PO_4^{3-}] = \sqrt[5]{(1.2 \times 10^{-26} \times 8/27)} = 5.13 \times 10^{-6}$ M
   molar concentration of $Ca_3(PO_4)_2 = [PO_4^{3-}]/2 = 2.6 \times 10^{-6}$ M

4. c. $K_{sp} = [Ca^{2+}]^3[PO_4^{3-}]^2$
   $= [Ca^{2+}]^3 \times (0.25)^2 = 1.2 \times 10^{-26}$
   $[Ca^{2+}]^3 = \sqrt{(1.2 \times 10^{-26} / 0.25^2)}$
   $[Ca^{2+}] = \sqrt[3]{(1.92 \times 10^{-25})}$
   $= 5.8 \times 10^{-9}$ M
   molar concentration of $Ca_3(PO_4)_2 = [Ca^{2+}]/3 = 1.9 \times 10^{-9}$ M

5. c. $PbCO_3(s) \rightarrow Pb^{2+}(aq) + CO_3^{2-}(aq)$ $\quad K_{sp1} = 3.3 \times 10^{-14}$
   $PbS(s) \rightarrow Pb^{2+}(aq) + S^{2-}(aq)$ $\quad K_{sp2} = 3.4 \times 10^{-28}$
   reverse equation 2 and add to obtain... *(the $Pb^{2+}(aq)$ cancel)*
   $PbCO_3(s) + S^{2-}(aq) \rightarrow PbS(s) + CO_3^{2-}(aq)$ $\quad K = K_{sp1} / K_{sp2} = 9.7 \times 10^{13}$

6. f. six (all except heat and work)

7. c.

8. d. $n = PV/RT = (8.00 \times 3.00)/(0.08206 \times 292.5) = 1.00$ mol

9. d. $w = -nRT \ln(V_f/V_i) = -1.00 \times 8.3145 \times 292.5 \times \ln(12/3) = -3371$ J

10. g. ΔS for the step A to B = $nR \ln(V_f/V_i) = +11.53$ JK^{-1}
    but ΔS for the step A to C is zero (adiabatic expansion where q = zero)
    and ΔS is path-independent, so ΔS for the step B to C = -11.53 JK^{-1}

11. f. first, no work done since the volume does not change
    second, work = $-P\Delta V = -2.00$ atm $\times (12.0 - 3.00)$ L $\times$ 101.325 J $L^{-1}atm^{-1} = -1824$ J

12. d. zero; ΔE is path-independent

13. f. eight molecules; number of arrangements with one molecule in left half = 8
total arrangements = 2^8 = 256
probability = 8 in 256, or 1 in 32

14. g. all are true

15. c. melting point = $\Delta H°/\Delta S°$ = 6120 J mol^{-1}/22.4 JK–1mol–1
= 273 K
= 0°C

16. a. oxidation, electrons move from anode to cathode in the external circuit

17. h. BrO_3^-; three O at –2 each, –1 for the charge on the ion, therefore bromine has an oxidation state +5, that is, 3(–2) + 5 = –1

18. c. $2\ e^- + 4\ H^+ + 2\ VO_2^+ \rightarrow 2\ VO^{2+} + 2\ H_2O$
$Zn \rightarrow Zn^{2+} + 2\ e^-$

$4\ H^+ + 2\ VO_2^+ + Zn \rightarrow 2\ VO^{2+} + 2\ H_2O + Zn^{2+}$

add 4 OH^- to both sides and cancell two H_2O:

$2\ H_2O + 2\ VO_2^+ + Zn \rightarrow 2\ VO^{2+} + 4\ OH^- + Zn^{2+}$

19. a. change the sign for the oxidation, and add
E° = +1.66 +0.337 V = +1.997 V

20. h. $2\ Al(s) + 3\ Cu^{2+}(aq) \rightarrow 3\ Cu(s) + 2\ Al^{3+}(aq)$
n = 6
lnK= nFE°/RT = 6 × 96485 C × 1.997 V / 8.3145 $JK^{-1}mol^{-1}$ × 298.15 K
= 466.36
K = 3.44×10^{202}

21. b.

22. g. 86 kg × (1000 g/1 kg) × (1 mol Na/22.99 g Na) × (1 mol e^-/1 mol Na) × (96485 C/1 mol e^-) × (1 A s/1 C) × (1/25 × 10^3 A) × (1 hour/3600 s) = 4.0 hours

FINAL EXAMINATION

CHEMISTRY 142 SPRING 2006

Solutions

1. d. the rate at which water is produced is 4/3 times the rate at which oxygen is consumed
   $= 2.4 \times 10^2 \times (4/3) = 3.2 \times 10^2$

2. d. three half-lives, to 1/2, to 1/4, then to 1/8, so 15 minutes

3. d. reactants are NO and H_2
   intermediates are N_2O_2 and N_2O
   products are H_2O and N_2

4. f. the half-lives of ^{13}N and ^{15}O are too short, essentially none will remain after 47 years
   the half-life of ^{14}C is too long for an accurate result, the radioactivity will hardly change in 47 years

5. e. all except e—the ultimate equilibrium is the same; it is merely reached more quickly

6. f. $K = 20 = [N_2H_4][HCl]^2 / [NH_3]^2[Cl_2] = (5/12)[HCl]^2 / (2/12)^2(1/12)$
   $[HCl]^2 = 16/144$
   $[HCl] = 1/3 \text{ mol L}^{-1}$
   number of moles in the 12 L flask $= 1/3 \text{ mol L}^{-1} \times 12 \text{ L} = 4$ mol

7. g. reverse the equation and divide by 2 so take the inverse and the square root of K = 0.22

8. h. HBr is a strong acid and will produce the most H_3O^+ ions in solution

9. d. assign the label 'base' to OH^- and then go from there

10. d. first calculate the pK_a: $pH = 2.57 \times 10^{-3}$
    $K_a = (2.57 \times 10^{-3})^2 / (0.10 - 2.57 \times 10^{-3}) = 6.78 \times 10^{-5}$
    therefore the $pK_a = 4.17$ and $pH = pK_a + \log([base]/[acid]) = 4.17 + \log 2 = 4.47$

11. d. HSO_3^- and SO_3^{2-}

12. e. $pH = ½(pK_{a1} + pK_{a2}) = 4.55$

13. c. $pH = pK_a + \log([base]/[acid]) = 3.52 - 0.52 = 3.0$

14. c. $pOH = 14 - pH = 11.9$
    $[OH-] = 1.3 \times 10^{-12}$

15. a. $K_{sp} = [Cu^{2+}][IO_3^-]^2 = 1.4 \times 10^{-7}$
    and $[IO_3^-] = 2 \times [Cu^{2+}]$
    $K_{sp} = 4 \times [Cu^{2+}]^3$
    $[Cu^{2+}] = 3.3 \times 10^{-3}$ M

16. d. $K_{sp} = [Cu^{2+}][IO_3^-]^2 = 1.4 \times 10^{-7}$
$= [Cu^{2+}](0.50)^2 = 1.4 \times 10^{-7}$
$[Cu^{2+}] = 5.6 \times 10^{-7}$ M

17. f. decrease in volume from B to C = 3.0L – 12.0 L = –9.0 L
$w = -P\Delta V = -8.0$ atm × –9.0 L × 101.325 J/L atm = 7.3 kJ

18. b. temperature at A and C = PV/nR = 2.0 × 12.0 / 0.50 × 0.08206 = 585 K
$w = -nRT \ln(V_f/V_i) = -0.50 \times 8.3145 \times 585 \times \ln(1/4) = 3.4$ kJ

19. a. must be zero; temperature is the same at A and C

20. e. vapor has the highest entropy

21. f. all of the reactions listed

22. f. H_2O_2

23. d. $E = E° - (RT/nF)\ln([\text{product}]/[\text{reactant}])$
$= 2.04 - (8.3145 \times 298.15/2 \times 96{,}485) \times \ln(1/4.5^4)$
$= 2.04 - (8.3145 \times 298.15/2 \times 96{,}485) \times -6.02 = 2.04 + 0.08 = 2.12$

24. d. leave equation 1 as it is; reverse equation 2 and divide by 2
$K = 2.3 \times 10^6 \times \sqrt{(1/1.8 \times 10^{37})} = 5.4 \times 10^{-13}$

25. d. $\Delta G° = -RT \ln K = -8.3145 \times 600 \times \ln(5.4 \times 10^{-13}) = 140{,}900$ J = 141 kJ

26. b. 2

27. d. Be B Xe

28. d. 0.67 g Ag × (1 mol Ag/107.87 g Ag) × (1 mol e^-/1 mol Ag) × (96485 C/1 mol e^-) = 600 C

29. b.

30. a.

31. b. 1 other isomer — *trans* isomer

32. f. 6

33. a. Fe^{2+} has a d^6 configuration and water is a weak ligand, so LFSE = –4Dq

34. c. atomic numbers must balance: 83 + 26 = 109 + x, so x = 0
mass numbers must balance: 209 + 58 = 266 + y, so y = 1
particle emitted = ${}^1_0n$

35. a. binding energy = mass defect × $(3.0 \times 10^8\ ms^{-1})^2$
$= 1.1555$ g × (1 kg/1000 g) × $(3.0 \times 10^8\ ms^{-1})^2 = 1.04 \times 10^{14}$ J

FINAL EXAMINATION

CHEMISTRY 142 FALL 2006

Solutions

1. f. if PH_3 is used at 12 mol L^{-1} s^{-1}, then P_4 is formed at 12 × (1/4) mol L^{-1} s^{-1} = 3 mol L^{-1} s^{-1}
after 5 seconds, the concentration will be 5 s × 3 mol L^{-1} s^{-1} = 15 mol L^{-1}

2. c. $\ln ([R]_t/[R]_0) = -k\ t$
$[R]_t = 0.25$ M, $[R]_0 = 1.25$ M, $k = 3.2 \times 10^{-4}\ s^{-1}$
$t = -\ln (0.25/1.25)/(3.2 \times 10^{-4}\ s^{-1}) = 5.03 \times 10^3$ s

3. a. $\ln (k_1/k_2) = -(E_A/R)(1/T_1 - 1/T_2)$
$\ln [(4 \times 10^{-2})/(4.24 \times 10^{6})] = -(E_A/8.314\ J\ K^{-1}\ mol^{-1})[(1/212\ K) - (1/945\ K)]$
$-18.48 = -E_A\ (4.4 \times 10^{-4}\ J\ mol^{-1})$ and $E_A = 42{,}000\ J\ mol^{-1} = 42\ kJ\ mol^{-1}$

4. b. $MoCl_5^-$ — an intermediate appears in the elementary steps of a reaction, but not in the overall chemical equation; it is formed in one step and used in a susequent step

5. h. $Rate = k[NO_3^-][MoCl_5^-]$ (rate depends on the slower elementary step)
$K = [MoCl_5^-][Cl^-]/[MoCl_6^{2-}]$ (equilibrium expression for second step)
$[MoCl_5^-] = K[MoCl_6^{2-}]/[Cl^-]$
$Rate = (k[NO_3^-]) \times (K[MoCl_6^{2-}]/[Cl^-]) = (kK[NO_3^-][MoCl_6^{2-}])/[Cl^-]$

6. d. 4 strong acids — HBr, HNO_3, HCl, HI

7. e. HPO_4^{2-} accepts H^+ — base
H_2O donates H^+ — acid
$H_2PO_4^-$ donates H^+ — acid
OH^- accepts H^+ — base

8. e. add a catalyst — no change
add S — no change (addition of solid does not change a gas phase equilibrium)
add H_2O — shift left
decrease V — shift right (shift to side with fewer moles of gas)

9. d. three — Fe, Ti, Mn

10. d. $K_p = (P_{CO2})^4/(P_{C2H2})^2(P_{O2})^5$ $K_p = (4)^4/[(2)^2(5)^5] = 2.05 \times 10^{-2}$

11. c. two — NH_4NO_3 and NH_4Cl — both contain the conjugate acid of a weak base (NH_3)

12. c. $C_6H_5OH + H_2O \rightleftharpoons C_6H_5O^- + H_3O^+$
$K_a = [C_6H_5O^-][H_3O^+]/[C_6H_5OH] = [H_3O^+]^2 / 3.0 = 1.6 \times 10^{-10}$
$[H_3O^+] = 2.19 \times 10^{-5}$ $pH = -\log[H_3O^+] = -\log(2.19 \times 10^{-5}) = 4.66$

13. e. $pH = pK_a + \log([\text{conj. base}]/[\text{acid}])$

$pH = -\log(1.6 \times 10^{-10}) + \log(0.75/3.0)$

$pH = 9.80 - 0.60 = 9.2$

14. e. $K_p = [N_2][HF]^6[Cl_2]/[ClF_3]^2[NH_3]^2 = (1.5)(9)^6(1.5)/(0.5)^2(0.5)^2 = 1.913 \times 10^7$

$K_p = K_c(RT)^{\Delta n}$

$K_c = K_p/[(RT)^{\Delta n}] = (1.913 \times 10^7)/(0.08206 \times 298)^4 = 53.5$

15. c. $pH = ½ (pK_{a1} + pK_{a2}) = ½ (1.187 + 4.215) = 2.70$

16. e. neutralization reaction $K_{neut} = K_a \times K_b / K_w = (1.35 \times 10^{-3})(5.6 \times 10^{-4})/(10^{-14}) = 7.56 \times 10^7$

17. b. $Ag_3PO_4(s) \rightarrow 3Ag^+(aq) + PO_4^{3-}(aq)$

let $[Ag^+] = 3x$ and $[PO_4^{3-}] = x$

$K_{sp} = [Ag^+]^3[PO_4^{3-}] = (3x)^3(x) = 27x^4 = 1.8 \times 10^{-18}$

$x = 1.61 \times 10^{-5}$ so $[Ag^+] = 3x = 3(1.61 \times 10^{-5}) = 4.8 \times 10^{-5}$

18. d. $K_{sp} = [Ag^+]^3[PO_4^{3-}]$

$[Ag^+] = (K_{sp}/[PO_4^{3-}])^{1/3} = [(1.8 \times 10^{-18})/(4.5 \times 10^{-2})]^{1/3} = 3.4 \times 10^{-6}$

19. c. $PV = nRT$ $P = nRT/V = (4 \times 0.08206 \times 300)/13 = 7.6$ atm

20. b. $PV = nRT$ $T = PV/nR = (12 \times 7.5)/(4 \times 0.08206) = 274$ K

or

new temperature = $300 \text{ K} \times (7.5/13) \times (12/7.6) = 274$ K

21. e. $w = -P\Delta V = -7.6 \times (7.5 - 13) = 41.8 \text{ L atm} \times 101.325 \text{ J L}^{-1} \text{ atm}^{-1} = 4240 \text{ J} = 4.24 \text{ kJ}$

22. f. temperature at state B = $300 \times (7.5/13) = 173$ K

$q = nC_p\Delta T = (4 \times (5/2) \times 8.314 \times (173 - 300)) = -10560 \text{ J} = -10.56 \text{ kJ}$

23. c. $\Delta E = q + w = -10.56 \text{ kJ} + 4.24 \text{ kJ} = -6.32 \text{ kJ}$

24. c. $\Delta S° = \Sigma S°_{(products)} - S°_{(reactants)}$

$\Delta S° = [(3 \times 33) + (2 \times 189)] - [(2 \times 206) + (1 \times 248)] = -183 \text{ JK}^{-1}$

25. c. $\Delta G° = \Delta H° - T\Delta S°$

$\Delta H° = [(2 \times -242)] - [(2 \times -21) + (1 \times -297)] = -145 \text{ k} = -145{,}000 \text{ J}$

$\Delta G° = (-145{,}000 \text{ J}) - (150 \text{ K} \times -183 \text{ JK}^{-1}) = -118{,}000 \text{ J} = -118 \text{ kJ}$

26. b. first find K at 150 K: $K = e^{-\Delta G°/RT} = e^{-[-118{,}000 / (8.314 \times 150)]} = 1.24 \times 10^{41}$

then find K at 770 K: $\ln(K_2/K_1) = \Delta H°/R [(1/T_1) - (1/T_2)]$

$\ln(K_2/1.24 \times 10^{41}) = -145{,}000/8.314[(1/150) - (1/777)]$

$K_2 = 2.72$

27. h. $4\,Ag(s) + 8\,CN^-(aq) + 2\,H_2O(l) + O_2(g) \rightarrow 4\,[Ag(CN)_2]^-(aq) + 4\,OH^-(aq)$

28. h. change the sign of the reduction potential for the oxidation at the anode, and then add the two potentials:

$$E° = 2.92\ V + 0.56\ V = 3.48\ V$$

29. e.

$$0.44\ g\ Mn \times \frac{1\ mole\ Mn}{54.94\ g\ Mn} \times \frac{4\ mole\ e^-}{1\ mole\ Mn} \times \frac{96{,}485\ C}{1\ mole\ e^-} = 3{,}091\ C$$

$$3{,}091\ C \times \frac{1\ mole\ e^-}{96{,}485\ C} \times \frac{1\ mole\ Ag}{1\ mole\ e^-} \times \frac{107.87\ g\ Ag}{1\ mole\ Ag} = 3.46\ g\ Ag$$

30. e. $[Co(NH_3)_5Cl]Cl_2$

31. b. two isomers – *cis* and *trans*

32. f. 6 —an octahedral complex has six ligands

33. b. In $[MnF_6]^{3-}$, the oxidation state of manganese is Mn^{3+}, therefore d^4
F^- is a weak ligand so the complex is high spin
LFSE = $(3 \times -4Dq) + (6Dq) = -6Dq$

34. d. balance the atomic and mass numbers on both sides

$${}^{239}_{93}Np \rightarrow {}^{239}_{94}Pu + ?$$

the missing particle is ${}^{0}_{-1}\beta$

35. b. total mass of the nucleons = $[(12 \times 1.007825\ u) + (12 \times 1.008665\ u)] = 24.19788\ u$

actual mass = 23.9784 u

defect = 0.21948 u
= $0.21948\ g\ mol^{-1}$
= $2.1948 \times 10^{-4}\ kg\ mol^{-1}$

binding energy = $E = mc^2 = 2.1948 \times 10^{-4}\ kg\ mol^{-1} \times (2.99792 \times 10^8\ m\ s^{-1})^2$
= $1.973 \times 10^{13}\ kg\ mol^{-1}\ m^2\ s^{-2}$
= $1.973 \times 10^{13}\ J\ mol^{-1}$

FINAL EXAMINATION

CHEMISTRY 142 SPRING 2007

Solutions

1. f. the rate at which aluminum is used is 10/9 times the rate at which water is produced
   $= 7200 \times (10/9) = 8000$ mol L^{-1} s^{-1}

2. j. 192 mol L^{-1} to 12 mol L^{-1} is four half-lives, to 96, to 48, to 24, then to 12, and takes 60 min
   60 min / 15 = 4 min. Successive half-lives are: 4 min; 8 min; 16 min; and 32 min
   two more half-lives are 64 min and 128 min = 192 min

3. c. A G B E

4. f. compare Experiments 1 and 3; rate and concentration of NO_2 increase by same amount

5. f. 6.3 *m*mol L^{-1} s^{-1} $\times (0.66/0.21) \times (0.18/0.70) = 5.1$ *m*mol L^{-1} s^{-1}

6. g. $\ln(k_1/k_2) = -(E_A/R)(1/T_1 - 1/T_2)$
   $\ln(12) = -(E_A/8.3145)(1/301 - 1/278)$
   $\ln(12) = (E_A/8.3145)(23/(301 \times 278))$
   $E_A = 75{,}000 \text{ J} = 75 \text{ kJ}$

7. d. H_2SO_3 (weak acid) NaOH (strong base) NH_4Br (acidic salt) NaCl (neutral salt)

8. i. $[OH^-] = 0.001$ M pOH = 3.0 and pH = 14 – 3.0 = 11.0

9. e. $K = 54 = [CO_2][H_2]^3 / [CH_4][H_2O] = (3.0)[H_2]^3 / (1.0)(1.5)$
   $2.0 \times [H_2]^3 = 54$
   $[H_2] = 3.0$ mol L^{-1}

10. c. assign the label 'base' to OH^- and then go from there

11. e. K_{neut} for the neutralization reaction $= K_a / K_w = 4.0 \times 10^4$

12. d. pH = pK_a + log([base]/[acid]); since in this case [base] = [acid], then pH = pK_a = 4.74

13. d. for $KHCO_3$, pH = ½($pK_{a1} + pK_{a2}$) = ½(6.37 + 10.25) = 8.31, i.e. basic
    so K_2CO_3 must be basic

14. c. reverse first equation and add to second, so pK = $pK_{a2} - pK_{a1}$ = 3.89

15. b. $K_{sp} = [Sb^{3+}]^2[S^{2-}]^3$

16. d. sulfide is more concentrated (more of them by a factor of 3/2)

17. a. $K_{sp} = [Sb^{3+}]^2[S^{2-}]^3 = ((2/3)[S^{2-}])^2[S^{2-}]^3 = (4/9)[S^{2-}]^5$ $[S^{2-}] = 3.285 \times 10^{-19}$ M
    solubility of 'Sb_2S_3' $= (3.285 \times 10^{-19} \text{ M}) / 3 = 1.1 \times 10^{-19}$ M

18. f. decrease in volume to one-half; the pressure must double to 6.0 atm

19. b. $PV = nRT$; $3.0 \times 10 = 1.5 \times 0.08206 \times T$; $T = 243.7K = -29.4°C$

20. e. $w = -nRT \ln(V_f/V_i) = -1.5 \times 8.3145 \times 243.7 \times \ln(1/2) = 2100$ J

21. a. must be zero; constant volume

22. d. $q_v = n\,C_v\,\Delta T = 1.5 \times (3/2) \times 8.3145 \times 243.7 = 4560$ J

23. d. increase in entropy

24. f. +5

25. g. $6e^- + 14H^+ + Cr_2O_7^{2-} \rightarrow 2Cr^{3+} + 7H_2O$ multiply by 2 and add
$3H_2O + C_2H_5OH \rightarrow 2CO_2 + 12H^+ + 12e^-$

$16H^+ + 2Cr_2O_7^{2-} + C_2H_5OH \rightarrow 4Cr^{3+} + 11H_2O + 2CO_2$

26. c. change the sign of 0.17 V (for oxidation) and add: 0.96 V – 0.17 V = 0.79 V

27. f. $\Delta G° = -nFE° = 6 \times 96{,}485 \times 0.79 = -457{,}000$ J = 457 kJ

28. d. 63.55 g × (1 mol Cu/63.55 g Cu) × (2 mol e^-/1 mol Cu) × (96485 As/1 mol e^-) × (1/10A)
= 19,297 s
= 322 min

29. c. 1 and 4 only (two ligands arranged *cis* or *trans*)

30. a. Fe^{2+} has a d^6 configuration and water is a weak ligand, so LFSE = –4Dq

31. j. N B F Be

32*a*. e.

32*b*. e. ^{15}O has the lowest n/p ratio (7/8)

33*a*. f.

33*b*. f. 28 – 2 for the 2+ charge + 8 from the four ligands = 34

34*a*. d.

34*b*. d. ΔG = zero at equilibrium

35*a*. a.

35*b*. a. disproportionation

FINAL EXAMINATION

CHEMISTRY 142 FALL 2007

Solutions

1. h. 36 mol L^{-1} min^{-1} ClO × (7 O_2 / 4 ClO) = 63 mol L^{-1} min^{-1} O_2

2. h. 64 mol L^{-1} to 8.0 mol L^{-1} is three half-lives, to 32, to 16, then to 8.0, and takes 28 min
   28 min / 7 = 4 min. Successive half-lives are: 4 min; 8 min; and 16 min; i.e. 28 min
   two more half-lives, to 4.0 mol and then 2.0 , are 32 min and 64 min = 96 min

3. g. compare Experiments 2 and 3; rate and concentration of CH_3I both double
   so the order with respect to CH_3I = 1
   compare Experiments 1 and 2; rate increases by a factor of 4
   and the concentrations of both C_5H_5N and CH_3I both double
   the effect of increasing the CH_3I is already known, so doubling C_5H_5N must also double the rate
   so the order with respect to C_5H_5N = 1
   overall order = 1 + 1 = 2

4. j. an intermediate is first formed and then used in a subsequent step, so O, ClO, Cl, and CF_2Cl are all intermediates

5. d. a catalyst changes the route or mechanism of a reaction

6. e. reverse equation 1 and add to equation 2
   $K_3 = K_2 / K_1$

7. c. KCN is weakly basic—the cyanide ion hydrolyses to produce HCN and OH^-

8. e.

9. e.

10. b. the neutralization constant = $K_a \times K_b / K_w = (1.8 \times 10^{-5})^2 / 10^{-14} = 3.24 \times 10^4$

11. e. the amount of acid added is one half of the base present
    the system is one-half of the way to the equivalence point, where $[NH_3] = [NH_4^+]$
    so, pOH = pK_b = 4.745
    pH = 14 – 4.745 = 9.26
    alternatively, use the Henderson-Hasselbalch equation, using the pK_a for NH_4^+
    pH = 9.26 + log([base]/[acid]) = 9.26 + log($[NH_3]/[NH_4^+]$) = 9.26

12. c. the pH stays the same, since the $[NH_3]/[NH_4^+]$ ratio stays the same, equal to 1

13. d.

14. a. at the equivalence point, the pH = ½($pK_{a1} + pK_{a2}$) = 2.71 —acidic

15. e. halfway to the second equivalent point, where $pH = pK_{a2} = 4.19$

16. d. moles of Ag_2CrO_4 per liter = 0.029 g / 331.73 g mol^{-1} = 8.74×10^{-5}
$[Ag^+] = 2 \times 8.74 \times 10^{-5}$
$[CrO_4^{2-}] = 8.74 \times 10^{-5}$
$K_{sp} = [Ag^+]^2[CrO_4^{2-}] = 4 \times (8.74 \times 10^{-5})^3 = 2.7 \times 10^{-12}$

17. a. b, c, d, and e have the same stoichiometry, and of these $CaCO_3$ has the highest molar solubility $= \sqrt{K_{sp}} = 9.3 \times 10^{-5}$ M
even though its K_{sp} is smaller, silver carbonate, stoichiometry Ag_2CO_3, has a higher solubility $= 1.2 \times 10^{-4}$ M

18. c. 5 atm × 4 L = 2.0 mol × 0.08206 L atm K^{-1} mol^{-1} × T, so T = 121.9 K

19. e. $w = -P\Delta V = -5 \times 8 \times 101.325$ J = –4050 J

20. g. $q_p = nCp\Delta T = 2 \times (5/2)R \times 243.7 = +10132$ J

21. a. constant temperature, so ΔE = zero

22. g. ΔS is path independent (S is a state function), so choose reversible isothermal expansion
$\Delta S = nR \ln(V_f/V_i) = 2 \times 8.3145 \times \ln(12/4) = 18.3\ JK^{-1}$

23. g. $\Delta G° = -RT \ln K = -8.3145 \times 298.15 \times \ln(4.5 \times 10^{-6}) = 30{,}500$ J = 30.5 kJ

24. a. very much reactant-favored

25. b.

26. e.

27. h. change the sign for the oxidation half-reaction and add: +0.76 + 1.49 = +2.25 V

28. f. $\Delta G° = -nFE° = -10 \times 96{,}485 \times 2.25 = -2171$ kJ

29. e. 1.45 A × 2.60 hr × (3600 s/1 hr) × (1C /1 As) × (1 mol e^-/96485 C) × (1 mol Cd / 2 mol e^-) × (112.41 g/1 mol Cd) = 7.9 g deposited at cathode (the Cd^{2+} ions are reduced at the cathode)

30. g. boron

31. c. 3

32. c.

33. e. 5 geometrical isomers

34. d.

35. h. configuration is d^6, all six electrons in the lower set, LFSE = 6 × –4 = –24 Dq

FINAL EXAMINATION

CHEMISTRY 142 SPRING 2008

Solutions

1. e. N_2O_5 is used at half the rate and oxygen is formed at one-fourth the rate.

2. d. the rate is independent of the concentration
   if it takes 60 seconds for the concentration to decrease by 36 mol L^{-1}
   it will take 1/4 as long to decrease by 9 mol L^{-1}

3. d. compare Experiments 2 and 3: rate and concentration of H_2 both double
   so the order with respect to $[H_2] = 1$
   compare Experiments 1 and 2: rate decreases by a factor of 4 when the [NO] decreases by 2
   so the order with respect to [NO] = 2
   Rate = $k[NO]^2[H_2]$

4. b. compare Experiments 1 and 4:
   Rate = $0.136 \times (0.105/0.420)^2 \times (0.488/0.122) = 0.0340$ mol $L^{-1}s^{-1}$

5. e. intermediate

6. f. P S R Q

7. g.

acidic:	NaH_2PO_4	CH_3CO_2H	HF	NH_4Cl
basic:	Na_2HPO_4	NaOH	CH_3NH_2	KCH_3CO_2
neutral:	LiBr	KCl	KBr	

8. c.

9. d. $K = 12 = (3.0)x^2 / (2.0)^2(1.0)$
   $x = 4.0$

10. b. H^+ concentration after dilution = 0.0010 M = 10^{-3} M
    pH = 3.0

11. c. $pH = ½(pK_{a1} + pK_{a2}) = 2.71$

12. b $K = K_{a2}/K_{a1} = 1.1 \times 10^{-3}$

13. g. $K_b = 1.8 \times 10^{-5} = [NH_4^+][OH^-] / [NH_3] = [OH^-]^2 / [NH_3] = [OH^-]^2 / 0.20$
    $[OH^-] = 1.90 \times 10^{-3}$
    pH = 11.28

14. e. conjugate base of the weakest acid, HCN

15. d. pH = 4.74 + log([base]/[acid]) = 4.74 + log($[CH_3CO_2^-]/[CH_3CO_2H]$) = 4.56

16. e.

17. c. $K_{sp} = [Ag^+][OH^-]$
$[Ag^+] = \sqrt{K_{sp}} = 1.2 \times 10^{-4}$

18. g. if the pH = 9.0, then $[OH^-] = 10^{-5}$
$K_{sp} = 1.5 \times 10^{-8} = [Ag^+][OH^-] = [Ag^+][10^{-5}]$
$[Ag^+] = 1.5 \times 10^{-3}$

19. a. ΔG must be negative if the reaction is spontaneous

20. f.

21. i. $4\,e^- + 4\,H^+ + Tl_2O_3 \rightarrow 2\,TlOH + H_2O$
$2\,NH_2OH \rightarrow N_2 + 2\,H_2O + 2\,H^+ + 2\,e^-$ $\times 2$
$4\,NH_2OH + Tl_2O_3 \rightarrow 2\,TlOH + 2\,N_2 + 5\,H_2O$

22. g. $q = -w = nRT\ln(V_f/V_i) = 1.0 \times 8.3145 \times 304.7 \times \ln(25/10) = 2320\text{ J} = 2.32\text{ kJ}$

23. d. A → B: w = zero
B → C: $w = -P\Delta V = -1.0 \times 15\text{ L atm} = -15 \times 101.325\text{ J} = -1.52\text{ kJ}$

24. e. $\Delta E = 0$, so $q = -w = 1.52$ kJ

25. b. pressure is greater, so more $P\Delta V$ work is done

26. b. neither process is spontaneous, $\Delta G = +$ for both

27. e. halve and reverse second equation and add to first, $K = \sqrt{(1/(1.8 \times 10^{37}))} \times (2.3 \times 10^6)$

28. e. $\Delta G° = -RT\ln K = 140{,}900$ J
$\Delta G° = \Delta H° - T\Delta S°$
$\Delta S° = -(\Delta G° - \Delta H°)/T = -(140{,}900 - 99{,}000)/600 = -70\text{ JK}^{-1}$

29. a. change the sign for the oxidation half-reaction and add: +0.403 – 0.257 = +0.146 V

30. b. 125 g Cr × (1 mol Cr/52 g Cr) × (3 mol e^-/1 mol Cr) × (96485 C/1 mol e^-) × (1 As/1 C) × (1/200A)
= 3479 s = 58 min

31. h.

32. f.

33. e. no unpaired electrons, bidentate, –24 Dq

34. d. dsp^2, square planar

35. e. positron; the mass numbers and the atomic numbers on the two sides must balance

FINAL EXAMINATION

CHEMISTRY 142 FALL 2008

Solutions

1. c. compare experiments 1 and 2; order with respect to I_2 is 1

2. g. $t_{1/2} = 0.693 / k \quad = 0.693 / 0.0606\ \text{day}^{-1} \quad = 11.4$ days

3. i. second-order decomposition, subsequent half-lives double
   100M to 50M takes 10 minutes, 50M to 25M takes 20 minutes, 25M to 12.5M takes 40 minutes, therefore 50M to 12.5M takes 20 + 40 = 60 minutes

4. c.

5. a. $K_c = [SO_3]^2 / ([SO_2]^2[O_2]) \quad = (0.056)^2 / (0.344^2 \times 0.172) \quad = 0.15$

6. a. reverse and multiply by 2 to obtain: $4\ HBr(g) \rightleftharpoons 2\ H_2(g) + 2\ Br_2(g)$
   take the inverse and square K: $K = 1 / (7.9 \times 10^{11})^2 = 1.6 \times 10^{-24}$

7. d. H_3PO_4, HF, and CH_3CO_2H are weak acids

8. h. $Sr(OH)_2 \rightarrow Sr^{2+} + 2\ OH^-$
   $[OH^-] = 2 \times 0.3M = 0.6M$
   $pOH = -\log(0.6) = 0.2 \qquad pH = 14 - 0.2 = 13.8$

9. d.

10. a. $HOCl^-$, acid

11. e. $HAsO_4^{2-}$ is the conjugate base of $H_2AsO_4^-$
    $K_b = K_w / K_{a2} = (1 \times 10^{-14}) / (5.6 \times 10^{-8}) = 1.8 \times 10^{-7}$

12. c.

	NH_4^+	+	H_2O	$\rightleftharpoons$	NH_3	+	H_3O^+	$K_a = K_w / K_b$
initial:	0.2M				0		0	
change:	–x				+x		+x	
equil:	0.2 – x				x		x	

$K_a = (1 \times 10^{-14}) / (1.8 \times 10^{-5}) = 5.6 \times 10^{-10} = x^2 / 0.2 - x \qquad x = 1.05 \times 10^{-5}$

$pH = -\log(1.05 \times 10^{-5}) = 4.98$

13. e. $pH = pK_a + \log([\text{conjugate base}]/[\text{acid}])$
    $= -\log(1.8 \times 10^{-4}) + \log(0.07 / 0.03) = 4.11$

14. g. $K_{neut} = K_a \times K_b / K_w = (2.0 \times 10^{-11} \times 1.8 \times 10^{-5}) / 1 \times 10^{-14} = 3.6 \times 10^{-2}$

15. d. $SnI_2 \rightarrow Sn^{2+} + 2\ I^-$ $K_{sp} = 1.0 \times 10^{-4} = [Sn^{2+}](2 \times [Sn^{2+}])^2 = 4[Sn^{2+}]^3$
$[Sn^{2+}] = 0.029$ M

16. a. add NaI; decrease solubility (addition of a common ion)
add HI; decrease solubility (addition of a common ion)
add HNO_3; no change

17. e. $Mg_3(PO_4)_2 \rightarrow 3\ Mg^{2+} + 2\ PO_4^{3-}$ $K_{sp} = 1.0 \times 10^{-24}$
dilution of Mg^{2+}: $(50\text{ mL})(1.0 \times 10^{-5}) = (100\text{ mL})[Mg^{2+}]$ $[Mg^{2+}] = 5.0 \times 10^{-6}$ M
dilution of PO_4^{3-}: $(50\text{ mL})(4.0 \times 10^{-5}) = (100\text{ mL})[PO_4^{3-}]$ $[PO_4^{3-}] = 2.0 \times 10^{-5}$ M
$Q_{sp} = [Mg^{2+}]^3[PO_4^{3-}]^2 = (5.0 \times 10^{-6})^3(2.0 \times 10^{-5})^2 = 5.0 \times 10^{-26}$
$Q_{sp} < K_{sp}$ therefore no precipitation

18. a. $w = -P\Delta V$ but $\Delta V = 0$ therefore $w = 0$

19. d. $w = -nRT \ln(V_f / V_i)$
$w = -2.0\text{ mol} \times 8.3145\text{ J K}^{-1}\text{ mol}^{-1} \times 365.5\text{ K} \times \ln(20/4) \times (1\text{kJ} / 1000\text{J}) = -9.8\text{ kJ}$
for an isothermal process $\Delta E = 0$ so $q = -w = 9.8$ kJ

20. e. $\Delta E = q + w$
$w = -P\Delta V = -3\text{ atm} \times (20\text{L} - 4\text{L}) = -48\text{ L atm} \times (101.325\text{ J} / 1\text{ L atm}) = -4864\text{ J}$
$q_p = n(5/2)R\Delta T = 2\text{ mol} \times (5/2) \times 8.3145\text{ J K}^{-1}\text{ mol}^{-1} \times (365.6\text{ K} - 73.12\text{K}) = +12159\text{ J}$
$\Delta E = -4864\text{ J} + 12159\text{ J} = +7300\text{ J}$

21. a. $q_{released} = q_{absorbed}$
$(317\text{ g}) \times (0.385\text{ J K}^{-1}\text{ g}^{-1}) \times (112 - T_f) = (359\text{ g}) \times (0.385\text{ J K}^{-1}\text{ g}^{-1}) \times (T_f - 42.6)$
$T_f = 75.16°C = 348.3$ K
$\Delta S_{cool} = m \times c \times \ln(T_f/T_i) = (317\text{ g}) \times (0.385\text{ J K}^{-1}\text{ g}^{-1}) \times \ln(348.3 / 385.15) = -12.27\text{ J K}^{-1}$
$\Delta S_{warm} = m \times c \times \ln(T_f/T_i) = (359\text{ g}) \times (0.385\text{ J K}^{-1}\text{ g}^{-1}) \times \ln(348.3 / 315.75) = +13.56\text{ J K}^{-1}$
$\Delta S_{process} = \Delta S_{cool} + \Delta S_{warm} = -12.27\text{ J K}^{-1} + 13.56\text{ J K}^{-1} = +1.29\text{ J K}^{-1}$

22. f. HN_3: H is +1, N is –3
N_2: N is 0 (an element by itself)
NO_2: O is –2, N is +4

23. h. half reactions: $(H_2O + SO_3^{2-} \rightarrow SO_4^{2-} + 2\ H^+ + 2\ e^-) \times 5$
$(5\ e^- + 8\ H^+ + MnO_4^- \rightarrow Mn^{2+} + 4\ H_2O) \times 2$
add: $5\ H_2O + 5\ SO_3^{2-} + 16\ H^+ + 2\ MnO_4^- \rightarrow 5\ SO_4^{2-} + 10\ H^+ + 2\ Mn^{2+} + 8\ H_2O$
simplify: $5\ SO_3^{2-} + 6\ H^+ + 2\ MnO_4^- \rightarrow 5\ SO_4^{2-} + 2\ Mn^{2+} + \underline{3\ H_2O}$

24. d. $E°_{anode} + E°_{cathode} = -0.337$ V (reverse sign for oxidation) + 0.799 V = +0.462 V

25. e. $Cu + 2\,Ag^{+} \rightarrow Cu^{2+} + 2\,Ag$
$\Delta G = -nFE° = -2 \text{ mol} \times 96458 \text{ C} \times 0.462 \text{ V} = -89152 \text{ J} = -89 \text{ kJ}$

26. e. $E° = (RT)/(nF) \times \ln K$
rearrange to solve for K: $\ln K = (E°nF)/(RT)$
$\ln K = (0.16 \text{ V} \times 2 \text{ mol} \times 96485 \text{ C}) / (8.3145 \text{ J K}^{-1} \text{ mol}^{-1} \times 298 \text{ K}) = 12.45$
$K = e^{12.45} = 2.6 \times 10^{5}$

27. h. $12.3 \text{ g Cu} \times (1 \text{ mol} / 63.55 \text{ g Cu}) \times (2 \text{ mol } e^{-} / 1 \text{ mol Cu}) \times (96485 \text{ C} / 1 \text{ mol } e^{-})$
$\times (1 \text{ As} / 1 \text{ C}) \times (1 / 1.89 \text{ A}) \times (1 \text{ hr} / 3600 \text{ s}) = 5.5 \text{ hr}$

28. f. carbon

29. c. aluminum — the diagonal relationship

30. g. all are true

31. b. 2 isomers — *cis* and *trans*

32. h. $[Ar]\ 3d^{1}$

33. c. Mn^{2+} EAN = 25 – 2 (for the 2+ charge) + 12 (from the six ligands) = 35

34. b. Fe^{2+} has a d^{6} configuration and water is a weak ligand
LFSE = 4(–4 Dq) + 2(6 Dq) = –4 Dq

35. b. ^{219}Rn

FINAL EXAMINATION

CHEMISTRY 142 SPRING 2009

Solutions

1. a. three I^- are used for every I_3^- made
   SO_4^{2-} is produced at twice the rate of I_3^-

2. e. reactants are A, B, and E
   intermediate (formed in step 1 and used in step 2) is C
   products are D (in step 1) and F(in step 2)

3. d. Rate of slowest step = k'[C][E]
   but from step 1: K = [C][D]/[A][B] so [C] = K[A][B][D]$^{-1}$
   and therefore Rate = k[A][B][E][D]$^{-1}$

4. d. rate of reaction = (96–24)/30 mol L^{-1} min^{-1} = 2.4 mol L^{-1} min^{-1}
   time required to reduce 96 mol L^{-1} to 6 mol L^{-1} = 90 mol L^{-1}/2.4 mol L^{-1} min^{-1} = 37.5 min

   alternatively, $t_{1/2}$ decreases by 2 every half-life, so

first $t_{1/2}$	= 20 minutes	96 to 48 mol L^{-1}
second $t_{1/2}$	= 10 minutes	48 to 24 mol L^{-1}
third $t_{1/2}$	= 5 minutes	24 to 12 mol L^{-1}
fourth $t_{1/2}$	= 2.5 minutes	12 to 6 mol L^{-1}
total time	= 20 + 10 + 5 + 2.5 min = 37.5 min	

5. c. $K_a = 3.5 \times 10^{-8} = [H_3O^+]^2 / 1.5 \times 10^{-2}$, $[H_3O^+] = 2.29 \times 10^{-5}$ and pH = 4.64

6. a. C_5H_5N is a base and $C_5H_5NH^+$ is its conjugate acid

7. d.

increase T	move to the left	exothermic reaction
remove PCl_5	move to the right	
decrease volume	move to the right	to the side of the system that occupies the smaller volume

8. c.

KNO_3 is neutral	salt of a strong acid and a strong base—no hydrolysis
LiCN is basic	salt of a weak acid and a strong base—hydrolysis of CN^- produces OH^-
NH_4Br is acidic	salt of a strong acid and a weak base—hydrolysis of NH_4^+ produces H_3O^+
NaF is basic	salt of a weak acid and a strong base—hydrolysis of F^- produces OH^-

9. e. choose an acid (H_2CO_3) with a pK_a (6.37) near to the pH required

10. b. H_3PO_4 is a weak acid

11. c. $K_b = K_w/K_a$

12. d. $K = [CO][H_2]^3 / [CH_4][H_2O] = (0.3)(0.8)^3 / (0.4)(0.1) = 3.8$

13. e. $K_{sp} = [Ba^{2+}][IO_3^-]^2$

14. b. equal amounts of acid and base, so first equivalence point is reached
pH is midway between 6.37 and 10.32, i.e. 8.35 —basic

15. g. half of the second H^+ removed; pH = 10.32

16. b. $[Pb^{2+}] = 3/2\ [PO_4^{3-}]$; there are more Pb^{2+} ions in solution than PO_4^{3-} ions

17. e.

$Ca(OH)_2(s) \rightleftharpoons Ca^{2+}(aq) + 2\ OH^-(aq)$	$K_{sp} = 6.5 \times 10^{-6}$
$2\ OH^-(aq) + 2\ H^+(aq) \rightleftharpoons 2\ H_2O(l)$	$K = 1/K_w^2 = 1.0 \times 10^{28}$
$Ca(OH)_2(s) + 2\ H^+(aq) \rightleftharpoons Ca^{2+}(aq) + 2\ H_2O(l)$	$K = 6.5 \times 10^{-6} \times 10^{28}$
the Cl^- ion is a spectator ion	$= 6.5 \times 10^{22}$

18. d. P V = n R T
$12 \times 20 = n \times 0.08206 \times 300$; n = 9.75 mol

19. f. ΔE = zero, $q = -w = nRT \ln(V_f/V_i) = 9.75\ \text{mol} \times 8.3145\ JK^{-1}mol^{-1} \times 300\ K \times \ln 4 = 33715\ J$

20. e. $q = n\ C_v \Delta T = 9.75\ \text{mol} \times 3/2 \times 8.3145\ JK^{-1}mol^{-1} \times (-225)\ K = -27360\ J$

21. g. $q = n\ C_p \Delta T = 9.75\ \text{mol} \times 5/2 \times 8.3145\ JK^{-1}mol^{-1} \times 225\ K = 45600\ J$

22. d. ΔE = zero, $w = -q = -(45600 - 27360) = -18240\ J$
alternatively, $w = -P\Delta V = -3.0\ \text{atm} \times 60\ L \times 101.325\ J\ L^{-1}\ atm^{-1} = -18240\ J$

23. b. +6

24. d.

$$e^- + 2H^+ + VO_2^+ \rightarrow VO^{2+} + H_2O \qquad \times 2$$
$$Zn \rightarrow Zn^{2+} + 2e^- \qquad \text{add}$$
$$Zn + 4H^+ + 2VO_2^+ \rightarrow Zn^{2+} + 2VO^{2+} + 2H_2O$$

25. d. nickel metal oxidized and oxygen reduced to water

26. e. $25.0\ g \times (1\ \text{mol Cu}/63.55\ g\ Cu) \times (2\ \text{mol}\ e^-/1\ \text{mol Cu}) \times (96485\ As/1\ \text{mol}\ e^-) \times (1/5.27\ A)$
= 14,405 s
$14{,}405\ s \times (1\ hr/3600\ s) = 4.00$ hour

27. d. energy liberated or absorbed as heat or work is path-dependent

28. d. heat lost by the hot water in cooling from 90.2°C to 0.0°C melts the 85 g of ice
$85\ g \times 333\ Jg^{-1}$ = mass of hot water $\times\ 4.184\ JK^{-1}g^{-1} \times 90.2\ K$
mass of hot water = 75 grams

29. c.

ice melting:	$\Delta S = 85\ g \times 333\ Jg^{-1}/273.15\ K = +103.63\ JK^{-1}$
hot water cooling:	$\Delta S = 75\ g \times 4.184\ JK^{-1}g^{-1} \times \ln(273.15/363.35) = -89.54\ JK^{-1}$
total:	$\Delta S = +103.63\ JK^{-1} - 89.54\ JK^{-1} = 14.08\ JK^{-1}$

30. e.

31. b. add the oxidation numbers: (n × 2+) + (2+) + (2 × 4+) + (7 × 2–) = 0, so n = 2
or, the pyrosilicate ion has a charge of 6–, which must be balanced by 6+

32. b. two, *cis*– and *trans*–

33. b. sp^3 tetrahedral

34. f. Mn^{3+} has a d^4 configuration and CN^- is a strong ligand; LFSE = 4 × –4Dq = –16 Dq

35. b. the mass numbers and the atomic numbers must balance

FINAL EXAMINATION

CHEMISTRY 142 FALL 2009

Solutions

1. b. second order reaction: $1/[R]_t - 1/[R]_0 = -kt$
   $1/0.3 - 1/0.6 = -(7.0 \times 10^9\ M^{-1}s^{-1})\ t$
   $t = 2.4 \times 10^{-10}$ s

2. g. B and D

3. b. start with the slow elementary step: Rate = $k[N_2O_2][H_2]$
   from the fast equilibrium: $K = [N_2O_2] / [NO]^2$
   rearrange: $[N_2O_2] = K[NO]^2$
   substitute into rate equation: Rate = $k[NO]^2[H_2]$

4. c. IO^-

5. a. Q = [products] / [reactants], therefore $2\ NO + Br_2 \rightleftharpoons 2\ NOBr$

6. b. $2\ NaHCO_3(s) \rightleftharpoons Na_2CO_3(s) + H_2O(g) + CO_2(g)$
   adding $NaHCO_3(s)$, a solid, will cause no shift in the equilibrium

7. a.

	$H_2(g)$ +	$Br_2(g)$	$\rightleftharpoons$	$2\ HBr(g)$	$K_c = 2.18 \times 10^6$
initial:	0	0		0.32 M	
change:	+x	+x		–2x	
equilibrium:	x	x		0.32 – 2x	

   $K = 2.18 \times 10^6 = (0.32 - 2x)^2 / x^2$
   $x = [H_2] = 2.2 \times 10^{-4}$ M

8. h. I, II, and IV are true

9. c. Co^{3+} and BF_3

10. f. NH_4Cl is acidic, $Mg(OH)_2$ is basic, $NaClO_4$ is neutral, and KCN is basic

11. f. at the second equivalence point, all of the second H+ has been removed; all HPO_4^{2-}
   $pH = (pK_{a2} + pK_{a3}) / 2 = (-\log(6.2 \times 10^{-8}) + -\log(4.2 \times 10^{-13})) / 2 = 9.79$

12. e.

	F^- +	H_2O	$\rightleftharpoons$ HF	+ OH^-	$K_b = K_w / K_a = 1.41 \times 10^{-11}$
initial:	0.5		0	0	
change:	–x		+x	+x	
equilibrium:	0.5 – x		x	x	

   $K_b = 1.41 \times 10^{-11} = x^2 / 0.5 - x$ $x = 2.66 \times 10^{-6}$ $pOH = -\log(2.66 \times 10^{-6}) = 5.575$
   $pH = 14 - pOH = 14 - 5.575 = 8.42$

13. d. multiply ($SO_3(g)$ $\rightarrow$ $SO_2(g)$ + 1/2 $O_2(g)$) by 2 so square K
$K = (1.32)^2 = 1.74$

14. b. $K_{neut} = K_a / K_w = (1.3 \times 10^{-10}) / (1.0 \times 10^{-14}) = 1.3 \times 10^4$

15. g. $Al(OH)_3(s) \rightleftharpoons Al^{3+}(aq) + 3\ OH^-(aq)$ $K_{sp} = 1.8 \times 10^{-33} = [Al^{3+}][OH^-]^3$
$[OH^-] = 3[Al^{3+}]$
$1.8 \times 10^{-33} = [Al^{3+}] \times (3[Al^{3+}])^3$
$[Al^{3+}] = 2.9 \times 10^{-9}$ M

16. b. $[Ca^{2+}]_{dilute} = [Ca^{2+}]_{conc} \times V_{conc} / V_{dilute} = (5.0 \times 10^{-3}\ M) \times 25\ mL / 50\ mL = 0.0025\ M$
$[F^-]_{dilute} = [F^-]_{conc} \times V_{conc} / V_{dilute} = (5.0 \times 10^{-3}\ M) \times 25\ mL / 50\ mL = 0.0025\ M$
$Q_{sp} = [Ca^{2+}][F^-]^2 = (0.0025) \times (0.0025)^2 = 1.56 \times 10^{-8}$
$Q_{sp} > K_{sp}$, therefore precipitation will occur

17. d. NH_3 will increase the solubility
HNO_3 will cause no change to the solubility
KBr will decrease the solubility because of the addition of the common ion Br^-

18. i. I, III, and IV are true

19. a. $w = -P\Delta V$; from A to B the volume is constant, therefore $\Delta V = 0$ and $w = 0$

20. e. find volume at state C
$P_aV_a = P_cV_c$; $V_c = (2.0\ atm \times 12.0\ L) / 8.0\ atm = 3.0\ L$
find temperature at State A = temperature at State C:
$T_a = PV/nR = (2.0\ atm \times 12.0\ L) / (0.50\ mol \times 0.08206\ L\ atm\ K^{-1}\ mol^{-1}) = 585\ K$
$w = -nRT \ln(V_f/V_i) = -(0.5mol) \times (8.3145\ J\ K^{-1}) \times (585\ K) \times \ln(3.0\ L / 12.0L) = 3371\ J = 3.4\ kJ$
isothermal, therefore $\Delta E = 0$ and $q = -w$
$q = -3.4$ kJ

21. g. total heat needed = heat to melt ice + heat to warm melted ice to 25°C
$q = (15.0\ g \times 333\ J\ g^{-1}) + (15.0\ g \times 4.184\ J\ g^{-1}\ K^{-1} \times 25K) = 6564\ J$

22. f. $\Delta S_{process} = \Delta S_{melt\ ice} + \Delta S_{warm\ water}$
$\Delta S_{melt\ ice} = (m \times \text{latent heat}) / T = (15.0\ g \times 333\ J\ g^{-1}) / 273.15K = +18.287\ J\ K^{-1}$
$\Delta S_{warm\ water} = m \times c \times \ln(T_f/T_i) = 15\ g \times 4.184\ J\ K^{-1} g^{-1} \times \ln(298.15\ K/273.15\ K) = 5.496\ J\ K^{-1}$
$\Delta S_{process} = +18.287\ J\ K^{-1} + 5.496\ J\ K^{-1} = +24\ J\ K^{-1}$

23. b. oxidation half-reaction: $H_2O + CN^- \rightarrow OCN^- + 2\ H^+ + 2\ e^-$ multiply by 3
reduction half-reaction: $3\ e^- + 4\ H^+ + MnO_4^- \rightarrow MnO_2 + 2\ H_2O$ multiply by 2
add: $3\ H_2O + 3\ CN^- + 6\ e^- + 8\ H^+ + 2\ MnO_4^- \rightarrow 3\ OCN^- + 6\ H^+ + 6\ e^- + 2\ MnO_2 + 4\ H_2O$
simplify: $3\ CN^- + 2\ H^+ + 2\ MnO_4^- \rightarrow 3\ OCN^- + 2\ MnO_2 + H_2O$

adjust for basic conditions by adding 2 OH^- to both sides:
$3\ CN^- + 2\ H_2O + 2\ MnO_4^- \rightarrow 3\ OCN^- + 2\ MnO_2 + H_2O + 2\ OH^-$
cancel the water: $3\ CN^- + H_2O + 2\ MnO_4^- \rightarrow 3\ OCN^- + 2\ MnO_2 + 2\ OH^-$

24. d. change the sign for the oxidation half reaction and add: $E° = +1.676\ V + 1.065\ V = +2.741\ V$

25. a. $\Delta G° = -nFE° = -6 \times 96485 \times 2.741 = -1.56 \times 10^6\ J = -1587\ kJ$

26. e. 75 min × (60 s / 1 min) × (2.15 A) × (1C / 1As) × (1 mol e^- / 96485 C) × (1 mol Zn / 2 mol e^-) × (65.39 g Zn / 1 mol Zn) = 3.3 g Zn

27. a. Be

28. i. I, III, and IV

29. h. N_2O_3 N is +3 and O is –2
$N_2O_2^{2-}$ N is +1 and O is –2
N_2O N is +1 and O is –2
N_2O_5 N is +5 and O is –2

30. b. F_2 is most difficult to oxidize and Xe can form compounds with other elements

31. d. [Ar] $3d^6$

32. c. H_2O, Cl^-, and OH^-

33. d. Fe^{3+} EAN = 26 – 3 (for the 3+ charge) + 12 (from the 6 ligands) = 35

34. b. Ni^{2+}: [Ar] $3d^8$
CN^- is a strong ligand so electrons will be paired, therefore dsp^2

35. a. Co^{3+} has a d^6 configuration and F^- is a weak ligand
LFSE = 4(–4 Dq) + 2(6 Dq) = –4 Dq

FINAL EXAMINATION

CHEMISTRY 142 SPRING 2010

Solutions

1. e. D: H_2O; produced 5/4 faster than CO_2
   C: CO_2
   B: O_2; used 13/2 faster than C_4H_{10}
   A: C_4H_{10}

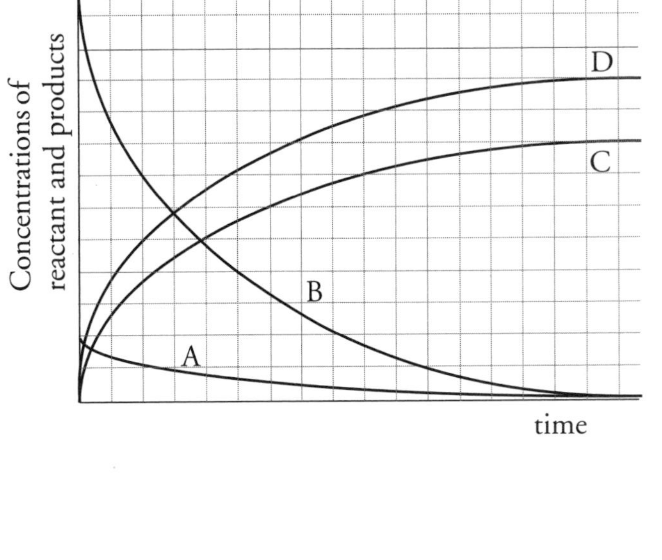

2. b. $t_{½}$ always the same = 12 min

3. b. order for [A] = 1
   order for [B] = 2
   overall order = 3

4. c. $\ln(1/3) = -k \times 12$
   $k = 0.09155$
   $t_{½} = 0.693/k = 7.57$ days

5. h. $K = 36^2 / (3.0 \times 2.0^2) = 108$

6. f.

7. e. add an H^+ to produce H_2SO_4

8. e.

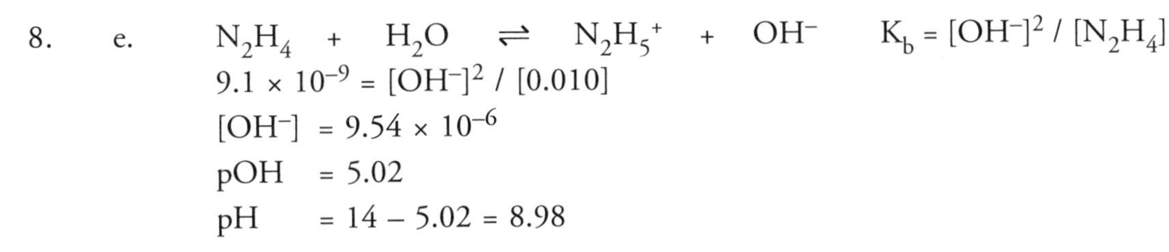

$N_2H_4 + H_2O \rightleftharpoons N_2H_5^+ + OH^-$ $K_b = [OH^-]^2 / [N_2H_4]$
$9.1 \times 10^{-9} = [OH^-]^2 / [0.010]$
$[OH^-] = 9.54 \times 10^{-6}$
$pOH = 5.02$
$pH = 14 - 5.02 = 8.98$

9. c. Neutral solution: $[H_3O^+] = [OH^-] = \sqrt{K_w} = 2.35 \times 10^{-7}$
   $pH = pOH = 6.63$

10. e.

11. b. $pH = pK_{a1} = 1.89$

12. d. $pH = ½(pK_{a1} + pK_{a2}) = 4.55$

13. h. $pH = pK_{a2} = 7.20$

14. g. $pH = pK_{a2} + \log_{10}([\text{base}]/[\text{acid}])$
    $= 7.20 + \log_{10}([1/4]/[3/4])$
    $= 7.20 - 0.477 = 6.72$

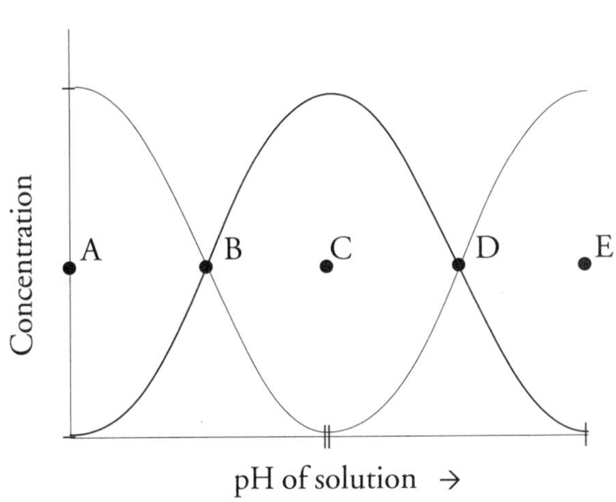

15. c. LiF and HF — a weak acid (HF) and its conjugate base (F^-)

16. f. $K_{neut} = K_a/K_w = 3.5 \times 10^{-4} / 1.0 \times 10^{-14} = 3.5 \times 10^{10}$

17. d. there are three Cu^{2+} ions per formula unit: $[Cu^{2+}] = 3 \times$ molar solubility $= 3 \times 1.6 \times 10^{-17}$
$= 4.8 \times 10^{-17}$

18. c.

19. f. the two solutions contain phosphate PO_4^{3-} and copper(II) Cu^{2+} in the correct stoichiometric ratio
what remains in solution after precipitation must also be in the correct ratio
$[Cu^{2+}] = (3/2)\ [PO_4^{3-}]$
$Ksp = 1.4 \times 10^{-37} = [Cu^{2+}]^3[PO_4^{3-}]^2 = ((3/2)\ [PO_4^{3-}])^3[PO_4^{3-}]^2 = (27/8)[PO_4^{3-}]^5$
$[PO_4^{3-}]^5 = 4.14810^{-38}$
$[PO_4^{3-}] = 3.34 \times 10^{-8}$
molar concentration = molar solubility $= [PO_4^{3-}]/2 = 1.67 \times 10^{-8}$

20. f. total arrangements $= 2^{10} = 1024$
of these, there are 10 with one molecule in the righthand half
probability = 10 in 1024 or 1 in 102

21. d.

22. e. +4 +5 +2 −1 = 10

23. h. $T = PV / nR = (16.0 \times 2.00)/(1.00 \times 0.08206) = 390$ K

24. d. $T = PV / nR = (1.59 \times 8.00)/(1.00 \times 0.08206) = 155$ K

25. g. $w = -n R T \ln(V_f/V_i) = -8.3145 \times 390 \times \ln 4 = -4500$ J

26. d. $w = n C_v \Delta T = (3/2) \times 8.3145 \times (155 - 390) = -2930$ J

27. a. w = zero (constant volume)

28. a. ΔE = zero (isothermal)

29. d. $q = 0$, so $\Delta E = w = -2930$ J (answer to 26)

30. g. $\Delta S = q/T = -w/T = nR \ln(V_f/V_i) = 8.3145 \times \ln 4 = 11.5\ JK^{-1}$

31. a. $\Delta S = q/T$ = zero ($q = 0$)

32. c. ΔS is path-independent
so the sum of ΔS for A to B and ΔS for B to C must = zero
ΔS for B to C is the negative of ΔS for A to B $= -11.5\ JK^{-1}$

33. e. $4\ Fe^{3+} + 2\ NH_3OH^+ + 5\ H_2O \rightarrow N_2O + 4\ Fe^{2+} + 6\ H_3O^+$

34. g. 10.5 g Ag × (1 mol Ag/ 107.87 g Ag) × (1 mol e^-/1 mol Ag) × (96485 C/1 mol e^-) × (1 As/1 C) × (1/5A) × (1 min/60 s) = 31.3 min

35. c. potassium

36. c. 3, so that +2 +4 –6 = 0

37. f. water, donor atom O

38. f. catenation of carbon—the ability to bond to itself

39. e. Xe

40. c. Fe atomic number 26
present as Fe^{2+}, so subtract 2 electrons; 24
each ligand supplies two electrons (Lewis base); 24 + (6 × 2) = 36

41. h. Fe^{2+} has the configuration d^6
in a strong field (cyanide ligands) the electrons will be paired in the lower level
LFSE = 6 × –4Dq = –24Dq

42. e. 5: all trans
all cis
and three with one pair trans and the other two cis

FINAL EXAMINATION

CHEMISTRY 142 FALL 2010

Solutions

1. a. O_3 used at 1/3 the rate at which SO_2 is used
   SO_3 is formed at the same rate as SO_2 is used

2. c. compare experiments 1 and 2 where $[Ce^{4+}]$ is the same:
   the rate increases in the same proportion as $[Fe^{2+}]$ — order 1 $[2.0 \times 10^{-7} \times (2.8/1.8) = 3.1 \times 10^{-7}]$

   compare experiments 2 and 3 where $[Fe^{2+}]$ is the same:
   the rate increases in the same proportion as $[Ce^{4+}]$ — order 1 $[3.1 \times 10^{-7} \times (3.4/1.1) = 9.6 \times 10^{-7}]$

   rate equation is: rate = $k[Ce^{4+}][Fe^{2+}]$

3. c. compare the two half-lives:
   0.080 to 0.040 M: time = 24.4 s
   0.060 to 0.030 M: time = 24.3 s
   half life is constant, therefore first order

4. g. $\ln(k_1/k_2) = -E_A/R(1/T_1 - 1/T_2)$
   $\ln(0.0315/k_2) = -260{,}000/8.3145 \times (1/800 - 1/850)$
   $= -260{,}000/8.3145 \times (50/(800 \times 850))$
   $= -2.30$
   $(0.0315/k_2) = 0.100$
   $k_2 = 0.315$

5. b. reactants: $(CH_3)_3CBr$, H_2O
   product: $(CH_3)_3COH$, HBr
   intermediates: $(CH_3)_3C^+$, Br^-, $(CH_3)_3COH_2^+$

6. f.

7. c. leave equation 1 as is
   reverse equation 2 and multiply by 2
   $K = K_1/K_2^2 = 14/1.3^2 = 8.3$

8. a.

9. b. approximately....$K_a = [H^+]^2 / [HF] = 6.6 \times 10^{-4} = [H^+]^2 / 0.80$
   $[H^+] = 0.0230$
   pH = 1.64

10. f. pH at first crossover = 2.97
    pH at second crossover = 13.44
    pH at first equivalence point = (2.97 + 13.44)/2 = 8.21
    species present at pH 10 (between 8.21 and 13.44) are $HC_7H_4O_3^-$ and $C_7H_4O_3^{2-}$

11. d. 8.21 (at first equivalence point)

12. e. strongest acid = chloroacetic acid (highest K_a)
acid with strongest conjugate base = weakest acid = hydrocyanic acid

13. a. weak base and its conjugate acid

14. g. $NaOH = 0.200\ L \times 3.0 \times 10^{-3}\ mol\ L^{-1} = 6.0 \times 10^{-4}\ mol$
$HNO_3 = 0.100\ L \times 5.0 \times 10^{-3}\ mol\ L^{-1} = 5.0 \times 10^{-4}\ mol$
at end, excess of 1.0×10^{-4} mol NaOH in 300 mL
$[OH^-] = 1.0 \times 10^{-4}\ mol / 0.300\ L = 3.33 \times 10^{-4}\ M$
$pOH = 3.48$
$pH = 10.52$

15. d. $K_{sp} = [Li^+]^2[CO_3^{2-}] = 8.2 \times 10^{-4}$
$[Li^+] = 2 \times [CO_3^{2-}]$
$K_{sp} = [2 \times [CO_3^{2-}]]^2[CO_3^{2-}] = 4 \times [CO_3^{2-}]^3 = 8.2 \times 10^{-4}$
$[CO_3^{2-}] = 0.059\ M$
$[Li^+] = 2 \times [CO_3^{2-}] = 0.12\ M$

16. e. II (addition of HCl) and III (addition of NH_3)

17. c. temp at point B $= 300\ K$
temp at point C $= (1.35\ atm \times 15.2\ L) \times 101.325\ JL^{-1}atm^{-1}/2.0\ mol \times 8.3145\ JK^{-1}\ mol^{-1}$
$= 125\ K$
$\Delta T = 125 - 300\ K = -175\ K$
$\Delta E = n\ C_v\ \Delta T$
$= 2.0\ mol \times 3/2 \times 8.3145\ JK^{-1}\ mol^{-1} \times (-175)\ K$
$= -\ 4365\ J$
$= -\ 4.37\ kJ$

18. c. A to B: ΔE = zero
B to C: $\Delta E = -\ 4.37\ kJ$
therefore A to C via B: $\Delta E = -\ 4.37\ kJ$
therefore A to C by any route: $\Delta E = -\ 4.37\ kJ$
since q = zero for the adiabatic expansion, $\Delta E = w$
and therefore $w = -\ 4.37\ kJ$ (the system does work)

19. e. A to C via B: $\Delta E = -\ 4.37\ kJ$
work done A to B $= -nRT\ln(V_f/V_i) = -\ 6.54\ kJ$
work done B to C = zero
$\Delta E = q + w$
$q = \Delta E - w = -\ 4.37\ kJ - (-6.54\ kJ) = +\ 2.17\ kJ$

alternatively, $q_{(ABC)} = q_{(AB)} + q_{(BC)} = +6.54\ kJ + (-4.37 kJ) = +2.17\ kJ$

20. g.

21. h. steam condenses: heat $= -20.0\text{ g} \times 2260\text{ J g}^{-1} = -45{,}200\text{ J}$
steam cools to T_f: heat $= 4.184\text{ J K}^{-1}\text{g}^{-1} \times 20.0\text{ g} \times (T_f - 100)\text{ K}$
water warms to T_f: heat $= 4.184\text{ J K}^{-1}\text{g}^{-1} \times 225.0\text{ g} \times (T_f - 35)\text{ K}$
sum to zero: $= -45{,}200 + 83.7T_f - 8368 + 941T_f - 32949\text{ J} = 0$
$1025T_f = 86517$
$T_f = 84.4°\text{C (357.6K)}$

22. c. steam condenses: $\Delta S° = \Delta H°/T = -45{,}200/373.15 = -121.13\text{ JK}^{-1}$
steam cools to T_f: ΔS = mass × specific heat × $\ln(T_f/T_i) = -3.56\text{ JK}^{-1}$
water warms to T_f: ΔS = mass × specific heat × $\ln(T_f/T_i) = 140.11\text{ JK}^{-1}$
sum: $\Delta S = 15.4\text{ JK}^{-1}$

23. e. I —q and w are not state functions
II —$\Delta S_{univ} > 0$ for a spontaneous process
III —it is G, the free energy of the system, that is minimized; $\Delta G°$ is constant at constant T

24. d. CH_3OH: C –2
$Mg_3(PO_4)_2$: P +5
CoF_3: Co +3
total = +6

25. g. $6\,Ag(s) + 3\,HS^- + 5\,H_2O(l) + 2\,CrO_4^{2-}(aq) \rightarrow 3\,Ag_2S(s) + 2\,Cr(OH)_3(s) + 7\,OH^-(aq)$

26. c. the titanium is oxidized to Ti^{2+} and the fluorine is reduced to F^-

27. h. Mn: the best reducing agent is the one most easily oxidized

28. e. 12 g Mn × (1 mol Mn/54.94 g) × (4 mol e^-/1 mol Mn) × (1 mol Al/3 mol e^-) × (26.98 g/1 mol Al) = 7.86 g Al

29. b. high ionization energy

30. a.

31. g.

32 - 36. b. Na_3P d. Mg (diagonal relationship) f. Be g. Xe h. O

37. c. $[Fe(NH_3)_5(OH)]Cl_2$

38. e. dsp^2

39. b. 2 (*cis*– and *trans*–)

40. b. H_2O is a weak ligand; Co^{3+} has a spin-free d^6 configuration; LFSE = –4 Dq

FINAL EXAMINATION

CHEMISTRY 142 SPRING 2011

Solutions

1. g. *change in $PH_3(g)$* = -12.0 mol L^{-1} min^{-1}
   change in $P_4(g)$ = 3.0 mol L^{-1} min^{-1}

2. f. compare 1 and 3:
   $[H_2]$ constant; [NO] halves; rate goes down a factor of 4
   order with respect to [NO] is 2
   compare 1 and 2:
   [NO] constant; $[H_2]$ halves; rate also halves
   order with respect to $[H_2]$ is 1
   overall order = 2 + 1 = 3

3. g. 224 to 56 mol L^{-1} takes 15 minutes—two half-lives.
   56 to 14 mol L^{-1} should take another 30 minutes if first order, but it takes longer probably second order then...

224 to 112 mol L^{-1}	5 min	
112 to 56 mol L^{-1}	10 min	15 minutes elapsed
56 to 28 mol L^{-1}	20 min	
28 to 14 mol L^{-1}	40 min	75 minutes elapsed
14 to 7 mol L^{-1}	80 min	
7 to 3.5 mol L^{-1}	160 min	315 minutes = 5 hr 15 min elapsed

4. f.

5. i. $K = [CO][H_2]^2 / [CH_4][O_2]^{½} = (3.0)(6.0)^2 / (1.0)(4.0)^{½} = 54$

6. e. add the equations and multiply K_{a1} and K_{a2} = 4.0×10^{-6}

7. f.

8. f.

Acidic solute	*Basic solute*	*Neutral solute*
H_2SO_4	NH_3	KNO_3

9. c. $HSO_3^- + HSO_3^- \rightleftharpoons H_2SO_3 + SO_3^{2-}$
   (SO_3^{2-}: base)

10. d. pH = 8.57
    pOH = 5.43
    $[OH^-] = 3.72 \times 10^{-6}$

    $F^- + H_2O \rightleftharpoons HF + OH^-$

    $K_b = [OH^-]^2 / [F^-] = (3.72 \times 10^{-6})^2 / 1.00 = 1.38 \times 10^{-11}$
    $pK_b = 10.86$ and therefore $pK_a = 3.14$

11. b. $ClO^-(aq)$ (base) + $H_2O(l)$ (acid) $\rightleftharpoons$ $HClO(aq)$ (acid) + $OH^-(aq)$ (base)

12. e. reverse equation 1 and add
$K = K_{a2}/K_{a1} = 9.4 \times 10^{-4}$

13. c, $pH = pK_{a1} + \log(base/acid) = 1.19 + \log((3/4)/(1/4)) = 1.67$

14. f. all of them are buffers solutions; all are weak electrolytes with their conjugate partners

15. b. two silver ions for each carbonate ion

16. e. K_{sp} for silver carbonate $Ag_2CO_3 = 8.1 \times 10^{-12} = [Ag^+]^2[CO_3^{2-}] = 4 \times [CO_3^{2-}]^3$
$[CO_3^{2-}] = 1.3 \times 10^{-4}$ M
which is the molar solubilty of silver carbonate—there is one CO_3^{2-} in each Ag_2CO_3 formula unit

17. g.
$Ag_2S(s) \rightleftharpoons 2Ag^+ + S^{2-}$ $K_{sp} = 6.0 \times 10^{-51}$
$Ag_2CO_3(s) \rightleftharpoons 2Ag^+ + CO_3^{2-}$ $K_{sp} = 8.1 \times 10^{-12}$
reverse equation 2 and add; divide K_{sp} for silver sulfide by K_{sp} for silver carbonate
$Ag_2S(s) + CO_3^{2-} \rightleftharpoons Ag_2CO_3(s) + S^{2-}$ $K = 7.4 \times 10^{-40}$

18. g. all are the same for the two paths except the heat and the work

19. f. use the ideal gas equation; n = 2.00 mol

20. h. constant volume change, w = zero

21. e. temperature at point B: $T = PV/nR = (29.24 \text{ atm} \times 5.0 \text{ L})/(2.00 \times 0.08206) = 890.81 \text{ K} = 617.7°C$
heat lost $q = nC_v\Delta T = 2.00 \times (3/2) \times 8.3145 \times (31.5 - 617.7) = -14621 \text{ J} = -14.6 \text{ kJ}$

22. g. $\Delta S = nC_v\ln(T_f/T_i) = 2.00 \times (3/2) \times 8.3145 \times \ln(304.7/890.8) = -26.8 \text{ JK}^{-1}$
alternatively, since the entropy change from B to A must be zero (adiabatic), then the entropy decrease (–) from B to C must equal numerically the entropy increase (+) from C to A:
$\Delta S = nR\ln(V_f/V_i) = 2.00 \times 8.3145 \times \ln(25/5) = +26.8 \text{ JK}^{-1}$

23. e. the internal energy change ΔE of the system in going from State B to State C = q + w = –14.6 kJ

24. b. 4 molecules; number of arrangements with one molecule in left half = 4
total arrangements = $2^4 = 16$
probability = 4 in 16, or 1 in 4

25. d. $\Delta S = k\ln(W_f/W_i) = k\ln(V_f^N/V_i^N) = k\ln(V_f/V_i)^N = Nk\ln(V_f/V_i)$
$= 4 \times 1.38 \times 10^{-23} \times \ln(2/1) = 3.83 \times 10^{-23} \text{ J K}^{-1}$

26. i. +6

27. b. 3 Cu(*s*) + 2 HNO_3(*aq*) + 6 H^+(*aq*) → 3 Cu^{2+}(*aq*) + 2 NO(*g*) + 4 H_2O(*l*)

28. f. E° = 0.74 V
$\ln K = nFE°/RT = 5 \times 96485 \times 0.74 / 8.3145 \times 298 = 144$
$K = e^{144} = 3.55 \times 10^{62}$

29. h. 100 min

30. c. potassium

31. f. Al

32. b. ($Al_2Si_2O_{10}$) has a charge of –6: (2×3 + 2×4 – 10×2)
charges must balance (+2) + (n×+3) + (2×–1) + (–6) = 0
2 + 3n – 2 – 6 = 0
n = 2

33. e. Xe

34. c. 3
the NH_3 can be *fac*– or *mer*–, and for the *mer*– isomer, the Br^- ligands can be *cis*– or *trans*–.

35. d. LFSE = –12 Dq